Contributions to Economics

More information about this series at http://www.springer.com/series/1262

Hansjörg Drewello • Bernd Scholl
Editors

Integrated Spatial and Transport Infrastructure Development

The Case of the European North-South Corridor Rotterdam-Genoa

Springer

Editors
Hansjörg Drewello
University of Applied Sciences Kehl
Kehl am Rhein
Germany

Bernd Scholl
ETH Zürich
Institute for Spatial and Landscape Development
Zurich
Switzerland

ISSN 1431-1933 ISSN 2197-7178 (electronic)
Contributions to Economics
ISBN 978-3-319-15707-8 ISBN 978-3-319-15708-5 (eBook)
DOI 10.1007/978-3-319-15708-5

Library of Congress Control Number: 2015948882

Springer Cham Heidelberg New York Dordrecht London

Printed on acid-free paper

Springer International Publishing AG Switzerland is part of Springer Science+Business Media (www.springer.com)

Foreword

Transport is at the crossroads of the economic, social and environmental goals of Europe, an essential driver for jobs and growth. It underpins both export and competitiveness. Against this background, Europe's transport infrastructure plays a key role to completing the internal market through the removal of physical barriers and the introduction of soft measures. Furthermore, in today's interconnected world, neither scope nor vision can be exclusively limited to the European internal market but has to incorporate the broader global context where economic gains, and likewise economic challenges, no longer stop at the border. It is, therefore, necessary to change the mental framework to understand that today's world no longer allows thinking in terms of a single mode or even country.

"*Connecting to complete*" and "*connecting to compete*" emerge as mottos for the creation of a single European transport area. Completing the transport network—in particular its core—interlinking modes and nodes—can provide a competitiveness boost and growth potential. This, however, will become reality only if all stakeholders act, and act together. In this context, three main strands emerge:

- balancing private and public interests, exploring potential areas of joint activity and collaboration, including through innovative financing schemes,
- reconciling short-term and long-term plans and vision, taking into account demand projections, and sustainability challenges within the larger European and global policy and economic framework, and
- above all, matching words with deeds, recognizing the need for the requisite firm political will, in the absence of which, even the best conceived project will flounder.

Transport, which has been part of one of the two founding policies present in the Treaty of Rome, holds a pre-eminent place in the European Union integration process. At a juncture when momentum for the European project is waning, it is the right time to deliver on political commitments, act on promises and progress with enforcement.

Against this backdrop, the CODE24 publication exploring the integrated spatial and transport infrastructure development on the Rotterdam Genoa corridor constitutes a vital instrument to analysing the transport challenges facing the continent as well as the specific needs of the Rhine–Alpine Corridor. As the publication highlights, recent developments and the emergence of the European transport corridors have added an additional layer of complexity when it comes to spatial planning, driven by the fact that corridors run not only across regional and national, physical and administrative borders but also cross-cut across issues related to economic development, environmental sustainability and social well-being.

As demonstrated in the opening part of the publication, the Rhine–Alpine Corridor stands out as an epitome of a mature, complex corridor that includes all modalities and which offers widely applicable insights. Altogether, more than 70 million people live, work and consume in the catchment area of the Rhine–Alpine Corridor. Leading manufacturing and trading companies, production plants and distribution centres are located within. The corridor runs through the so-called "Blue banana", which includes major EU economic centres. Along its course, over 1 billion tonnes of freight are transported annually, resulting in a corridor GDP of more than 2700 billion euros. With 13 % of EU's population, the Corridor regions generate 19 % of the EU's GDP.

The Rhine–Alpine Corridor is a "forerunner" for other corridors with initiatives on innovations and new technologies paving the way into the future. The CODE24 publication offers an all-encompassing overview of the corridor, which can serve as a useful template for other corridors as well as for the larger questions confronting transport in the European Union. Part II provides a comprehensive overview of the freight sector, exploring bottlenecks in railway infrastructure, their impact on regional transport costs and analyses scenarios towards achieving an integrated railway network. Logistical processes, clusters and the connection of terminal ports to hinterland are addressed in detail in Part III, which also explores rail competitiveness, inland ports and urban development. Fundamentally, the publication addresses critical environmental aspects of railway infrastructure development along the corridor, such as noise reduction and the management of ecological measures for major projects, while its penultimate chapter factors in local preferences and closes in on the vital need for a buy-in from the European citizens as a requirement for political legitimacy.

Looking towards the future evolution of the corridor and transport in Europe at large, effective coordination emerges crucial for all European traffic flows. A project influences the performance of the entire Corridor, in every country involved, stressing the need to look beyond single initiatives. The work and involvement of all stakeholders remain vital. The comprehensive information about the state of play as well as the needs and challenges of the corridor must be translated into reality. In this context, Code24 is a fitting example of the type of collaboration needed in order to provide crucial momentum and implement this corridor whose centrality cannot be overstated.

Brussels, Belgium Ana Palacio

Ana Palacio was the European Coordinator for the Rhine-Alpine Core Network Corridor until May 2015. She is an international lawyer and founding partner of Palacio y Asociados, a law firm with offices in Madrid, Brussels and Washington, DC. Between 1994 and 2006, she was Member of the European Parliament, Minister of Foreign Affairs of Spain and Member of the Spanish Parliament. In 2006, Ms Palacio became Senior Vice President and General Counsel of the World Bank Group; in 2008 she was member of the Executive Committee of AREVA. In 2012, Ms Palacio became member of the *Consejo de Estado de Reino de España* (Council of State). She takes part of various boards, think tanks and public institutions and is distinguished professor of multiple universities.

Foreword

It was an ambitious intent when the project partners of CODE24 first applied for this seminal project. The project application for EU funding has been approved at the first attempt, and the project has also officially been awarded as a "strategic initiative".

The project CODE24 with its 18 project partners from five countries has produced a number of tangible results culminating in the joint strategy for the development of the Rhine–Alpine Corridor. Partners from different professional backgrounds have cooperated seamlessly and have successfully delivered the intended outcome.

This publication provides a comprehensive insight into the concrete achievements and results of more than 4 years of cooperation. Obviously, a very broad field of issues has been tackled. This is due to the ambitious goal of the project to perform an interdisciplinary approach, looking at a variety of issues which are equally important for a reasonable regional development.

Looking ahead, the coordination of development activities and measures in the Rhine–Alpine Corridor is an enduring task and a continuing challenge. Thus, the intended creation of the European Grouping of Territorial Cooperation "Interregional Alliance for the Rhine–Alpine Corridor EGTC" plays a crucial role and will enable its members to perpetuate the fruitful collaboration for the sake of the development of this corridor.

The achievements of CODE24, which are presented in this publication, will serve as an excellent groundwork for this prospective cooperation.

I'd like to thank the editors for having attended to this complex endeavor!

Mannheim, Germany — Ralph Schlusche

Ralph Schlusche After a degree in Jurisprudence at the University Mannheim and a post-graduate course in Management Sciences, he entered into the higher public service of the Land Baden-Württemberg.

In 2010 he was elected as Director of the Regional Planning Association Rhein-Neckar. He also acts as Chief Executive Officer at the Metropolregion Rhein-Neckar GmbH and is Vice-Speaker of the "Initiativkreis Europäische Metropolregionen in Deutschland". (450 Zeichen)

Contents

Introduction

From Sea to Sea: Changing a Trade Route and Transit Area into a Common European Living Space

Hansjörg Drewello and Bernd Scholl

Abstract The Rhine axis and the adjoining trade routes southwards over the Alps to the Mediterranean are among the most important transport routes in Europe. The development and significance of this trading route has inspired the sciences as well as current policy-making. The French geographer Roger Brunet explains that this link has arisen from a fundamental 'north-south dissymmetry' of the European cultural landscape, starting as early as Roman times: the evil barbarians in the north facing the civilized cultures of the Mediterranean states. In economic terms, this inequality was manifested in the range of different resources: in the north amber, wool and wood and in the south spices, silk and precious stones, but also bronze weapons, often imported from Asia or the Near East. The resulting exchange of goods, technologies and culture began in the Bronze Age and was intensified up until the twelfth century CE on the shortest route between the Mediterranean and the North Sea. Between today's North Sea ports of Antwerp, Rotterdam and Amsterdam and the ports of the Ligurian coast in northern Italy, one finds the highest concentrations of settlement activity and population, wealth, infrastructure and traffic in Europe, today called the Rhine–Alpine Corridor. This article illustrates the approach of the INTERREG project CODE24 dealing with direct negative consequences of the economic strength of the corridor, such as rising land prices, increased pollution, formidable traffic problems and further urban sprawl outside outside of the core cities.

The research done on the development opportunities of European cities by a group of French geographers under the direction of Brunet, the Réseau d'étude des changements dans les localisations et les unités spatiales (RECLUS), placed the core area of European economic power at the end of the 1980s in the centre of scientific discourse (RECLUS 1989). The study warned of growth deficits in

H. Drewello (✉)
University of Applied Sciences Kehl, Kinzigallee 1, 77694 Kehl am Rhein, Germany
e-mail: drewello@hs-kehl.de

B. Scholl
ETH Zürich, Institute for Spatial and Landscape Development, Zurich, Switzerland
e-mail: bscholl@ethz.ch

H. Drewello, B. Scholl (eds.), *Integrated Spatial and Transport Infrastructure Development*, Contributions to Economics, DOI 10.1007/978-3-319-15708-5_1

France, as it is not connected to this central growth axis. In the discussion of the results of the study in the French press and the political arena, the term *Blue Banana* began to be used for this area, which extends from London via Randstad Holland, Brussels, Rhine-Ruhr, Rhine-Main, Rhine-Neckar, the Upper Rhine, Basel and Zurich to northern Italy and the cities of Milan, Turin und Genoa. Brunet explains that this important trading link has arisen from a fundamental 'north-south dissymmetry' of the European cultural landscape, starting as early as Roman times: the evil barbarians in the north facing the civilized cultures or the Mediterranean states (Brunet 2002: 15). Other studies as well acknowledged that this region, with names such as European Megalopolis, Manchester-Milan Axis or European Backbone, as the region with Europe's greatest economic potential (Schätzl 1993; Delamaide 1994). Then another point of discussion was whether, in future, the Blue Banana could possibly lose its dominant position in Europe to the European Mediterranean coast (European Sunbelt) or to the regional axis Cologne-Berlin-Warsaw (yellow banana) (Lambooy 1994; Erzner 1999).

More than a fifth of EU citizens, about 110 million people, live in the Blue Banana region. Up to today, a loss of economic prosperity has not been discernible. A comparison of the countries involved reveals that in each case the regions lying in the blue banana exhibit high productivity and low unemployment.

In addition to these economic strengths, one must also acknowledge some direct negative consequences, such as rising land prices, increased pollution, formidable traffic problems and further urban sprawl outside of the core cities. All of this has sparked diverse political activities and the construction of new infrastructures. In 1992, the citizens of Switzerland voted for the NRLA concept (New Railway Link through the Alps) and the corresponding financing plan by a clear majority. In 1994, the first measures were taken to construct the 34.6 km long Lötschberg Base Tunnel in Switzerland, which became operational in 2007. The opening of the 57 km long Gotthard Base Tunnel is planned for December 2016 and the 15.4 km long Ceneri Base Tunnel is scheduled to open at the end of 2019. Thus, capacities for transalpine goods traffic have been created, but up to now neither the German nor the Italian side have been able to utilise this possibility with their existing access routes. In 1996, the Federal Republic of Germany and Switzerland signed a bilateral agreement, the Lugano Convention, to expand the capacity of the northern railway connection to NRLA between Karlsruhe and Basel to four tracks, to keep pace with traffic demand.

In this context, the upgrading of the northern part of the traffic corridor in the Netherlands is worth mentioning. In 2007, a strictly freight route was put into operation in order to improve transport of the increasing quantities of goods handled in the Port of Rotterdam to the European hinterland. This Betuwe line connects Rotterdam with the city of Emmerich on the Dutch-German border, a distance of 160 km. It increases the capacity pressure on the corridor from the north.

Against a background of existing traffic problems, the Association of Chambers of Industry and Commerce on the North-South Railway Link Rotterdam–Genoa, (Communauté des Chambres de commerce de l'axe ferroviaire nord-sud Rotterdam–Gênes) was established in 2004 with Italian, Swiss and German chambers of industry and commerce. The association is committed to the expansion of the rail link between Genoa and Rotterdam. At the same time, informal meetings took place at Karlsruhe University, Dept. of Urban and Regional Planning, with experts in regional and railway planning in Germany and Switzerland.

In 2007, the Association hosted a conference: The AlpTransit Rotterdam–Genoa: Its Significance for Europe, in Karlsruhe. There the Karlsruhe Declaration was adopted, in which 87 European chambers of industry and commerce called for the rapid expansion of the rail link, especially the access routes to both of the large Swiss tunnels. The first contact between the chambers of industry and commerce and the Swiss Federal Institute of Technology Zurich (ETH) took place at this conference, which ultimately resulted in the Corridor Development Rotterdam–Genoa INTERREG Project (CODE24). The determining factor for this decision was a visit by members of the Rhine-Neckar Regional Association to the construction site of the Gotthard Base Tunnel in Sedrun in 2007 and a subsequent visit to ETH Zurich.

Parallel to the activities of economic players, planning associations along the traffic corridor also developed diverse initiatives (Saalbach 2012: 441). In 2007, in a joint position paper, they called for a harmonisation and coordination of national requirement planning, the acceleration of the planning process, a reduction in traffic noise as well as regular and systematic assessments of the situation (Rhine-Neckar Regional Association VRRN 2008), among other things. A further impetus for the CODE24 project came from the position paper by the Regional Studies and Planning Academy of the ARL (2009) on the future challenges of large-scale traffic development. Here the north-south railway link for Europe was used as an example to show that 'these problems can only be solved by an infrastructure policy which is carried out across regions and modes of transport, and which serves to integrate traffic and regional development.'

In 2010, the CODE24 project began under the auspices of the Rhine-Neckar Regional Association, starting with 15 partners, which later grew to 18 partners. The partnership comprises four organisation types: (1) regional planning bodies and local authorities, (2) universities and research institutes, (3) port authorities, and (4) private engineering firms. The funding came from the EU programme INTERREG IVB Northwestern Europe.

Three project partners from Italy have also been involved, although the country is not in the programme region. This exception, involving the expansion of the region, was possible due to the flexibility rule, which is provided for by EU regulation and the funding programme if good reasons can be shown. In particular, this applies to designated corridors throughout the EU that extend beyond the individual funding areas. ETH Zurich is also involved as a partner from Switzerland. Co-financing for the ETH was ensured through the participation of the Swiss Federal Office of Spatial Development (ARE) and the Gotthard Committee (Fig. 1).

The main objective of this EU project is to coordinate regional, national and international traffic development of the most important European north-south corridor between the North Sea and the Mediterranean (Rhine–Alpine Corridor). It can be aptly described with the slogan: One corridor–one strategy (Saalbach

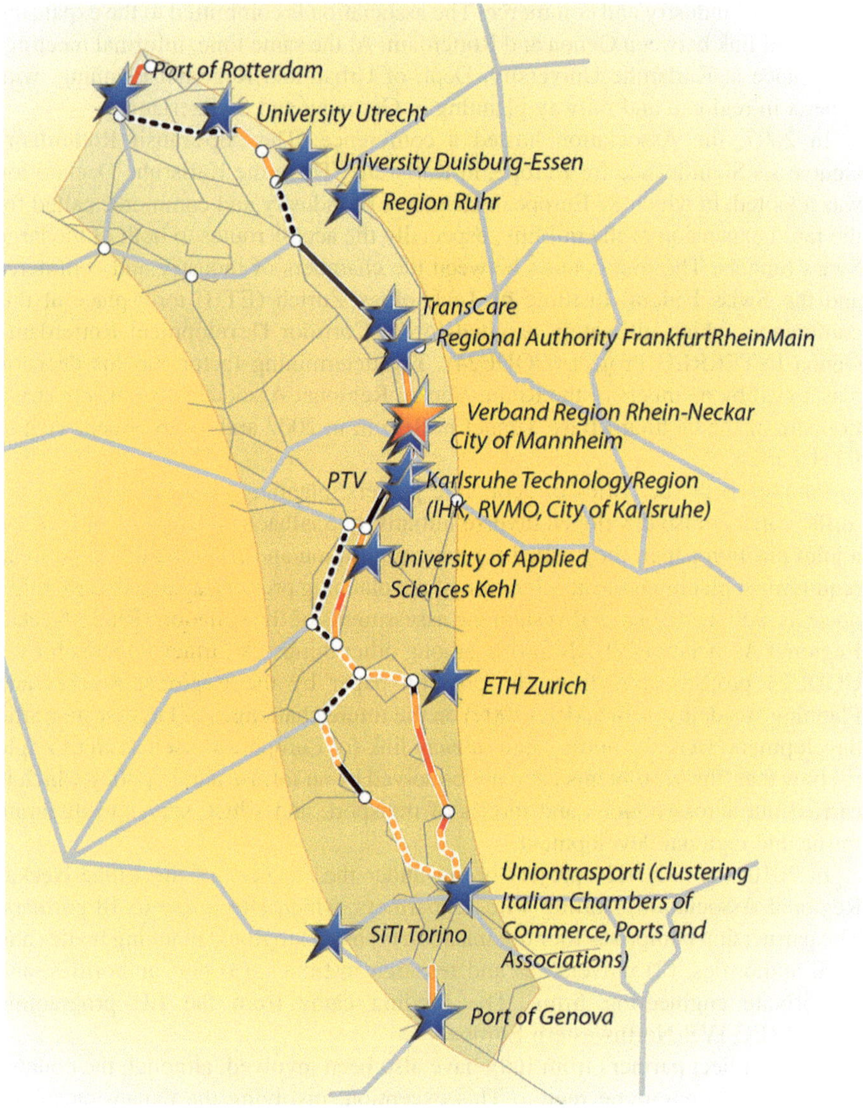

Fig. 1 Location of the partners. *Source:* Project map CODE 24

2011: 34–37). With the project, the partners would like to help to further intensify the discussion about the expansion of the entire railway link, especially the northern and southern access routes leading to the Alpine crossing, as well as examine the specifics in bottleneck areas. Regions were examined that are expected to experience increasing traffic bottlenecks, such as the German-Dutch border region at Emmerich, the Rhine-Main region, the Upper Rhine and northern Italy. With the interdisciplinary project approach, it has become clear that what is important for the partner nations, in addition to the increase in economic performance, is the minimisation of the negative effects on the environment and the population, above all with regard to goods transport and logistics networks. The various studies want to establish a traffic and noise management system on the railway that takes the development of the railway system as well as sustainable regional planning into account.

The studies were carried out in four study segments:

- Study segment 1: Coordinated regional and infrastructure development
- Study segment 2: Environmental aspects and noise reduction
- Study segment 3: Goods transport and logistics
- Study segment 4: Communication, acceptance and continuation of interregional cooperation

In addition, the project was used as a platform to bring together the partners most affected by this topic. Thus, processes and joint initiatives could be set in motion. Through its practical approach, the project contributed to a better integration of economic promotion, logistics, traffic and regional planning. Another important goal of the project was the development of sustainable cooperation in the corridor area. The basis for this has been created with the continuation of the cooperation within the framework of a European Association for Territorial Cooperation (EVTZ), which was established in November 2014.

The editors of this book have attempted to explain and illustrate the issues, proposals and progress made, as well as the interdisciplinary and cross-cultural approach necessary for a sustainable, efficient and generally accepted design and development of the Rhine–Alpine traffic corridor and living space.

Part I illustrates the framework and challenges of developing the Rhine–Alpine Corridor from a heavily used transit area to a sustainably designable living space of national and European significance. In his article, Bernd Scholl shows that regional planning and development must prove to be of value on a European scale in order to achieve integrated solutions. Approaches and experiences with this topic exist in the individual countries. Large-scale cooperation could be tested within the framework of CODE24, just as in a laboratory. At the same time, the ability to present arguments from a regional planning perspective plays a decisive role. In his article, Rolf Signer presents important basic principles and insights into this aspect, as logically consistent argumentation leads to important foundations for decision-making. In her article, Isabella Lami deals with decision-making in conflict situations in regional planning. Peter Endemann and Birgit Simon present the challenges of the Rhine–Main area, which has always been a first-rate European transportation hub due to its central location. Finally, Tejo Spit and Patrick Witt examine the challenges from a Dutch perspective.

Part II focuses on articles dealing with the topic of regional accessibility. With the CODE24 model developed during the project, Hansjörg Drewello, Marcel Huschebeck and Norbert Schick simulate the effects of a shift from rail-to-road caused by bottlenecks in goods transport. The article by Alessandro Africani et al. presents a qualitative method of regional analysis for identifying deficits in regional accessibility, using the example of the 'Technology Region' of Karlsruhe and the region of Liguria. In their article, Mauricio Arnone et al. show how regional accessibility can be increased in regions not directly connected to a high-speed network by an intelligent link-up with existing light rail lines, without building new infrastructures.

Efficiently organised logistics processes create new transport capacities. This idea is the focus of the articles in Part III. With the help of case studies, Julian Brenienek and Rudolf Juchelka describe the development of logistics clusters along the Rhine–Alpine Corridor. The article 'Linking the ports to the hinterland' by Alessandro Africani et al. shows, with the help of a newly developed port-performance indicator, the strengths and weaknesses of seaports in supplying their hinterland with goods. In his article, Peter Endemann describes a possible way of better utilising unused transport capacity in freight trains. An online exchange for freight shipping services, which was developed in the project, can improve the coordination of the supply and demand of freight shipped by rail. The inland ports play an important role in the Rhine–Alpine Corridor. Using two case studies, Nina Marzioch, Manfred Rausch and Kerstin Ruppenthal analyse synergies arising from cooperation among the ports. The last article of the chapter by Markus Ibert and Daniel Halter provides information about the results of a regional economic feasibility study for a logistics service centre at the Black Forest Airport in Lahr.

Increasing freight traffic creates disruptive effects on the environment. Integrated, sustainable traffic planning must take these effects into account. In Part IV, Sebastian Wilske analyses different approaches to noise protection with the aim of increasing the acceptance of traffic infrastructure projects in the population. Regulations on conservation and environmental protection often seem to be contrary to the goals of traffic planning. Claus Peinemann uses examples to show how development and infrastructure projects in the Rhine–Alpine Corridor can be brought in line with national and European regulations.

The focus of the last chapter is how to establish acceptance of European infrastructure projects through participation. Using a case study, Frank Joneit describes the involvement of regional lobbyists in an informal planning process in the Wesel area. With the help of the economic theory of federalism, Hansjörg Drewello shows that one can achieve better economic efficiency through more effective involvement of regional players in planning processes for a European infrastructure. The European Association of Territorial Cooperation is still a new legal form of cooperation for regional authorities in Europe. Finally, Jörg Saalbach portrays how this construct should ensure the future representation of the interests of regional authorities in the Rhine–Alpine Corridor.

As editors, we hope that the new ideas discussed in this book can make a small contribution towards coping with the emerging challenges in European regional and traffic infrastructure policy. On this note, we hope the book will give the reader much inspiration and food for thought. And as the American composer John Cage once said: 'I can't understand why people are frightened of new ideas. I'm frightened of the old ones.' (Kostelanetz 1988).

References

Brunet R (2002) Lignes de force de l'espace européen. Lappemonde 66(2):14–19
Delamaide D (1994) The new superregions of Europe. Dutton, New York, NY
Erzner F (ed) (1999) Wirtschaftsgeographie. Cornelsen, Berlin
Kostelanetz R (1988) Conversing with Cage. Limelight, New York, NY
Lambooy JG (1994) Structure and transformation: regional capitals. In: Kooij P, Pellenbarg P (eds) Regional capitals: past, present, prospects. Van Gorcum, Assen
ARL Akademie für Raumforschung und Landesplanung (2009) Künftige Herausforderungen der großräumigen Verkehrsentwicklung. ARL Position Paper, No. 49, Hannover
RECLUS (1989) Les villes europeénnes. In: Rapport pour la DATAR. RECLUS, Montpellier
Saalbach J (2011) CODE24: one corridor–one strategy! Eur Railway Rev 2011(2):34–37
Saalbach J (2012) Die europäische Entwicklungsachse Rotterdam—Mannheim—Genua. Informationen zur Raumentwicklung Rev 2012(7/8):439–450
Schätzl L (ed) (1993) Wirtschaftsgeographie der Europäischen Gemeinschaft. UTB, Paderborn
VRRN Verband Region Rhein-Neckar (2008) Positionspapier der Regionalverbände zur Nord-Süd-Transversale Rotterdam–Genua. Mannheim

Since 2009 **Hansjörg Drewello** has been a professor of economics, first at the University of Applied Sciences Göttingen and now at the University of Applied Sciences Kehl. He is the German director of the European Competence and Research Centre Cluster Management. His teaching and research deal with regional economics, in particular infrastructure development and the regional economic impact of cluster initiatives in a cross-border context. Until 2009 Hansjörg Drewello was the director of the Department of International Commerce, Transportation, Commerce and Tourism at the Chamber of Commerce and Industry Karlsruhe and a member of its management board. He was the lead partner of the study segment "Freight transport and logistics" of CODE24.

Bernd Scholl is since 2006 a full professor for Spatial Planning and Development at the ETH Zürich. His professorship is part of the Institute for Spatial and Landscape Development at the ETH Zurich. He is director of the institute and Delegate (Dean) of the department for the Master of Advanced Studies in Spatial Planning (MAS RP). His teaching and research focal points are on land and spatial management in the local and regional development, space and infrastructure development, transnational tasks as well as development and organization of innovative planning processes and methods in spatial planning and regional development. In 2010 he initiated the Interreg-Project Corridor development Rotterdam/Antwerp-Genova (CODE24).

Part I
Developing the Transport Corridor Rotterdam-Genoa Within Spaces of National and European Importance

Spatial Planning and Development in a European and Macro-regional Context

Bernd Scholl

Abstract Throughout the entire history of Europe and its spatial development, the north-south corridor from Rotterdam to Genoa has formed a quasi 'transport backbone'. The most densely settled areas of Europe are to be found here, together with diverse landscapes and urban regions of various sizes. It is also the area where the highest economic value is created. To a considerable extent, the economic power of this axis determines many European developments. Major investments of all kinds: airports, railway systems, distribution centres, industries, etc., are concentrated here.

The renovation and development of the regions along this corridor are a challenge for those involved. Global, European, national, regional, and local processes overlap each other and changes are plain to see. The serious conflicts within spatial development in this transit area also inevitably lead to addressing Europe's cultural and political history, as well as its future and the future of its nations. What are the central questions, and what solutions are possible? In recent years, stakeholders from all the countries involved have formed, for the first time, a group for developing a spatial strategy in a European and macro-regional context. This contribution deals with the challenges and perspectives of one of the most important corridors in Europe and the need for an action-oriented European spatial programme.

1 Introduction

Up to now, spatial planning and development in Europe and the European Union has been seen as a national task. This contribution will not question this task distribution, conducted on the basis of the subsidiarity[1] principle. On the contrary,

[1] The principle of subsidiarity is defined in Article 5 of the Treaty on European Union. It ensures that decisions are taken as closely as possible to the citizen and that constant checks are made to

B. Scholl (✉)
ETH Zürich, Institute for Spatial and Landscape Development, Zurich, Switzerland
e-mail: bscholl@ethz.ch

H. Drewello, B. Scholl (eds.), *Integrated Spatial and Transport Infrastructure Development*, Contributions to Economics, DOI 10.1007/978-3-319-15708-5_2

this contribution wants to promote a strengthened collaboration between the various governance levels and spatially relevant actors in a European context.

1.1 European Spatial Development from Concept to Territorial Agenda

This is in keeping with the Territorial Agenda 2020 (TA 2020), published by the EU, in which it states: '. . .to provide strategic orientation for territorial development, to foster integration of territorial dimensions within different policies at all governance levels and to ensure the implementation of the Europe 2020 Strategy according to the territorial cohesion principle' (European Union 2011).

The global economic crisis triggered by the USA in 2008 has affected the countries of Europe to various degrees. Crisis times are also times of opportunity. They should be used to strengthen the solidarity in Europe through special initiatives and make economic prosperity a possibility for European nations and their inhabitants, thereby contributing to the protection of peace. These goals could be achieved, especially if the national and European institutions can manage to inspire the greatest number of actors possible to embrace European thinking and to motivate them to take up their own initiatives. This has succeeded, specifically through solving problems that lay in the common interest, meaning tasks in which it has been recognised that they were not simply in the interests of nearby local, regional or national levels, but also in the European interest.

Spatial planning has a special responsibility here: planning's central task is to bring a frugal approach to the use of land, a limited resource. The resulting mandate, in coordination with sector planning, is to effect an integrated sustainable spatial development—a challenging task. This is being practised in different European countries in different ways and with various levels of success, because the approach to land, and therefore to spatial planning, is connected to a considerable extent with the language and culture of the country.[2]

---

verify that action at Union level is justified in light of the possibilities available at national, regional or local level. Specifically, it is the principle whereby the Union does not take action (except in the areas that fall within its exclusive competence), unless it is more effective than action taken at national, regional or local level. It is closely bound up with the principle of proportionality, which requires that any action by the Union should not go beyond what is necessary to achieve the objectives of the Treaties. See: europa.eu/legislation_summaries/glossary/subsidiarity; Accessed 14.11.22.

[2] The newest publication on the topic, *Spatial Planning Systems and Practice in Europe,* also makes this clear. The publication is the outcome of an ARL European Working Group that devoted their time to this topic over a 3-year period. 'Although the emergence of the EU spatial planning agenda played an influential role as a driving force in domestic planning reforms, this did not lead to a complete 'harmonisation' of spatial planning systems and practises across Europe,' (Reimer et al. 2014). As Switzerland was not included in this analysis it may be of interest to be aware of the differences. Its understanding of planning is clearly differentiated from its neighbours. One of the

The Territorial Agenda 2020 clarified that the approach to land issues should be an action-oriented[3] political framework. One conclusion is that spatial planning must also be action-oriented.[4] In an action-oriented process, important tasks are revealed through mutual treatments and decisions, and solutions are introduced within the framework of well-considered and well thought-out processes. Deviations of these spatial development tasks from their usual form will also be required informally in the completion of the formal spatial planning procedures and instruments. Time-limited, customised processes are suitable for identifying strategic tasks and for the application of solutions that can then be implemented with the usual methods and legislative foundations of the respective countries and institutions.

The results of spatial planning seldom materialise in the short term because the results of spatial planning activities can often take many years or even decades to complete, not only due to the length of time needed for actors and planners to reach a decision, but also for the measures to pass through all the administrative levels of implementation. Unfortunately, undesirable elements can creep in or are only recognised as the process unfolds. The space itself is 'inert.' Many developments proceed slowly, for example, sprawl spreading into more and more agricultural land. 'Wrong' results are mostly, if at all, noticed after a long time period. The corrections that follow likewise stretch over a longer timeframe (see Sect. 2 on the time aspects of spatial planning) and can also produce unwanted effects.

One burden of long-term planning is that it seems to be heavier and more cumbersome in comparison to results effected by the electronic media, especially the Internet, where the wheel appears to turn faster and the flood of information it brings make it more and more difficult to distinguish the essential from the non-essential. In addition, many political actors, who are subject to the rhythm of election periods, do not see spatial planning as an attractive sphere of activity.

---

main differences is that Swiss planning is understood as a 'rolling' process as indicated by the *Richtplan* (Guidelines) in which the categories of Pre-orientation (Vorientierung), Interim Results (Zwischenergebnisse) and Final Assessment (Festsetzung) of the various plans must be represented, including a final representation of the desired state. The pragmatic approach used in Switzerland recognises that spatial planning must also be intensely concerned with what could be important for the coordination (Pre-orientation) and the representation which, until now, has shown what solutions have been achieved through planning (Interim Report) (Zwischenergebnisse) and, finally, what direction the solution should follow (Final Assessment). These categories deliver an overview that is important for the coordination of an effective process and its decisions.

[3] The ministers responsible for spatial organisation and territorial development state that: 'The TA 2020 is our action-oriented policy framework to promote territorial cohesion in Europe as a new goal of the European Union introduced by the Treaty of Lisbon (Art 3. TEU).'

[4] In my work on action-planning, the use of the essential principles and methods of action-oriented spatial planning to solve complex core tasks was presented. The knowledge is based on the evaluation of comprehensive complex tasks in which the author was also involved.

This should not discourage actors active in spatial planning, however, spatial planning can be strengthened, especially through well-founded initiatives[5] and successful examples extending to other European connections as well. In this way, initiatives of the various EU presidencies and of the minister responsible for spatial planning have brought the topic of territorial cohesion to a dialogue between cities, which led to the passage of the European Spatial Development Concept (ESDP) in Potsdam in 1999.[6] Further initiatives came from this foundation, which led to the first generation of the transnational INTERREG Cooperation Programme, among others, and the establishment of a European Spatial Observation Network (ESPON) (European Commission 2008).

In 2004, an intensive debate on European spatial development started under the Dutch EU presidency against the background of EU expansion and the contributions of spatial development to the Lisbon Strategy. Results of this, as well as the Rotterdam processes, were on the Territorial Agenda in 2007 '. . . for a competitive sustainable Europe of diverse multi-facetted regions'. A Green Paper appeared in 2008 on the appraisal and intensification of the debate on territorial cohesion: *Turning Territorial Diversity into Strength* (European Commission 2008).

In order to convey the importance of an expanded Europe and the states' noteworthy contributions to this, the German EU presidency passed the Leipzig Charter on the sustainable state in 2007. Under the Hungarian presidency, the Territorial Agenda was finally further developed and passed in 2011 with the title *Territorial Agenda 2020: For an Integrated, Intelligent and Sustainable Europe of Diverse Regions* by the ministers responsible for spatial organisation and territorial development. This was followed in mid-2014 by the *Sixth Status Report on Economic, Social and Territorial Cohesion* (European Commission 2014a).

[5] Worth mentioning in this connection is the *Landschaftsinitiative* of Switzerland. It led to an intensive public debate about spatial planning and to the revision of the spatial planning law. A central goal is limiting sprawl through the implementation of the basic strategy: Redevelopment before New Development. The law was approved by over 60 % of the voting population in Spring 2013.

[6] ESDP 1999: 'With the aim to provide an integrated, multi-sector and indicative strategy for spatial development, the key ideas of ESDP are: To achieve an integrated approach, i.e. not just look at specific sectors of development activity, e.g. environment, economic development, or transport, but to recognise that they all affect each other; an understanding of spatial development, a much wider view of sector development vital for an integrative approach; to include strategic actions, which means also to set priorities.' Of interest in this connection are also the spatial policy guidelines, summarised here:

- Development of a polycentric and balanced urban system, and strengthening of the partnership between urban and rural areas, in order to create a new urban-rural relationship.
- Promotion of integrated transport and communication concepts, which support the polycentric development of the EU territory, so that there is gradual progress towards parity of access to infrastructure and knowledge.
- Wise management of the natural and cultural heritage, which will help conserve regional identities and cultural diversity in the face of globalisation.

The policy and guidelines appear in a different form in the Territorial Agendas of 2007 and 2020.

The Territorial Agenda mainly follows the goal of using European regions potentially capable of economic growth. Less in focus is a large-scale regional cooperation, which is desirable for transit spaces of European importance. The macro-regional strategies of the EU have moved in this direction since 2011, similar to that for the Danube area or the Baltic region. The collaborations along the Main East-West Railway (Paris–Budapest), which has merged into the TEN-T core network, have also delivered important experiences.

1.2 The Need for Action- and *Problem-Oriented Spatial Development*

All in all, the last decade has seen an intense debate take place about the basic concepts and programmes of European spatial development. Anyone in the respective countries involved with the practice of spatial planning will also discover that spatial planning argumentation in concrete local, regional, or national plans hardly plays a role in the European connection. This also relates to the fact that many plans of European importance involve sectors and small spaces as well as less problem-oriented areas. In a real sense, an action- *and* problem-oriented European spatial development does not exist.[7] This was confirmed by a multi-year exchange among high-ranking European experts from the science and practice of regional spatial development.[8] In the opinion of the author, this is required in order to attain a sustainable European spatial development.

To bring the coordination of large-scale infrastructures, e.g. in the branches of energy and transport, in line with the desired spatial development of affected cities and regions is one of the central tasks (METREX 2013). In particular, narrow

[7] I use 'difficult, challenging tasks' as synonymous with complex problems. They differentiate themselves from routine tasks in that such tasks don't fit into commonly used solution patterns and sometimes it is not at all clear at the beginning of the clarification process, which tasks are to be solved in general.

[8] Scholl 2012: In the summary it was stated: 'Once more, it has been confirmed that spatial planning is a discipline that is linked to language, culture and paradigms of thinking. For these reasons alone, mutual understanding is already a major challenge. Nevertheless, the cross-border dialogue is important in order to recognise new developments, attain new knowledge and stimulate and encourage through discussions—perhaps even formulate new initiatives. Through the exchange that took place during the symposia and the discussions of the case studies, the knowledge that a pan-European spatial planning does not exist has become a consensus. The increased development of trans-European infrastructural networks, for example, the energy sector (such as cable and pipeline routes) and the transport sector already call for this development. This shows that sector development is very often neglected during the customary development of the living space. A compilation of the spaces and projects of national importance in the individual countries would be a very good foundation for the spaces and projects that lie within mutual European interests'.

reciprocal effects persist between transport development and settlement development, even on a large-scale (Initiativkreis Europäische Metropolregionen 2007). Spatial and transport development are two sides of the same coin in the effort to bring about the sustainable development of common living spaces.

An action-oriented European spatial development should start from tasks that are concrete, strategically important and lie in the common European interest. According to the subsidiarity[9] principle, it is incumbent upon spatial planning and the responsible actors to implement the decisions taken and the appropriate processes.

On the regional and national levels, tasks bound up with spatial development are already challenging enough, and, without a clear strategy for spatial and transport development that sets authentic priorities and provides planning security through secure financing, it becomes nearly unmanageable.[10]

However, this deficiency cannot be used as a reason to ignore the European dimension of a required collaboration across several national boundaries. When, for example, major investments in the harbours of northwest Europe are active in order to increase their handling capacity, then the transport level in the nearby hinterlands, both inside and outside of the settlement spaces, will need to be managed. If in the areas of the transport infrastructures, the corresponding provisions are not met, then increasing traffic jams on the highways and freeways will be the result. If the quality level for maintaining connections for entire regions grows worse, this will also have negative consequences for the economic development of the affected regions. Even the welcome shift of any amount of goods transport from the roads to the rails, as stipulated in the EU's *White Paper on Transport* (2011) and most of the national planning documents, hide some risks: insufficient rail capacity could displace railway passenger transport, thus once again reducing the quality of public transport railway connections, which in turn plays a central role in the containment of sprawl. The catchment areas of the railway stations and stops are a crystallisation point for the philosophy of redeveloping settlements first.

A capable, attractive public transport system is of great importance for a redevelopment-oriented settlement strategy because it would allow the basic strategy of sustainable spatial development to be realised. It builds upon using the reserves of the settlement's building stock for development before any new 'green fields' are re-zoned. This basic strategy is not only standard for countries such as Switzerland or Germany, but for all European countries that want to make sustainable spatial development a reality. This argumentation makes it clear that an

[9] IRL 2011: 'According to the subsidiarity principle, tasks should only be transferred to a superordinate state level when it can demonstrate that it can fulfil the task as well as the subordinated state levels.'

[10] ARL *Position Paper No. 79*, 2009: In this paper on the 'future challenges of the large-scale transport development in Germany', it was stated: 'In addition, it needs a clear commitment to the prioritisation of planning preparation as well as the implementation of concepts on the main axes. Correspondingly, strategic approaches and considerations have not been, however, on hand for a long time. And not on the side of the government either.'

effective interplay between transport development and spatial development on a European scale, namely for the entire European space, is of extraordinary, as well as general, European interest. Obviously, if value creation declines in strongly performing regions because of uncoordinated and non-sustainable spatial development, then there will necessarily be fewer means available for the economically weaker regions of Europe.

Moreover, the argument that joint European spatial development would first require harmonising European legislation is not convincing. A European comparison of spatial planning laws and their connected instruments and processes would certainly show just how different these are from one another. This also supports the idea that the ways and means of how spatial planning is conceived and conducted are shaped, as already mentioned, by pattern of thinking, culture and language. Given that an action-oriented European spatial development should start from the tasks to be jointly solved, a common understanding of what tasks this actually concerns, what sort of time sequence is needed for their solution and how the financial means for it should be raised, then taking steps in this direction is urgent. One could call it setting priorities.

The necessary, current and future-compatible overviews are mostly missing. Generating them does not require a legally regulated formal process. Using informal, time-limited processes to achieve an expansion or completion of an overview should be sufficient (see 'Die Methode der Testplanung' in *Grundriss der Raumentwicklung,* ARL, Hannover). Within the framework of such processes, for example, agreements can be made and multi-lateral contracts can be prepared. Informal processes can also be used to find cross-border solutions for known tasks. The actors responsible can also bring potential solutions before the respective national and regional legislatures, much as this is carried out in current practice for spatially important tasks in the 'Triangle' area of Germany, France and Switzerland.[11]

For an effective collaboration and the further development of large-scale structures of European importance, jointly prepared ideas about the spatial development of the affected spaces are of central importance. Because the construction of such infrastructures takes years or decades, the timeframe for such ideas must also range over 30 or 40 years.[12] This is not an exceptional demand, because the life span of

[11] In connection with a 'bypass discussion' about a railway in the course of the planned four-track construction on the Upper Rhine River, the regions from Switzerland, Germany and France have found their way to cooperating on a plan for an integrated spatial and railway development. The results were published in Scholl (2007).

[12] At present, a very promising test trial is taking place in the Limmat Valley, an approximately 20 km long space of national importance between Zurich and West Baden. The report has recommendations from an international *Ideenkonkurrenz* (idea competition) that could be important for similar tasks. Published by ETH Zürich, Fall 2013.

such infrastructures stretches over a much longer timeframe, usually 60-to-80 years—or even longer.

An invaluable advantage for such important European collaboration projects is to be able to gain a common background on the problems to be solved through cooperation. Because of the cultural polymorphism of Europe, which is also its wealth, cooperation depends on creating projects of European importance in spaces of European importance. Through the collaboration on spatially important projects of European importance, the cohesion of Europe can also be strengthened. Discovering and disclosing the connections of spatially important activities in various areas of Europe are therefore the first step. Clarification of the tasks to be solved together, its financing and realisation are further steps. Of pivotal importance is that the connections, tasks, and solutions should be clearly communicated to the inhabitants of the affected regions.

In an enlightened pluralistic society, projects of European importance can only be realised with—and not against—its inhabitants. Spatial planning can play an important role when its actors manage to make the connections between transport development and spatial development generally understandable on both a large scale and in European standards, i.e. in everyday language, illustrated and defined for the search for integrated solutions, as the EU's Territorial Agenda 2020 requires.

1.3 TEN-T Projects as Pioneering Works

The projects of the Trans-European Network of the European Union (TEN-T) could serve as concrete examples of the above-mentioned types of problems. The intention is to produce high-capacity transport and communication infrastructures for a polycentric European network, as proclaimed in TA 2020 (Fig. 1).[13]

By 2030, the current gaps should be closed and bottlenecks, with co-financing from the TEN-T facilities (European Commission 2014b), should be eliminated.[14] To achieve this, the integration of the respective countries and regions in spatial development and spatial planning is urgently required. Many countries stipulate that the spatial coordination for projects, such as eliminating bottlenecks and closing gaps in the infrastructural network, must follow legal and formal processes.

[13] Under point 24 of the TA 2020, it emphasises that '...a polycentric and balanced spatial development in the EU is a key element for the achievement of territorial cohesion.'

[14] European Commission 2014a, p. 43. 'The TEN-T consists of two layers: a core network to be completed by 2030 and a comprehensive network feeding into the core network, to be completed by 2050. The core network will provide essential support for the Single Market by facilitating the flow of goods and people around the EU, including in the less developed Member States (Map 1.22). It involves connecting 94 main European ports to rail and road links, 38 key airports with rail connections into major cities, 15,000 km of railway lines upgraded to high-speed and 35 cross-border projects to reduce bottlenecks.'

Fig. 1 TEN-T Network. *Source*: TENtec, European Commission, Mobility and Transport, 2013

Not only is the agency of public issues important, so is the inclusion of non-governmental organisations (NGOs) and the population. Especially on questions of the environment and spatial planning, sensitive inhabitants reject large-scale infrastructure plans with great resistance, mainly in the areas of transport and energy transfer. For example, the planned four-track construction for the railway section from Offenburg to Basel was met with resistance from the population, which did not accept the planned route. Likewise, the construction of the section between Frankfurt and Mannheim (beginning of 2000) was not accepted because the needs of the region and the cities were not sufficiently respected.

In Switzerland as well, a new solution had to be found for the approach routes to the Gotthard Base Tunnel, which forms the core of the most important European north–south railway connection. With a planned opening at the end of 2016, there was a distinct lack of clarity about the construction of the approach routes in the northern and southern areas. For example, in Felderboden, another tunnel was under consideration between the Axis Tunnel and the planned Urmiberg Tunnel, possibly to avoid a challenging coordination of the spatial development on the surface. However, it turned out that another tunnel was not feasible for a variety of reasons and a new solution was found (see chapter on Integrated Spatial and Railway Development: Case Study of One of the Hot Spots in the Canton of Schywz—Felderboden (Scholl 2014).

1.4 Integrated Corridor Development Approach

The Felderboden 'crisis' also provided an opportunity for the author to get intensively involved with the corridor from Rotterdam/Antwerp to Genoa, formerly designated as Corridor 24, it is now in operation as the Rhine–Alpine Corridor (Fig. 2).

Using this example of spatial and transport development in the most important north-south corridor of the EU, it can be exemplarily demonstrated which connections and conflicts between spatial and transport development are to be observed on a large-regional scale and how the tasks to be solved can be identified and solved in steps. In a 4-year process, important actors along the Rhine–Alpine Corridor worked within the framework of a EU INTERREG project. Results and interim results were reported in detail in several publications (see Scholl 2011a, d, 2012, 2014).

1.5 Structure of this Contribution

This contribution concentrates on the underlying paradigms and starting points of spatial development, as well as its applied methods. Section 2 goes deeper into action- and problem-oriented spatial development to find a basic underlying understanding of spatial planning, while Sect. 3 presents considerations and deliberations on the importance and necessity of spatial planning in a European and large-scale context. This involves the trade routes of European importance, especially within the framework of an intensive goods exchange. As a result of technical advances, such spaces can mutate into transit spaces, which then causes the resident population to lose control over the increasing amount of traffic through their living space with all its negative after-effects.

The newest example is the resistance to the planned tunnel connection under Fehmarn Sound on the southern approach from Germany to Denmark (*Neue Zürcher Zeitung* 2014). Selected findings on the focal points of spatial development in the Rhine–Alpine Corridor will be introduced in Sect. 4 of this contribution and conclusions concerning its transferability are drawn in Sect. 5.

Continental transit spaces have come into being over the course of the century and they are per se 'Spaces of European and National Importance' (see chapter "Introduction", Scholl 2012). These have proven to be of very special importance for both a cultural exchange and economic growth within Europe and beyond and they will continue to be so in the future. A major challenge exists therein to bring the anticipated growth in traffic, mainly in goods transport, in balance with the needs of the people in the settlement areas. This is especially clear in the Rhine–Alpine Corridor, which is in the most densely settled area of Europe.

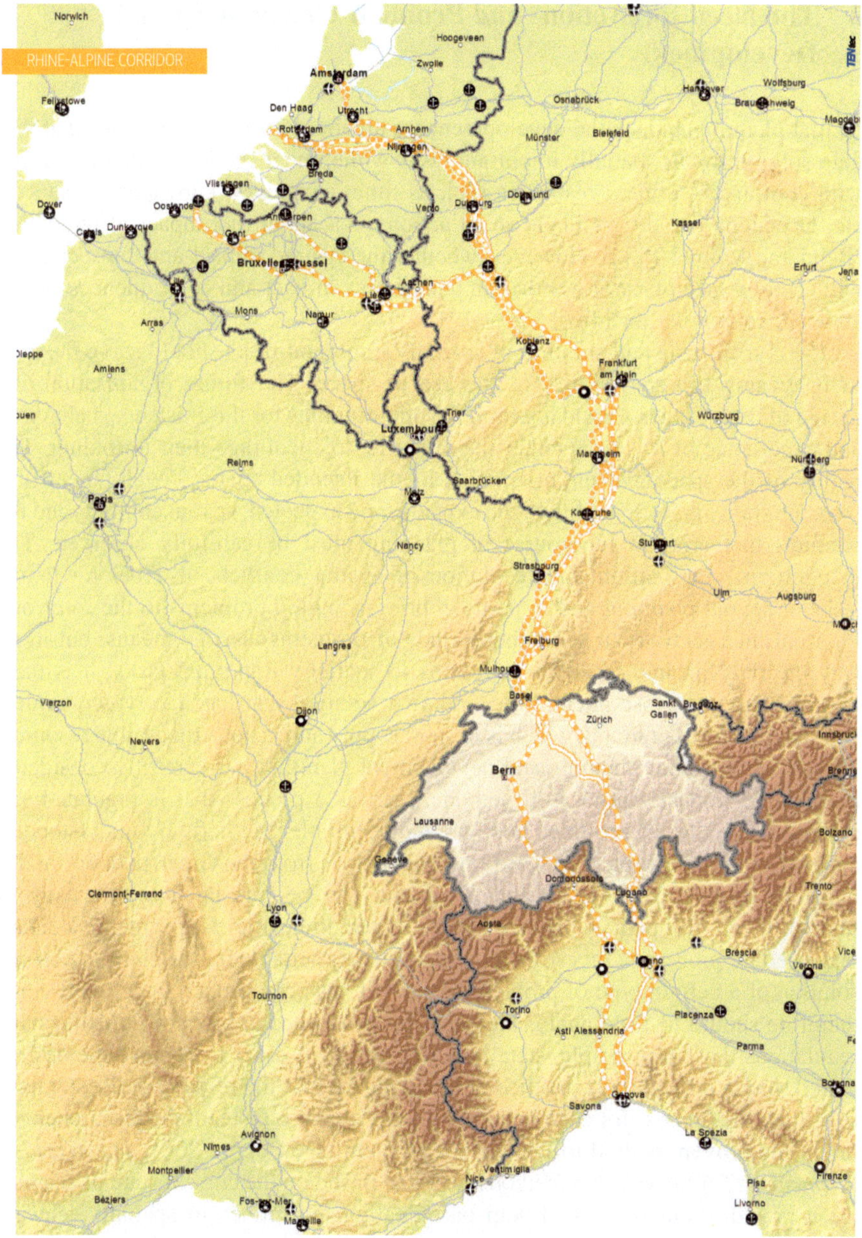

Fig. 2 Rhine-Alpine Corridor/former CODE24. *Source*: European Commission, Mobility and Transport, 2014

2 The Need for Action- *and* Problem-Oriented Spatial Development

Spatial planning and spatial development are all about initiating decisions that will help solve difficult spatially important tasks while under the restraints of working with continually limited resources such as time, personnel, knowledge, finances, etc. Decisions are always made about actions, not about the impact of the action (Behn and Vaupel 1982). However, whether the desired impact and consequences will actually take place or whether an undesirable impact and consequences can be avoided, cannot be 'decided'.

The clarification and solution of spatially important tasks should serve the goal of increasing the possibilities of the use of space. The future organisation and design of living space should not limit the opportunities for the existence, behaviour and experience of its users, rather the opposite, to encourage their unfolding. The reality of the space and the orientation of the intended spatial development thus form the framework. In addition, the existing, and expected, spatial conflicts and the conflicts that arise in the course of planning must be carefully observed. The recognition, clarification and resolution of spatial conflicts also belong to the tasks of spatial planning and in part can bring complex problems to the forefront. Physical interventions to solve these kinds of problems are one means, but in no way the only means. In all the solutions to spatially important tasks, it is most important to pay close attention to finding a solution that has a very economical approach to the use of land, which is in short supply and is not a renewable resource. To achieve a spatial planning and development plan that has carefully considered its possible impact, purposeful methods, such as a process that is regulated and committed to a pre-determined goal, are indispensable. Methods are understood as a means of investigation. The word 'method' comes from the Greek μέθοδος and is made up of two parts: μετά and ὀδός, and can be translated as 'the way of searching' or 'the path of investigation'. In our linguistic usage, method means that a certain process of investigation is particularly suitable for the clarification and solution of a certain type of problem because it has been proven in actual use and therefore should be used again. Methods can be defined as a set of guidelines and regulations that, in principle, state how to proceed in order to achieve the desired goal. Using a proven 'best practice' method means that the researcher doesn't have to rethink the process for each new individual situation. Instead, the researcher can orient the proven method to the current problem (see Popper 2001, *The World of Parmenides* for his view on methods, p. 190). Based on the current state of science, these proven methods should help participating disciplines in spatially relevant planning and development to achieve the best possible results. Because spatial planning is a per se interdisciplinary area, it follows that the spatially relevant methods of all the disciplines involved in spatial planning could be drawn in. Tasks of European spatial and transport development are strategic tasks because the solutions for the various national levels are of great importance. This can be graphically represented using railway development as an example. The reorganisation of infrastructure nodes will give the affected city and regional development considerable more economic stimulation and can contribute to a more sustainable spatial development. Efficient train stations and railway connections strengthen the network of cities and sites of the respective regions and countries and require the desired polycentric European development. And, they are a necessary precondition for stopping the incessant advance of sprawl.

Because bundled tasks are not only complex but also strategic, generating them requires a main focus, and, therefore, they are called complex strategic tasks.

2.1 Starting Points are Complex Strategic Tasks

The clarification and solution of this type of problem is usually of strategic importance for spatial development and defies—at least at the beginning of a clarification process—any kind of routine handling. This is also one of the reasons why the legally required, formal processes for the clarification and solution of difficult problems of spatial planning often do not lead to the desired goal and therefore need informal supplementation. In the beginning phase of a clarification process, informal processes can help overcome the difficulties of differentiating the key tasks to be surmounted.

The work of the planner will be influenced by the fact that he sees the world through his own set of lenses, i.e. he follows a particular planning approach. To this also belong the current leading ideas on how to approach the problem and what sort of aids or tools one can purposefully apply—and which rather not. The following tool, the Cloud-Tree Metaphor, represents such a pair of glasses and starts with the following premise (Fig. 3).

- At the beginning, there is a disorderly, confusing situation (the cloud) that must be dealt with.
- At the end of the clarification process, a specific option should be selected that will either solve the problem or avoid it. Therefore, a decision is to be found (the tree) in the metaphor.

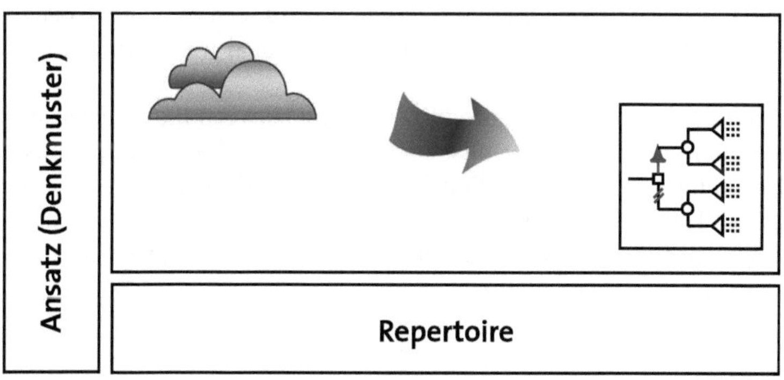

Fig. 3 The Cloud-Three-Methaphor. *Source*: Grundriss der Raumordnung und Raumentwicklung, ARL, 2011

- Because an intricate problem cannot be handled as a routine problem, the clarification process needs some custom tailoring for the content. For the organisation of the process and the application of certain methods, specific managerial ideas are helpful, which will be presented in the following paragraphs using a method that will serve as an illustrative example.
- There are many compilations of aids and tools available to support the clarification process and these compilations can serve as a useful 'repertoire'.

2.2 Paradigm of Spatial Planning

Spatial planning is a task-oriented discipline, and at the core, most problems can be traced back to problems of decision-making. The task is to clarify the main possibilities open to spatially relevant decisions and actions.

What are the basic options for action? What conditions should be observed? For example, if one assumes there are sufficient financial resources available for specific plans and then it turns out that no money has been reserved for them, the reason for this oversight is always the first question. What side effects and consequences might arise as a result? Once an option has been selected, the question of how to implement it brings up operational questions. Who should do what with which resources and by when?

The answers to difficult problems, which, of course, are not limited to spatial planning, are not immediately obvious and require, at least in part, longer processes to work them out. Intended improvements are often implemented years and sometimes even decades later. Extensive infrastructure plans in the area of roads and railways, airport expansions and the transformation of settlement areas in urban and regional contexts can deliver clear descriptive examples. In Switzerland, for example, the Gotthard Base Tunnel, which is the core of the north-south link from Rotterdam to Genoa, is scheduled to go into operation at the end of 2016, nearly 20 years after the start of construction. Whether intended effects, such as the extensive transfer of goods transport to the railways, can be achieved and whether undesirable effects, such as a displacement of regional transport, can be avoided, cannot be decided in advance.

Because the lead time, understood as the time between the actual selection of a specific option and the final results, stretches out over long time periods in many spatial planning processes, functional procedures and methods take on increased importance. These should, and it's actually required in democratic countries, also include transparent and sustainable decisions and actions that contribute to the causative planning argumentation.

2.3 Consequences of the Methodology Applied

If one follows this paradigm, then information gathering procedures must also be adjusted to avoid superfluous, time-consuming and expensive data collection. The method should contribute to gradually penetrating the core of the difficult task, thus helping to differentiate important data or facts from the less important. When less information is needed, then the information is also less specific, and therefore less risky because it is not tied to a specific outcome. Or, from another perspective, decisions are more robust when they are flexible enough to deal with possible changes (De Jouvenel 1967).

When less information is needed, communication is also lighter because the required common knowledge of the participants remains manageable. New information and communication technology can lead to a 'flash flood' of information and thus building 'dams' against them is an important task. The uninhibited application of new technology can bring about the opposite from what was actually intended.

Put more broadly, methods in spatial planning can be understood as a regulated approach to the systematic discovery, clarification and solution of difficult spatially relevant problems; likewise, appropriate methods of inspecting the effects and consequences of already realised solutions should be supported.

To be functional, the various methods have to meet certain requirements, which are listed here and can be found in detailed form in the chapter 'Die Methode der Testplanung' in the *ARL Handbuch: Grundriss der Raumordnung und Raumentwicklung* (Scholl 2011c):

1. Insecurity, risks and surprises
2. Time constraints and timing mismatches
3. Limited resources
4. Abstraction and concretisation
5. Robust decisions
6. Strategies and tactics
7. Task differentiation

3 Sustainable Development in Spaces of European Importance: Transit Spaces

It has been said that spatial development in a European context can be studied along the historic trade routes of European importance. The contributions of spatial planning will influence the future development of the transport carriers (rails, roads, water and air), and should be coordinated with the concerns and interests of the respective regions and cities in the sense of an integrated spatial development.

Transit spaces are areas that have an intensive exchange of goods, services and people. Many of these spaces started ages ago as footpaths that became trade routes and developed into important sites, regions, and landscapes that are now connected to each other by both land and water routes. Who hasn't heard of the Silk Road, the Salt Road or the Amber Road? Traders had to undertake long and arduous journeys through rough terrain and unknown regions in order to acquire exotic wares and bring them back home. They not only brought tea, coffee, silk and spices with them, but also exciting stories from their travels. Many myths and legends have grown up around these old trade routes that still fascinate us today.

And, naturally, such routes also brought with them an intensive cultural exchange.[15] Buildings and cultural artefacts of quite different types bear witness to this along the ancient trade routes. Trade routes have strongly influenced mankind's behaviour and, in the best cases, they also brought prosperity.

In the beginning, trading was all about overcoming obstacles, while the next step was establishing a useable transport chain, including people and beasts of burden as well as that free infrastructure: water. It was not by chance that the availability of water for transport was a central location factor for the founding of most cities.

In some areas of Switzerland, one also hears about '*Susten und Nauen*'—toll stations and small freighters—where the horses were changed, cargo delivered, passengers transferred and the piers where the first boats, or even rafts, were the 'modern' form of transport. Later, these places provided the first postmasters and captains for the ever-larger boats. Finally, in a quantum leap, the first railways arrived and then motorised transport, the lorry.

Since the purpose of these routes was trade, there has always been the desire to increase the capacity of such routes in order to make the highest possible profit by bringing as many goods as possible from the place of production to a place where the demand is high. Establishments of various kinds sprang up along these routes and with the increasing exchange, larger harbours and fairs, markets and bazaars followed. Later, at the important crossroads of such trade routes, railway stations, highway intersections and airports took over the transport function. This international commerce laid the groundwork for today's global economy based on the division of labour with its corresponding exchange of goods, services and people.

In today's economic point of view, people say: Time is money. The reduction of travel and transport time is an important matter—as the development of the travel time between Zurich and Milan clearly shows. The first private postal service at the end of the sixteenth century required 4 days, the opening of the Gotthard railway in 1882 reduced it to 10 h and with the start of the operation of the two base tunnels, Gotthard (end of 2016) and Ceneri (end of 2019) it will be just a little over 2 h. The

[15] The example of the Amber Road shows how the Mediterranean societies (Phoenicians, Egyptian, Greek) valued the stones of fossilised resin iron available mainly in the Baltic and brought from the northern lands. Already in the Bronze Age, Mycene, Bernstorf and Osterrada had developed into trading centres for the popular material (Graichen and Hesse 2012).

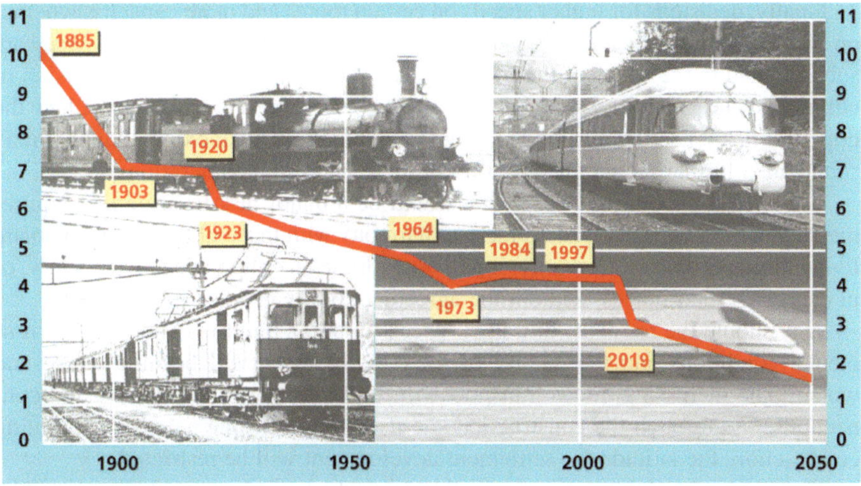

Fig. 4 Time–space relation between Zurich and Milano. *Source*: Hans-Peter Vetsch, Alptransit, 2006

space-time ratio has changed dramatically; northern and southern Europe are lurching ever closer together (Fig. 4).

In the beginning, trade routes were close to settlements, in fact, the settlements were sometimes right on the crossroads, which were a decisive factor in their founding. Along with the exchange, or perhaps because of the exchange, went a building culture of high quality. Many trade cities here in Switzerland also bear witness to this development.

Of course, it cannot be denied that there were also shadow sides to this trade and, since industrialisation, we can observe many legacies of these misguided developments. More and more, the infrastructures required for transport and financial turnover have become the drivers of urban and regional development. At first, it was the river and seaports, then the main route to the railway station, and in the twentieth century, the road to the airport. These became the areas of the most dynamic development with settlements going up around these points without any direction or plan. Today, as a result of far-reaching economic changes, these infrastructures are once again the drivers. However, new possibilities, not only in spatial planning, but also in redesigning infrastructures and settlements have emerged from the 'redevelopment before new development' maxim, i.e., a transformation of the building stock of settlements, infrastructures and public spaces before using agricultural land. We will come back to this subject later.

With the advances in technical innovations, many infrastructures are now outside the city, e.g., airports, or are separated from the city centre by commercial development, e.g. harbours. These new alignments mean that former areas of dynamic development have fallen prey to sprawl or decay, while settlements are further away, often pushed to the periphery or into housing developments. With this dislocation, settlements now need to be reconnected in order to take part in the new opportunities for mobility and transport.

Finally, decisions for higher speeds on certain routes and more capacity, whether on rails or on roads, leads to a situation where the needs of the living spaces are subordinated to the needs of the transit spaces. The physical and visual division of cultural landscapes by non-standard infrastructures, noise, and emissions reduce the opportunities for settlement development, and, possibly, paradoxical though it may sound, also further reduce the mobility of these and future generations.

The central hypothesis here is that it will be a major task to bring the requirements of a high-capacity trade route into harmony with those of the people who are living in transit spaces. Certainly, it will not work in an enlightened civil society, or in densely settled spaces, without the agreement of the people.

Why should the residents of transit spaces support a build-up when the environmental conditions and the quality of their living space will become worse and worse? The increasing goods transport will reduce accessibility, quality and connections, and through the construction site itself, and the long years of continuing construction, the latitude for settlement development will be restricted.

The agreement of the population will be strongly dependent on whether it will be possible at every step to cancel or compensate for the disadvantages and to eliminate possible 'sins' of the past. That will, in particular, require (1) that all sides have agreed upon a long-term perspective for spatial development and (2) continuous efforts in persuasion, because many years—and even decades—can go by between the first idea and its completion.

4 CODE24 Initiative: An Approach to Integrated Space and Transport Planning

Using the example of the route from Rotterdam to Genoa, the more general aspects of the challenges and conflicts of the transit corridor will be differentiated in this section. Part of this transit area is the Rhine axis between the North Sea and Basel and the bordering regions of the Netherlands, Germany, Switzerland and Italy (Fig. 5).

In the collective history of Europe and its spatial development, this axis is the equivalent of a backbone.[16] The most densely populated areas of Europe are situated here with the most diverse landscapes and city regions of all sizes. The economic strength of this axis is decisive for European life. Major investments of all kinds such as airports, railway stations, distribution centres, industrial areas, etc.

[16] In a programme on main focus points given by the German Research Society, it was recommended that the historical connections should be more closely examined. In this situation, we find: 'The Rhine, as a large river flowing through the centre of Europe, was a significant waterway already in prehistoric times. With the arrival of the Romans, it became even more important. In order to develop the Germanic provinces, including the construction and maintenance of military and civilian infrastructure, large quantities of raw materials and other goods were required, which were primarily transported by boat. Consequently, the new settlements on the Rhine had harbours with loading and mooring facilities, which are often known but seldom well researched' (DFG 2014).

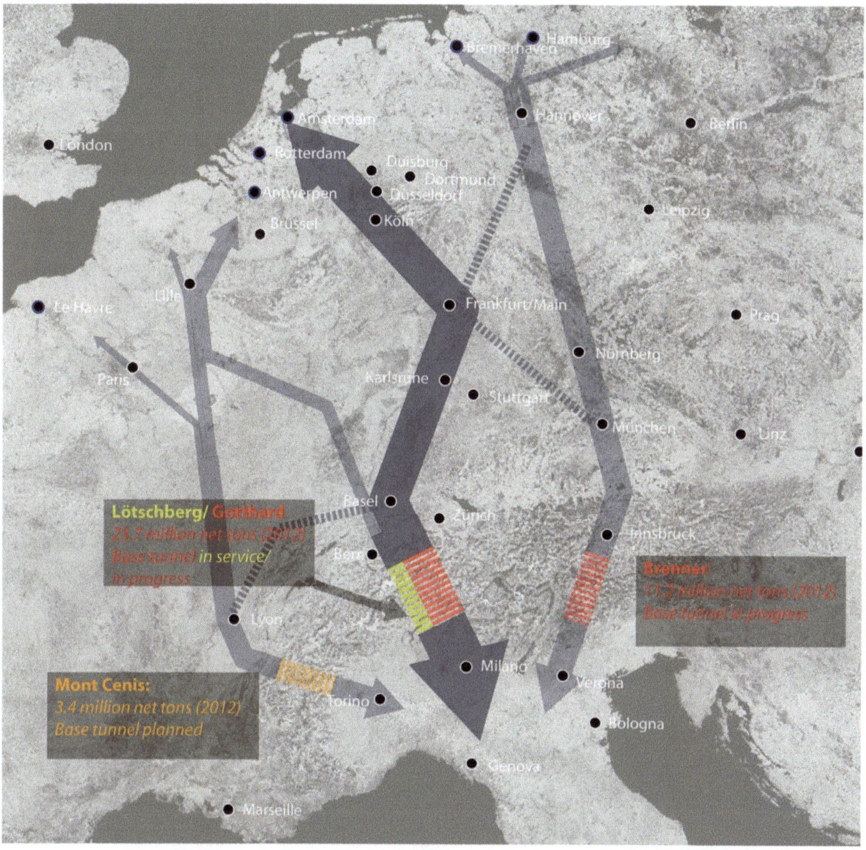

Fig. 5 The Lötschberg–Gotthard Axis, central for the alpcrossing transportation. *Source*: F. Guenther, CODE24, IRL, 2014

are situated here. The renewal and development of the regions are both a challenge. Global, European, national, regional and local processes intersect here. The changes are more than evident. The serious engagement with the spatial development of this transit area will inevitably lead to the need to deal with the cultural and political history and the future of Europe and its nations. What are the central issues? What are possible solutions?

Using an analogy to the laboratories and field experiments of natural science, there is a similar need for experimental space in spatial planning. One needs to deal with actual present and future problems in order to experience the value of using hypotheses and proven methods, as well as to communicate and cooperate effectively across the limits of disciplines and institutions. Meeting people who struggle with the current problems, who break out of their familiar surroundings in order to participate helps planners gain insight into the relationships between international, national and local procedures.

Learning how to sift through a mass of information, to order, clarify and evaluate in such a way that the core problem becomes visible, and finally, to present the experience and the discoveries in a comprehensive way is indispensable for learning how to handle complex situations. In an actual sense, this means learning from actual problems through testing various approaches and solutions. And, this brings considerable advantages with it—also in practice.

Starting from the idea of creating such an experimental space, in 2010 the Chair of Spatial Development of ETH Zurich initiated the INTERREG Project CODE24, to bring together the major actors from along the transit route to create a collective strategy of spatial- and infrastructure planning. (This strategy was approved at the Corridor 24 Conference in Mannheim in 2014). After much preliminary work and the authorisation from the EU as a strategic project, the work began in Germany under the auspices of the Regional Association of Rhein-Neckar.

4.1 Corridor Implementation

The European railway's north-south connection from Rotterdam to Genoa is a main artery of the European railway network and until recently was known as Corridor 24. It is now called the Rhine–Alpine Corridor in the EU's TEN-T network, which consists of nine lines (Fig. 6).

This connection has grown over the last few centuries to be more than 1200 km long. It connects the large North Sea harbours in Belgium and the Netherlands to

Fig. 6 TEN-T Network with Rhine–Alpine Corridor. *Source*: TENtec, European Commission, Mobility and Transport, 2013

Germany, Switzerland and the prospering economic centres of northern Italy as well as the Italian Mediterranean area. The main purpose of the planned extension is to improve the international passenger and goods transport and to divert goods transport from the roads onto the railway tracks. This is particularly relevant from the Swiss point of view, as according to the Swiss Constitution, all transit of goods transport (and Alp crossings) going from border to border must use the railways.

The flat railways (NEAT) through the Alps form the core of this plan. Alone in Switzerland, investments have reached over 15 billion CHF (including noise reduction). The launch of the first Alpine base tunnel in Switzerland in 2007 on the Lötschberg (35 km) is the result of a quantum leap in railway development, and not just for the Alpine area. The Gotthard (57 km), also part of the planned Alpine base tunnels, is scheduled to open at the end of 2016 and the Ceneri (15 km) in 2019.

The Rhine–Alpine Corridor travels through European regions that have the highest economic value, with a strong increase in transport on the top-ranking railway and road networks and a high population density. Almost 70 million people, roughly a fifth of the entire population of the EU, live in the catchment area of this important European north-south connection. 700 million tonnes of freight is transported along the Corridor; thus, over 50 % of the entire collective freight movement in the north-south direction.

4.2 *Overview of the Alpine Arc and Other Flat Railways*

Two other tunnels are planned on EU territory for the Alpine Arc, namely the Brenner Base Tunnel and the Mont Cenis Base Tunnel, which are also part of the TransEuropean Transport Network (TEN). The test soundings at the two base tunnels of Mont-d'Ambin (52 km) and Bussoleno (12 km) at Mont Cenis between Lyon and Turin have already started. In 2006/2007, the test soundings phase also started for the pilot tunnels on the 55 km Brenner Base Tunnel.

In spring 2011, the starting shot was given for the building of the main tunnel. The launch of the tunnels will be in 2030, approximately.

This means that the base tunnel of the Gotthard stretch will open more than a decade earlier than the Brenner Base Tunnel. It can be expected that the logistics and transport industries will anticipate and adjust to this situation. A steady increase can be expected in freight transport in terms of time and space on the fastest railway connection from the northern harbours to the prospering northern Italian economic region. The diversion of freight transport to the railways was the main reason for building the series of tunnels and, following the acceptance of the law on the Protect the Alps initiative, this will be pursued as before (at that time about 1.2 million journeys could be reduced to 600,000 journeys per year in lorry transport crossing the Alps.)

How much transport will be transferred to the rails after the launch of the Brenner Base Tunnel cannot be said in advance because of certain routine operations and previous investments; even though the favourable geo-strategic position of the NEAT to economic centres in northern Italy makes the Gotthard route the shortest connection to the North Sea harbours. From a European point of view, however, it is advantageous if redundant north-south tunnel systems are available.

4.3 *Importance of the Harbours for Hinterland Transport*

A strong increase in transport in the north-south direction can be expected for 2030 as globalisation continues to grow. The world is changing slowly but steadily into one single large market. In particular, the economic growth in Asia and the increase of goods exchange within the EU will play a vital role. The Mediterranean Sea has taken over from the Atlantic as the main shipping route for maritime exchange. In the long term, it will be important to take the opportunity to use the harbour potential situated on the Mediterranean and Adriatic coasts in order to achieve a balanced European goods exchange. Highly productive hinterland connections, in particular, those on railway systems, like the northwest European harbours, will, as before, play a decisive role (Fig. 7).

At present, the main parties dealing with sea transport in Europe are the ZARA harbours (Zeebrugge, Antwerp, Rotterdam, Amsterdam) and the German harbours (Willemshafen, Hamburg, Bremen/Bremerhaven). Antwerp and Rotterdam alone had a turnover of more than 16 million TEU in 2010. By 2020, a redoubling of capacities to over 36 million TEU is expected (Fig. 8).[17]

The preference for northern harbours over Mediterranean harbours is indicated by the financial capacities of these regions and the much larger and more effective cargo handling facilities, as well as better hinterland connections, e.g. waterways, railways and high capacity roads. The container capacities of Rotterdam are five times larger then those of the Ligurian harbours (Savona, Genoa and la Spezia) put together. The harbours in Northern Europe, particularly in Rotterdam, have invested massively in the extension of their harbour infrastructures. Rotterdam will increase its cargo handling capacity with a harbour extension 'Maasvlakte' from about 450 million tonnes at present to about 700 million tonnes in 2035. However, the harbour extension will make it increasingly difficult to guarantee a smooth hinterland transport. In the Netherlands, the launch of the Betuwe line was achieved in 2007 under considerable pressure. Starting in 2015, a full use of this line between Oberhausen and Emmerich will mean the need for a third track and a block concentration, which will not be available because the German railway network has other priorities. From now on, commissioning and costs of around 2 billion euros will be expected for the year 2023.[18]

---

[17] TEU is the abbreviation for 20-foot equivalent unit, a measurement of the capacity of container ships and harbour cargo handling amounts.

[18] Only Rotterdam and the deep-sea harbour of Wilhelmshaven are equipped for container ships with a large cargo handling capacity, ca. 18 k TEU per ship). At present, ships with 12 k TEU translate to a cargo of 120 trains or 6000 lorries.

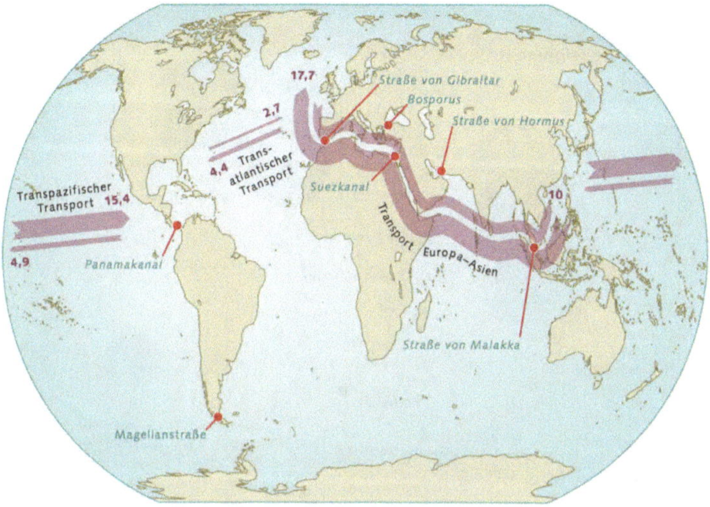

Fig. 7 Global sea routes. The numbers are representing the quantities of transported TEU in mio. *Source*: World Ocean Review 1, Hamburg 2010, Hrsg: maribus gGmbH

4.4 The Problem of Bottlenecks: Separation of Goods Transport from Passenger Transport

More bottlenecks are to be expected. One of these will be the strategic railway section of Frankfurt-Mannheim, which is also highly significant for the approach to the Brenner Tunnel and the Main East-West Railway as well as the missing continuation of the four-track extension of the Rhine line from Offenburg to the area south of Freiburg. It should be mentioned that, despite all the financial difficulties, the 17.6 km Katzenberg Tunnel was launched in time for the timetable change in December 2012. The building work for the Raststatter Tunnel could also begin in 2013 after a nearly 10-year postponement due to missing financial support. The launch is now expected to be in 2022, with construction costs of around 700 million euros. According to prognoses, the finalisation of the extension of the Rhine Valley route will not take place until after 2030, although according to the contract, a timely launch with the Gotthard Base Tunnel had been secured.[19] In this

[19] Traffic jams are already happening daily in the areas surrounding harbours and their hinterlands. One request from the Rotterdam Harbour is increasingly about transferring the share of the container hinterland traffic from 2009, a total of 4 million TEU (with 18 million TEU expected in 2035), to the waterways and railways. Percentage-wise, the share on the waterways will rise from 40 % to 49 % and the share of the railway from 14 % to 20 % TEU. The share of road transport should be reduced from 46 % to 35 %. Measured by the possible capacity of the harbour by 2035, a ship could carry five times more TEU, the railway six times more and lorries would have to take over four times more cargo. (See also Braun 2014). In the above-mentioned briefing from the German government, no date has been given for an implementation of the Rhine Valley extension (Deutscher Bundestag 2014).

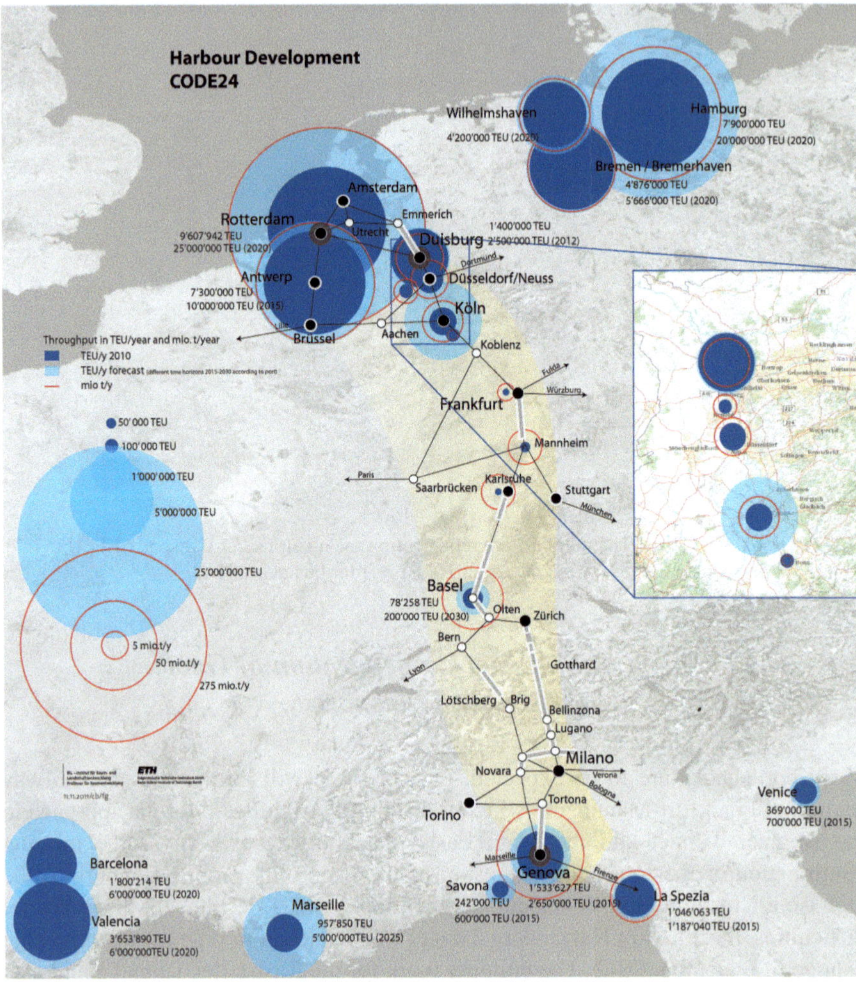

Fig. 8 Harbour Development on the Rhine-Alpine Corridor. *Source*: C. Braun, CODE24, IRL, 2011

connection, the development of a multimodal freight transport centre at Lahr will be of major importance.[20]

Whether the core network that the Rhine–Alpine Corridor belongs to can be realised by 2030, as announced by the EU, is questionable. There is around

[20] The actors of the Goods Transport Centre in Lahr were pushed into the CODE24 Project during its extension phase, 2013/2014, in order to be able to test possibilities of integrated development. A parallel situation of placing the tracks along the highway is a precondition for a direct transition from roads to rails.

23 billion euros[21] available on behalf of the EU for the TEN-T network up until 2020. Based on this discrepancy, it becomes clear that in the long term a large number of bottlenecks will continue to exist, with the result of not being able to offer enough capacity for all the desired railway transport. It will be necessary therefore to create sensible priorities, together with the national and regional authorities, and find integrated solutions through innovative forms of cooperation. This can only happen on the basis of an assessment of the entire situation. CODE24 has created an important foundation for these.

4.5 The Potential of Inner Development

It has been mentioned a few times that eliminating bottlenecks is of major importance for the redevelopment of cities and regions. The catchment areas of the railway stations are the crystallisation points of redevelopment. This encourages the need to keep, or actually improve the availability of public transport. With redevelopment, public transport's share of the overall traffic can be increased when relevant attractive offers for public transport are available. Its decrease or reduction through increasing goods traffic, which would compete with the passenger transport on rails, is counterproductive. Of all these reasons, the integrated development in the triangle of passenger transport, goods transport and redevelopment is of central importance.

The catchment areas of nearly all mayor railway stations offer special opportunities for redevelopment. The city of Frankfurt is a good example: The transfer of freight stations and switching stations here offer a particularly clear picture. Near the central railway station and several metro stations, a new city area, the 'Europaviertal', is being created on a 100-ha site where about 30,000 people will have the unique opportunity to modernise and expand the Messe Frankfurt in its inner city location.[22]

Estimates of the available settlement potential along the north-south axis show a grand total of around 150,000 ha. Some of these potential areas, e.g. in the German state of Rheinlandpfalz, in the regions of Rhein-Neckar and Upper Rhine-Bodensee, and in the Swiss cantons of Basel-Landschaft, Schwyz, Uri and Tessin, were assessed using the Raum+ method (Scholl 2008) in individual communities

[21] According to CEF (Connecting European Facility), this was decided in February 2013 by the EU Presidency.

[22] The project was developed using several test planning processes and the Consilium Europaviertel Frankfurt, which was led by the author. (See Scholl 2011b, ISOCARP).

of the participating areas (Nebel 2014). When these key statistics were applied to the entire area, the resulting potential settlement would be around 100 million users today, 'space users' of the 150,000 ha mentioned earlier.[23] Speaking from experience, a conservative estimate would mean that around 50 % of the site would be available immediately. An average gross floor ratio of about 0.5 would result in 37,500 ha of floor space. If one would fill these exclusively with apartments at a rate of 50 m^2 of floor space per inhabitant,[24] around seven million additional inhabitants could be housed along the Corridor. With an average gross floor ratio of 1, which would probably be expected in the catchment area of high-capacity railway stations, this would mean around 14 million inhabitants along the Corridor. These are significant figures for a potential settlement area, which can be actively implemented in the coming decades.

Furthermore, the economic impulse for real estate development is significant. If we assume that 300 euros is needed per cubic meter of converted space (based on the assumed 37,500 ha of floor space), then with one billion square meters (at 2.65 m average storey height) of converted space, this would result in an investment of around 300 billion euros (not including possible investments in the connecting infrastructures) (Fig. 9).

In addition, the opportunity arises, supported by new planning approaches, to achieve the separation between goods transport and passenger transport in some densely populated areas, for instance, by using goods transport bypasses. Route sections where this should definitely be tested are: between Frankfurt and Mannheim, from Offenburg to just south of Freiburg, Basel to Central Switzerland, a bypass for Bellinzona, and south of Ceneri to the Italian border. Transferring dangerous goods transport not only decreases railway noise and eliminates unsatisfactory noise reduction installations, it also plays a significant role in creating relevant planning considerations. This transfer would also create additional experimentation space for local and regional development where smaller distances to the railways are possible. Such elements could play an important role in the discussion around the approval of additional railways.

[23] Space users includes inhabitants and workers in the catchment area of the corridor. Per user there is a potential of ca. 30 m^2, of which half can be counted as part of the reserves (Nebel 2014, p. 23).

[24] This value is for Switzerland. In the surrounding countries, the value is lower, which could result in more inhabitants being counted as potential users, always under the prerequisite that the surface area used remains approximately the same.

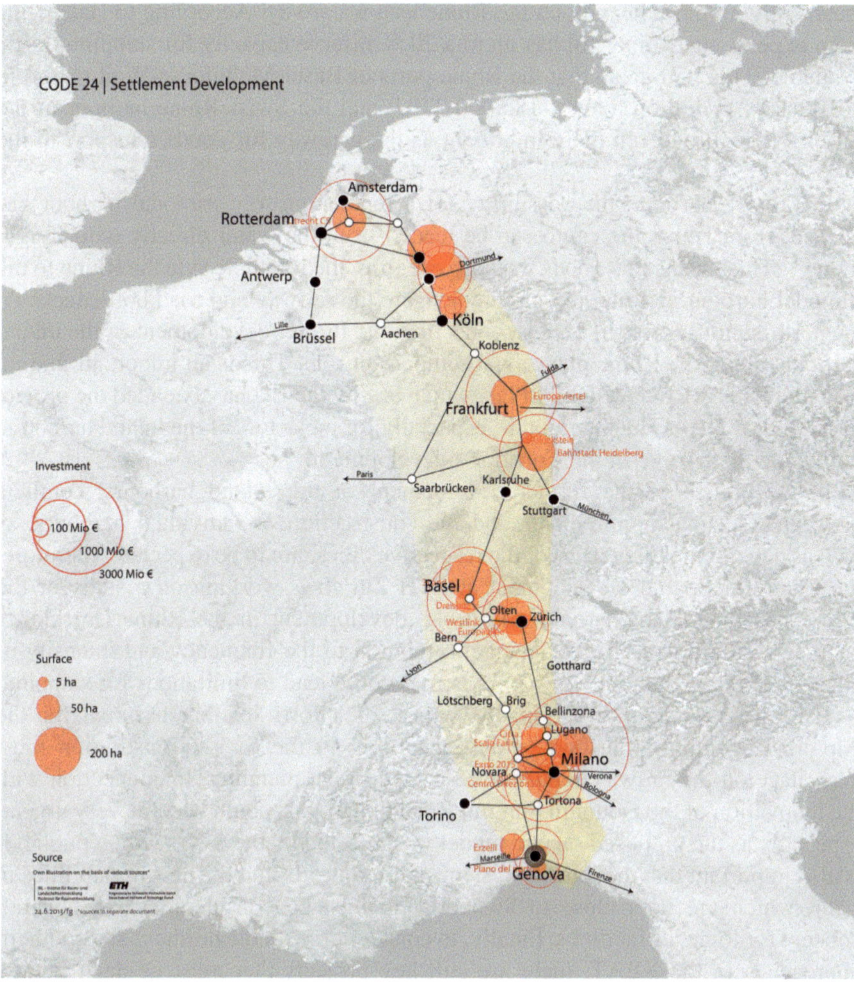

Fig. 9 A representation of settlement potential. *Source*: C. Braun, CODE24, IRL, 2011

4.6 The Importance of the Rhine River as an Inland Waterway

Thus, in future, it will still be about the further development of the railway network, but mainly, it will be about sufficient, reliable and environmentally compatible capacities along the heavily burdened main routes.

For example, Switzerland has to face a starting situation where, because of the shortage of money in its European surroundings, despite contractual agreements, the partner countries of Germany and Italy will not be able to fulfil their obligations (Schweizerische Eidgenossenschaft 1998). For the northern approaches, this means relying more heavily on the Rhine as a waterway. According to statements from experts, the Rhine still has up to a 20 % reserve capacity for shipping traffic available. This will mean that the Rhine ports of Basel-Muhlhouse-Weil (cargo in 2010 ca. 12.5 million tonnes, 185,000 TEU) and the Swiss Rhine harbour in the Basel area will be even more important as the gateway for goods transport to the north.

For Basel Harbour, the possibility for a renovation to a trimodal harbour has opened, from which the cargo can be loaded from the water directly onto freight trains.[25] This means that goods would come from the North Sea via the Rhine to the trimodal harbour and literally circumnavigate the railway and road bottlenecks on the north-south approach. In this connection, the further development of the inland harbours along the Rhine plays a very important role. This is an important finding from the INTERREG CODE24 Plan, which is why the EU has extended the project to the end of 2014. The intention is, especially for the actors of the inland harbours, to include them more in the Corridor's development.

In the course of the further development of the inland harbours, conflicts between logistical requirements and the construction of individual harbours, as well as development fantasies of the respective cities, are to be expected. Therefore, the Chair for Spatial Development at ETH Zürich is working on a study of the future possibilities of harbour and urban development in the Rhine Corridor,[26] keeping in mind that climate change in relation to the frequency and intensity of changes to the high and low water extremes could lead to limitations for shipping.

For Switzerland, in addition, the option of a lorry-loading terminal for the Alp-crossing goods transport, which according to the Swiss Constitution must take place on the railways, must be respected in the planning. In connection with the renovation of the Gotthard Summit Road tunnel, this could become very urgent, especially as the proposed second tunnel was rejected by the Swiss voting public. It is also important that the loading systems in Freiburg/D and the new possibilities in Lahr, which are both close to Switzerland, must be considered in any overall concept for logistics facilities. Finally, in connection with the northern and southern approaches to the NEAT, there are still key projects that must be dealt with—without fail. These include planning and securing space for a new Jura crossing (Nordwestschweizer Regierungskonferenz 2012), the realisation of the Axis Tunnels, the ring road around Bellinzona and securing space for the construction of the southern approaches. With the acceptance of the FABI decision in early 2014, financing for the planning has been made available.

The southern access to the Gotthard Base Tunnel must reckon with the reverse situation: a major part of the goods transport must be reloaded onto lorries to continue to Italy, because the infrastructure systems there will not be ready in time. The loading equipment in the southern border area therefore takes on special importance. If Italy's infrastructures become operational very much later than the

[25] The earliest this new essential harbour basin would be available is 2017. In addition, the project opens far-reaching possibilities for the city's development (Braun 2014).

[26] An introductory report is in preparation (Braun 2014).

commissioning of the Gotthard Route, i.e. after 2019, the Greater Milan Area will have to reckon with a massive capacity overload on their high-capacity roads.

4.7 Intermediate Results

1. As a result of the continuing advance of globalisation, a continued increase in goods transport must be expected. The relocation policy of the EU will lead, in cases of limited capacity on the high-capacity roads, to a (desired) increase of goods transport on the rails, also the routes crossing the Alps. This can lead to conflicts with other rail transport, especially if it is in the same speed limit areas as the locally operating regional transport.
2. It must be taken into consideration that bottlenecks along the Rhine–Alpine Corridor could lead to a displacement in regional transport operating in the same speed category or could lead to a situation where the railway potential for goods transport cannot be utilised.
3. The Rhine waterway wins by a (temporary) circumnavigation of bottlenecks and leads to a steady decline in the importance of road and rail in Germany. For Switzerland, the Swiss Rhine harbours move back into focus, namely the construction of the harbour in Kleinhüningen with a trimodalen rotary platform loading terminal for Alp-crossing transport.
4. All efforts are, however, on promptly adjusting the construction around the expected bottlenecks in all countries and coordinating with one another to bring this forward so that the displacement of regional transport is not allowed nor is the transfer of goods transport to the rails endangered.
5. In the spaces identified as having bottlenecks, special time-limited planning processes should be brought in immediately in order to manage an integration of solutions for spatial and transport development. With the test planning method, for example, a nationally and internationally tested method is available (Scholl 2006, 2011c; Scholl and Staub 2013).
6. A separation of the goods transport from the passenger transport in the densely settled areas should be a main goal.
7. Integrated spatial and transport development creates a large potential settlement area and its activation will bring large sustainable economic investments.

4.8 Achievements

1. A stable foundation for a further collaboration was created within the framework of the CODE24 Plan, as requested in the Territorial Agenda 2020.[27]

[27] TA 2020 pt. 56: '. . .building on recent experiences, where appropriate full advantage could be taken of the new opportunities offered by the legal instrument of the European Grouping for

2. This involved working jointly on problem screening and identifying the complex points of future actions. In several situation assessments and in individual thematic reports, the advances were documented.
3. In numerous regional conferences and in the annual Corridor 24 conferences, the results and findings were introduced to the public and the representatives of all governmental levels as well as political committees.
4. The findings, together with the outcome from the debates with the public and political representatives, were consolidated into a common strategy for the Corridor's development through the leadership of the Rhein-Neckar region. A conference was suggested for the close of the INTERREG Project CODE24 in Mannheim at the end of 2014.
5. Finally, under the leadership of the Rhein-Neckar Regional Association, a European special purpose association was formed, the European Grouping for Territorial Cooperation (EGTC), which will be the organisational arm for further work (Fig. 10).

4.9 *What to Do Next*

The following tasks should be approached within the framework of the EGTC and the principles of the Joint Strategy—available now as a basis for the mutual 'Will to Do'.[28]

1. Conduct a start-up event with the EGTC to announce an action programme.
2. Form task forces to identify complex strategic tasks with the goal to achieve integrated solutions through special planning processes, e.g. test planning. Some possibilities for an upcoming collaboration on the affected sections of the Corridor, for example, the section Frankfurt to Mannheim and the section Lahr–Basel–Central Switzerland.
3. Conduct conferences with the actors of the relevant railway and harbour companies along the Corridor.
4. Update and record the basic principles and the work progress that are seen as functional, advisable, advantageous, and practical.
5. Prepare a status report that summarises the results of 3 year's work on the project and formulates further steps to be completed by 2030.

Territorial Cooperation (EGTC). Furthermore, voluntary coordination of planning activities along borders within the framework of the EGTC should be explored where it is perceived as an added value.'

[28] According to the official bulletin of the TEN-T Priority Programme, the bottlenecks in the core network of the EU should be eliminated by 2030.

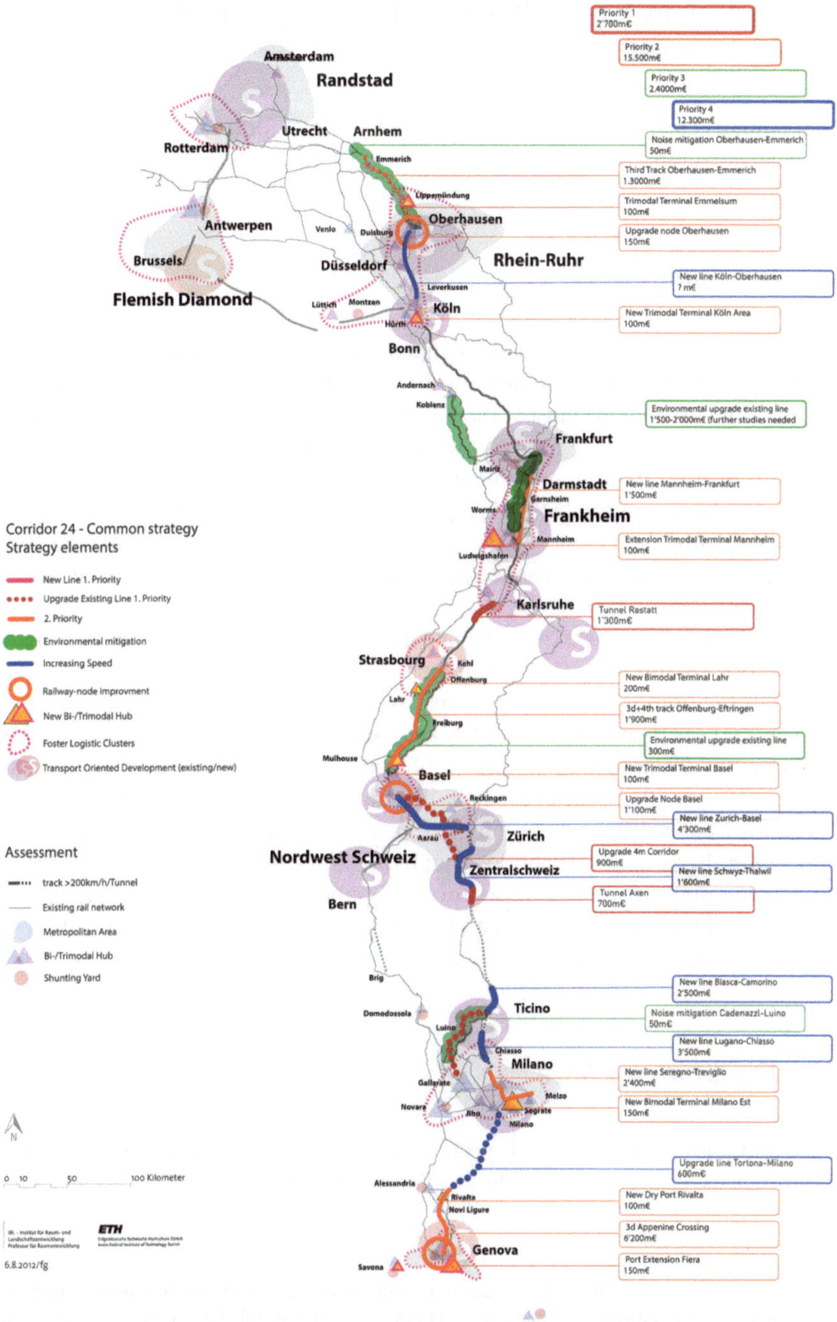

Fig. 10 Joint Strategy of CODE24. *Source*: F. Günther, CODE24, IRL, 2014

5 Conclusions for the Practice of an Integrated Corridor Development

What findings can be drawn from the CODE24 project for the development of other European corridors? The following sections present an answer in a brief form.

5.1 *Initiation Phase*

The experiences from CODE24 show that a multi-year approach can be critical in organising a smooth collaboration for the development of an integrated corridor. The initiative of an actor must come out of his experience. As actors for initiating new project ideas, all the responsible figures from the region and the cities are possibilities, as well as representatives of science, culture and politics. Actors should also be aware that initiating also means investing. This means first, that capable specialists and modest financial means are needed for the initiative to even be able to start. In this connection, universities and colleges have special importance.[29]

Difficult unsolved tasks with their corresponding spatial conflicts, especially those whose solution lies in the common interest, should therefore be a starting point. Information gained from the first overview and the status assessment can make up a set of arguments (Signer 1994) to be used for a potentially more intensive collaboration. Compiled in a variety of coding types, e.g. words, images and numbers (Signer 2012), these can help illustrate the difficult circumstances.

The trust created during the initiative phase is important capital for continued cooperation. In case the partner does not have sufficient capital of his own, the EU can provide start-up money (also for the application phase), for example, the INTERREG Programmes Office can be approached. There is also a possibility of receiving financial support from TEN-T.

5.2 *Exploration Phase*

The exploration phase, about 1 year, is about studying and assessing the joint circumstances in order to be able to identify the focal point of the work. For a

[29] It was in this connection that representatives of numerous regional organisations and cantons, as well as industry associations and chambers of commerce in the middle Upper Rhine came together to first experience an exchange on an informal basis. An invitation from my then Chair for Urban Design and National Planning at the University of Karlsruhe (today KIT) did bring many actors together. After the visit of a delegation from the Regional Rhein-Neckar Association in Autumn 2008 to a construction site of the Alptransit (Sedrun), an idea for an INTERREG project arose—as well as the readiness to take over the general management.

macro-regional space, based on our experience, regional workshops are indispensable. Over the course of the 4-year cooperation on CODE24, there were almost 30 regional workshops presented: Italy, Switzerland, Germany and the Netherlands. Of these, eight workshops were prepared in Switzerland. Together with the semi-annual, 2-day meetings of the entire group, there were also about 40 coordination meetings across the 4 years where tasks were prepared, information exchanged and findings evaluated. In the times between phases, work contracts were carried out according to the agreed upon work programme.

5.3 Development of a Common Strategy Phase

The information gained in the exploration phase should be combined into a common strategy step-by-step and through several review cycles over several passes. In order to be able to find a version acceptable to all participating actors, the first drafts should be presented and discussed at the latest at the halfway point of the cooperative agreement. The common strategy is the common directional orientation between the 'will to do' and the future 'to do'. Based on our experience, intensive feedback allows the individual participating institutions to also create a broadly supported and well-founded basis for further processes. All these activities need time.

One important point is that the actors need to internalise the essential elements of the strategy, so they can plausibly communicate with those opposite in their own institution and in other participating institutions, as well as to important political committees and interested members of the population. At this point, it is very important to have the strategy worked out collaboratively, including work assignments.

5.4 Joint In-Depth Analysis Phase

Important questions often arise out of the exploration phase and the first strategy proposals. These can be clarified through activities, such as workshops and special processes, designed to take the participants deeper into the topics. For example, one problem that surfaced in the CODE24 project was bottlenecks in the German and Netherlands border (area around Oberhausen/Emmerich). A modified test planning process was used to work out the best way to approach the problem.[30]

[30] A detailed representation of the result of the integrated space and infrastructure development can be found in: IRL (2013a, b). The dissertations by Günther (2015) and Tosoni (2015): 'The methodical aspect of large-scale and integrated spatial and infrastructure development will be evaluated and should deliver scientific principles for the macro-regional cooperation.'

The section Frankfurt–Mannheim is especially important for the European network and both of the participating regional associations, Rhein–Main and Rhein–Neckar, found an approach over several workshops that provided a good basis for further cooperation with the management of the Deutsche Bahn.

Finally, in Switzerland, more workshops were devoted to the question of railway development in the area around Basel Mittelland. One result was a discussion about a new Jura crossing, which sought additional financial resources for planning as part of the FABI Decision,[31] which was subsequently accepted by a strong majority of the population of Switzerland and could therefore be arranged.[32] In Canton Schwyz, a test-planning process was conducted in the area of Felderboden to achieve an integrated development of the transport, space and landscape plans for a strategic important space (Scholl 2014). The work led to a new organisation of the transport tracks (Bundesamt für Verkehr 2014). This example makes it clear that consistent cooperation with the responsible actors as part of an informal process can lead to desirable developments and finally find their precipitation into the formally planned process and instruments.

5.5 Implementation Phase

Exploration, in-depth analysis and strategy deliver sound basic principles for decision-making, on whether a long-term cooperation with a set organisational framework is necessary in order to follow up tasks of mutual interest. A European special purpose association can function as such a framework. A part of its activities could be contacting important state levels and bringing additional important actors together to accompany the realisation process of strategic spaces. It can come back to setbacks and surprises again and again with comprehensive complex plans of infrastructure and spatial development. Through its network and the experience of background cooperation, a European special purpose association can activate strategic reserves in order to be able to offer assistance in special situations.

The implementation of the individual plans must use the already mentioned subsidiarity principles of the individual countries and must be carried out with valid formal processes and instruments. Nonetheless, a continual update as part of the informal process has achieved advances and knowledge of extraordinary importance. This allows the set of arguments, once organised, to be further developed and

---

[31] Federal decision on the financing and construction of a railway infrastructure (Finanzierung und Ausbau der Bahninfrastruktur (FABI)), 9.2.2014.

[32] The results have also led to rethinking the Mixed Transport Strategy (joint use of the rails by passengers and goods). It could be shown that the partial separation of the goods transport from the passenger transport can thickly settled settlement spaces unburden and lead to limited costs in the realisation. The findings were published in a trilogy in the professional journal of the Swiss Architects and Engineers Association, *TEC21* 2013.

refined for accuracy and, if necessary, to be revised. Plans such as those introduced here, might take several decades to proceed from planning to realisation. All the actors should be aware of this.

The CODE24 actors are organising a group called the European Grouping for Territorial Cohesion. From my view, the important points for such a work programme are listed in the previous paragraph and will need, naturally, some adjustment and updating.

6 Closing Remarks

At the beginning of this text, the commitment to European spatial development was mentioned as important for creating strong regions and emphasising a strong Europe. Naturally, this is something many can agree on.

When it concerns an integrated spatial and infrastructure development at the European and macro-regional standard, then the action- and problem-oriented cooperation between many cities and regions is truly in demand. Actors of the Rhine–Alpine Corridor have seized this opportunity and taken the initiative. They have contributed in many ways, but also for Europe.

When the contribution in your hands can awaken an interest in a new fruitful cooperation around language, culture and ways of thinking, then one of the goals of this contribution has been achieved.

References

Akademie für Raumforschung und Landesplanung (2009) Position Paper No. 79: The future challenges of large-scale transport development in Germany. Akademie für Raumforschung und Landesplanung, Hannover

Behn RD, Vaupel W (1982) Quick analysis for busy decision makers. Basic Books, New York, NY

Braun C (2014) Rhein, Raum und Logistik. Exposé im Rahmen MAS Raumplanung. ETH Zürich

De Jouvenel B (1967) Die Kunst der Vorausschau. Luchterhand, S 234

Deutsche Forschungsgemeinschaft (2014) Special Research Programme SPP 1630: Harbours from the Roman period to the middle ages. DFG. http://www.spp-haefen.de/en/projects/der-rhein-als-europaeische-verkehrsachse/. Accessed 24 Aug 2014

Deutscher Bundestag (ed) (2014) Unterrichtung durch die Bundesregierung Bericht über die Projektfortschritte beim Ausbau der grenzüberschreitenden Schienenverkehrsachse. Drucksache 18/357, 22 January 2014, S2

European Commission (2008) Territorial cohesion: turning territorial diversity into strength. Brussels

European Commission (2014a) Investment for jobs and growth: sixth report on economic, social and territorial cohesion. European Commission, Brussels

European Commission (2014b) Infrastructure – TEN-T – connecting Europe. http://ec.europa.eu/transport/themes/infrastructure/ten-t-guidelines/project-funding/cef_en.htm. Accessed 24 Aug 2014

European Union (2011) Territorial Agenda of the European Union 2020: towards an inclusive, smart and sustainable Europe of diverse regions. European Union, Gödöllö, Hungary
Graichen G, Hesse A (2012) Die Bernstein Strasse: Verborgene Handelswege zwischen Ostsee und Nil. 2. Auflage, Ulm
Günther F (2015) Raumplanerische Exploration von Makroregionen. Unpublished Dissertation, IRL, ETH Zurich
Initiativkreis Europäische Metropolregionen (2007) Verbindungsqualitäten zwischen Metropolregionen. Positionspapier, Institut für Konfliktmanagement, April 2007
Institute für Raumplanung und Landschaftsentwicklung (2011) Teaching module on the subsidiarity principle. Chair for Spatial Development, ETH Zürich, Zurich
Institute für Raumplanung und Landschaftsentwicklung (2013a) CODE24: a common strategy for the corridor Rotterdam–Genoa. ETH Zurich, Zurich
Institute für Raumplanung und Landschaftsentwicklung (2013b) Perspektiven der räumlichen Entwicklung Limmattal. Empfehlungen des Begleitgremiums. ETH Zürich, Zurich
METREX (2013) Territorial visions and scenarios for Europe. ESPON 2013 Programme Project. European Union, Brussels
Nebel R (2014) Flächenmanagement Schweiz. Dissertation, ETH Zürich, vdf, Zurich
Neue Zürcher Zeitung (2014) Streit um Dänemarks Prestigeprojekt. 16 August 2014
Nordwestschweizer Regierungskonferenz (2012) Schweizer Beitrag zu CODE24. ORR Resolution. November 2012
Popper K (2001) The world of Parmenides. Routledge/Chapman & Hall, New York, NY, p 190
Reimer M, Panagotis G, Boltevogel HH (2014) Spatial planning systems and practices in Europe. Akademie für Raumforschung und Landesplanung, Hannover, p 5
Scholl B (2006) Test planning procedures as a method for supporting decision-making in complex planning projects. Milan
Scholl B (2007) Strategies for spatial and railway development for Europe in the North-South link. In: Perspektiven der Raum- und Eisenbahnentwicklung am Hochrhein Schriftenreihe. Institute for Urban Development and Spatial Planning, University of Karlsruhe, Karlsruhe
Scholl B (2008) Nachhaltiges grenzüberschreitendes Siedlungsflächenmanagement, Regions- und Nationalgrenzen überschreitendes Kooperationsprojekt. Zusammenfassende Einführung. In: Raum+ Abschlussbericht. Umweltministerium und Wirtschaftsministerium von Baden-Württemberg, Stuttgart
Scholl B (ed) (2011a) SAPONI, spaces and projects of national importance. vdf Verlag, Zurich
Scholl B (2011b) Strategies for integrated spatial development along the European North-South Railway Link. In: ISOCARP Review 07, The Hague
Scholl B (2011c) Die Methode der Testplanung. Exemplarische Veranschaulichung für die Auswahl und den Einsatz von Methoden in Klärungsprozessen. In: Grundriss der Raumentwicklung. Akademie für Raumforschung und Landesplanung, Hannover
Scholl B (2011d) Die Methode der Testplanung. In: Grundriss der Raumentwicklung. Akademie für Raumforschung und Landesplanung, Hannover
Scholl B (2012) Gedanken zur Nord-Süd-Transversale. In: SIA Tec21, 17 December 2012
Scholl B (2014) Integrated spatial and infrastructural development: the need for adequate methods and spatial strategies for collaborative action and decision-making. In: Lami I (ed) Analytical decision-making methods for evaluating sustainable transport in European corridors. Springer, Cham
Scholl B, Staub B (2013) Test planning: a new method with a future. ETH Zürich, Canton of Solothurn
Schweizerische Bundesamt für Verkehr (2014) Sachplan Verkehr, Teil Infrastruktur Schiene. 30 April 2014
Schweizerische Eidgenossenschaft (1998) Vereinbarung zwischen dem Vorsteher des Eidgenössischen Verkehrs- und Energiewirtschaftsdepartements und dem Bundesminister für Verkehr der Bundesrepublik Deutschland zur Sicherung der Leistungsfähigkeit des Zulaufes

zur neuen Eisenbahn-Alpentransversale (NEAT) in der Schweiz. Abgeschlossen am 6 September 1996

Signer R (1994) Argumentieren in der Raumplanung. ETH Dissertation, Zurich

Signer R (2012) The image proceeds the idea. In: International Doctoral College Space, The Logbook. Spatial Research Lab, Jovis, p 51

Tosoni I (2015) Shared spatial strategies and actions design: approaches and instruments enabling collaborative design processes at the large, regional and macro-regional scales. ETH Dissertation, Zurich

Bernd Scholl is since 2006 a full professor for Spatial Planning and Development at the ETH Zürich. His professorship is part of the Institute for Spatial and Landscape Development at the ETH Zürich. He is director of the institute and Delegate (Dean) of the department for the Master of Advanced Studies in Spatial Planning (MAS RP). His teaching and research focal points are on land and spatial management in the local and regional development, space and infrastructure development, transnational tasks as well as development and organization of innovative planning processes and methods in spatial planning and regional development. In 2010 he initiated the Interreg-Project Corridor development Rotterdam/Antwerp-Genova (CODE24).

Reasoning in a Macro Level Spatial Context

Rolf Signer

Abstract Dealing with problems in a macro level spatial context, such as the transnational corridor from Rotterdam to Genoa, requires a careful and thorough approach. Sheer size and an international composition means that the project cannot be grasped in its entirety all at once, that many different actors are involved, that most of the available information has to be translated, and since different countries often treat material fields differently, some form of consolidation will have to be arranged for the whole to operate. In such situations, it is important to be aware of an overall paradigm that can help with recognising and deciphering a series of possible difficulties (traps), which are often hidden or emerge only upon investigation. The paradigm includes a collection of helpful guidelines (maxims) for dealing with such large-scale and, at the same time, fragmented difficulties. Some of these traps and maxims will be explained and illustrated in this contribution.

1 Introduction

Many problems in a spatial context are of a local character, while others are regional. Adequate means of treatment are available for most problems, though more for the local problems than regional ones. Other problems are of a supra-regional character, e.g. national, such as important harbours or airports, or even of transnational relevance, such as the railway corridor from Rotterdam to Genoa, which is part of a macro level context. This contribution is about dealing with problems in a macro level context. As such, these problems require an especially thorough approach for the following reasons:

- It is not possible to grasp the entire corridor as a whole all at once, which also applies to other non-local problems.
- Whatever written, graphic or drawn material is available, due to the different kinds of codification in different countries, translations are inevitable; this in turn often leads to misunderstandings (obvious and nonobvious) and gaps

R. Signer (✉)
ETH Zürich, Institute for Spatial and Landscape Development, Stefano-Franscini-Platz 5, 8093 Zurich, Switzerland
e-mail: signerr@ethz.ch

H. Drewello, B. Scholl (eds.), *Integrated Spatial and Transport Infrastructure Development*, Contributions to Economics, DOI 10.1007/978-3-319-15708-5_3

(obvious and nonobvious). Furthermore, countries often treat material fields differently, such as how to set priorities for the development of infrastructures or settlements. In Switzerland, for example, the importance of speed in the passenger railway network is considered the interaction among infrastructures, timetables and rolling stock. The priorities in railway infrastructure development do not include achieving the fastest connections possible, rather it is to create a system of travel possibilities throughout the country by providing reliable connections between trains and/or buses.

While misunderstandings and gaps that have been identified can be clarified (though not always very easily), precautionary measures have to be taken into account for hidden misunderstandings. This chapter concentrates on hidden problems or misunderstandings by outlining the traps in certain processes and by presenting maxims that help to reveal hidden issues and to avoid spending time and money on non-critical aspects.

2 Framework

Of course, the problems or possible solutions that come up for discussion are dependent on what approach or paradigm is used (Signer 2013). Different 'lenses', so to speak, lead to different judgements of the situation, as well as different evaluations of the possibilities. This framework focuses on 'problems first' (Schönwandt 2011) and has as its foundation the 'cloud-tree metaphor' developed by Signer (2011). The main points of this approach are outlined in Fig. 1.

The 'cloud-tree metaphor' is useful in situations that focuses on a certain problem situation often labelled a 'mess' or a '*Schlamassel*' (for the origin of the word, see Althaus 2010). The main characteristic of a 'mess' is that the problem

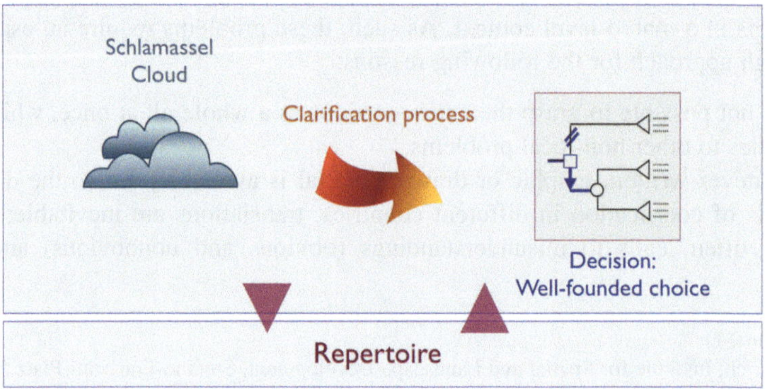

Fig. 1 The cloud-tree metaphor: cloud, clarification process, decision and repertoire (Signer 2011)

cannot be described clearly in the early stages, so it is like a cloud in that you cannot see into it and each attempt at description gives it a different shape and form.

Since the approach is action-oriented, we look for measures to help us see into this cloud. This part of the task is called the clarification process and leads to the collection of possible and feasible measures and, finally, to a decision, hence the name: decision tree. The picture of a tree is just one way to visualise a decision problem; others are written descriptions or the use of a matrix.

The tree has proven useful for exploring the options an actor may choose from; it works like an exploration device [*Erkundungsraster*]. The example used here is called a 'basic decision dilemma', where the actor has a choice between two options: one leads to known results, here the upper option, while the other option leads to a node, called 'circumstances', which have a definitive effect on the results. Real decision situations in spatial planning normally have more than two options to evaluate. The 'basic decision dilemma' in these cases is a more didactically oriented type of decision problem (for more about the different types, see Behn and Vaupel 1982).

Some knowledge is needed in order to get out of the cloud, e.g. What exactly is the problem? What kind of measures could be adequate and successful? In complex problem situations (Scholl 2011), this gathering of knowledge is often a process of trial and error, of putting aside some ideas and improving others: a tailor-made process. We could call this phase the reasoning phase, because reasoning addresses the exploratory and preliminary character of the knowledge at a certain moment. According to Bunge (2003), reasoning is 'a thought process triggered by some problem and which ideally reaches a conclusion in real time'.

This is where one's repertoire is called into action. An individual's or a groups' collection of instruments, tools and experiences that can be accessed to assist in the clarification process forms the repertoire (Signer 1995). It consists of methods, texts, key figures and graphics, such as principle sketches, that can be adapted or applied to the current situation.

One of the most important elements in making a decision is the circumstances, i.e. the situational conditions [*Gegebenheiten*] and occurrences and events [*Begebenheiten*] that the actor cannot influence. The actor has to take into account that the chosen option may not lead to the desired results, but, in fact, may fail. There are two ways to deal with that danger: either the actor improves the option during evaluation, e.g. with what may be called accompanying measures [*Flankierende Massnahmen*] or the actor prepares measures to apply in case the results are bad [*Korrekturdispositiv*], in which case, he is looking at decisions as a sequence of decisions and actions (Signer 2011). The concept of circumstance (often called a 'state of the nature') and the concept of the sequence of decisions and actions are therefore of great importance. Both are based on the assumption that neither knowledge of actions nor future results are complete and perfect and in the, often, long delay between a decision and the results of the chosen option [*Verzugszeit*] many circumstances can change.

As already stated, the aim of the clarification process is to come to a number of possible measures to solve a problem. The actor (person or organisation) then has to decide which of these options have the best chance of solving the problem. After deliberating and debating, he chooses the option he thinks is best. The actor will have to argue in favour of his preferred option, even though he may have reservations about it. However, any reasons against the preferred option must be acceptable with regard to the expected results or correctable later in the process, otherwise, they become exclusion criteria and the option has to be eliminated. The proposer will also have to present and defend the chosen option if the desired results do not materialise.

This can be called the argumentation phase in which the pros and cons are debated in order to draw out as many aspects or factors that might be hidden or misunderstood in the search for the best option possible. In contrast to the reasoning phase of trial and error exploration, this phase is characterised by its attention to the fitness and feasibility of the option chosen. The actor will thus have three collections of arguments at the end of the clarification process:

1. Arguments in favour of the chosen option
2. Arguments against the chosen option
3. Arguments that are no longer of any value, although they might have been useful during the reasoning phase (can be a huge number)

3 Traps

During the clarification process, a series of traps can arise at any point. I will concentrate on certain of these, beginning with the danger of overlooking some important aspects during the clarification process.

3.1 Hidden Gaps

The main reason why being aware of gaps is so important lies in the fact that clarification processes are tailor-made in relation to the material field. Physical or societal laws are seldom applicable, though knowledge about regularities may be available. New pieces of information can thus lead to a different decision. Let us assume that a specific, mountainous part of a railway network in an important international corridor needs to be remodelled within the next few years. The company has decided to carry out the needed measures in the 5-year period from 2020 to 2025. In the course of an observation campaign, experts come to the conclusion that the probability of a heavy landslide in the near future is very high, and would require that the corridor be closed for weeks. The company reconsiders their decision and reschedules the necessary measures to an earlier date.

In summary, avoiding information gaps in these tailor-made clarification processes is the first consideration when organising a process. Scholl (1995) partitions knowledge into 'knowing that you know', 'knowing that you don't know', and 'not knowing that you don't know'. Of course, the last type is what we have to focus on. What is a favourable framework to reduce the risk of gaps or misunderstandings? This will be covered in Sect. 4.

3.2 *Hidden Misunderstandings*

As mentioned in Sect. 2, circumstances play a crucial role in the phases of reasoning and argumentation. Circumstances may come into being with a certain probability, which has to be expressed in terms that are clear and specific. As part of a lecture series at ETH Zurich, called *Planungsmethodik* (Planning Methodologies) up to 2013 and since then *Theorien und Methoden der Planung* (Theories and Methods of Planning), with Markus Nollert as co-lecturer, I repeatedly carried out an experiment using German probability phrases, such as *Es ist wahrscheinlich, dass …* (It is probable that …), in which students were invited to give a spontaneous response for the meaning of the phrase using a figure between 0 and 100 %. The experiment has its origin in Behn and Vaupel (1982).

As the experiments were carried out using German phrases, I have included an English translation here:

Es ist wahrscheinlich, dass …	It is probable that …
Es besteht eine Chance, dass …	There is a chance that …
Es ist ziemlich unwahrscheinlich, dass …	It is rather improbable that …
Die Chance ist grösser denn je, dass …	Chances are higher than ever that …

Figure 2 shows a typical example of the results, in this case, a group of 17 students from autumn semester 2013. The orange horizontal bars indicate the total range of answers and the red vertical mark stands for the median. Note the wide range of answers, e.g. for the phrase '*Es besteht eine Chance dass …*' (There is a chance that…). This means that within a given group, the sender of the message cannot be sure what the audience's interpretation of the phrase will be. This is especially relevant when the basic decision paradigm of the audience is Baysian, i.e., the actor's decision is based on the expected values of the options, where the

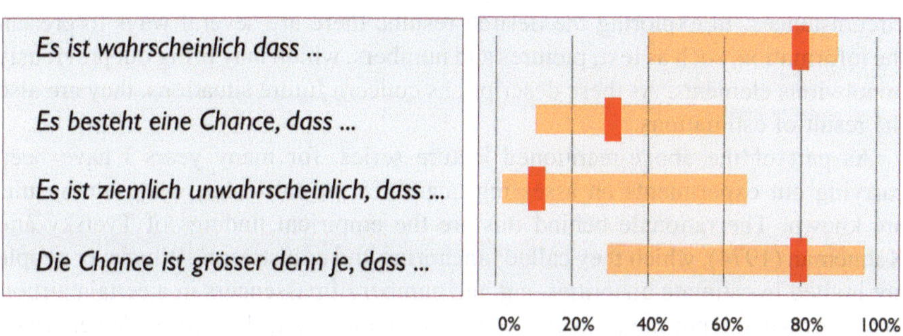

Fig. 2 Misunderstandings (Signer)

probability of its occurrence is one factor. In this case, decisions may be taken without being aware of sit. Behn and Vaupel (1982) illustrate this danger with the Bay of Pigs fiasco:

'In early 1961, President Kennedy ordered the Joint Chiefs of Staff to study the Central Intelligence Agency's plan for an invasion of Cuba by expatriates. The general in charge of the evaluation concluded that its chances of overall success were 'fair', by which he meant that they were 30 %. Yet, when the Joint Chiefs sent their report to the president, no probabilities were included; instead the report stated, 'This plan has a fair chance of ultimate success.' The rest is history. Years later the general felt that the misinterpretation of the word 'fair' had been one of the central misunderstandings of the Bay of Pigs fiasco, and he was still unhappy with himself for not insisting that a specific, numerical assessment be used. Recalled the general, 'We thought other people would think that 'a fair chance' would mean 'not too good''.

Of special interest is the phrase '*Es ist ziemlich unwahrscheinlich, dass …*' (It is very improbable that…; see Fig. 2). Among the answers, we typically get figures higher than 50 %. The same thing happened with Behn and Vaupel's group (1982) with the English equivalent ($n = 163$). Their comment: 'This phrase, however, seems to be particularly ambiguous. Responses included 80, 70, 55 and 35 %.' This example of misunderstanding shows the necessity of being aware of vagueness in certain expressions. Again, what would be a favourable framework for lessening the risk of misunderstandings? We will discuss this later in Sect. 4.

3.3 Overestimating One's Knowledge: Anchoring and Adjustment

In addition to the discussion on circumstances, the discussion about the results of a chosen option is important. What kind of results should we expect if the option actually leads to the desired result, and what might we expect if the results are not what we wanted? One may point out here that decision-making is about preferring one option—and not about choosing results, since these are a function of the circumstances. In exploring the desired results, there are several ways to present the information, such as text, pictures and numbers, which may bring out previously nonobvious elements. As these descriptions concern future situations, they are also the result of estimations.

As part of the above-mentioned lecture series, for many years I have been carrying out experiments on assessing quantities, in particular, where the results are known. The rationale behind this are the empirical findings of Tversky and Kahneman (1974), which they called 'anchoring and adjustment'. Whenever people are invited to estimate quantities, e.g. the number of passengers in a certain airport in a given year, a normal answer is a specific number, e.g. 12 million. When asked how much they trust their knowledge, they typically change their answer to a range, e.g. between 10 and 14 million. This, in fact, may better reflect their knowledge, but will typically still be insufficient, because the first expressed number is like an anchor that is hard to move away from. This experiment appeared originally in Behn and Vaupel (1982).

They suggest starting by assessing quantities from the extremes in order to not overestimate one's knowledge. Besides avoiding that particular trap, it is often not necessary to be so precise, since the decision may be the same for a certain range of the relevant quantity (see Sect. 4.4, Modigliani and Cohen 1961).

In my experiments, the students had to answer a series of such specific questions, being aware of the danger of anchoring. Figure 3 shows the example question on the number of flight movements in three airports for the year 2007. The second column (so-called Almanac Value) is the value taken from a statistical yearbook, which would, of course, be filled in afterwards. The right side shows five large rectangles for each airport, starting from the lower extreme (on the left) to the upper extreme (on the right). By convention, students' answers should come out so that only in 1 % of the answers the almanac value is below the lower extreme; the same applies for the upper extreme. We call the ratio of the answers outside the extremes the Surprise Index, which would be 2 % if the students' ability to assess quantities were perfect.

After each experiment, students were presented with the results of both the individuals and the group as a whole. The focus was to avoid having the almanac value outside the range, i.e. achieving a low Surprise Index. Not all groups showed the same performance (see Fig. 4): the first group of 17–19 students was very cautious from the beginning; the Surprise Index was not higher than 12 %. The second group, with 12–19 students, dramatically overestimated their knowledge at the beginning, despite my warnings, so the Surprise Index was 33 %, meaning that the almanac value was outside the range of the estimated extremes for one-third of the answers. However, they did improve their performance afterwards. The third group (my most recent experience) had 19–24 participants who again started with overconfidence and a Surprise Index of 24 %, but showed good development during the series, thus avoiding overconfidence. The other side of this performance is that the group did not really exploit their knowledge, but rather underestimated it.

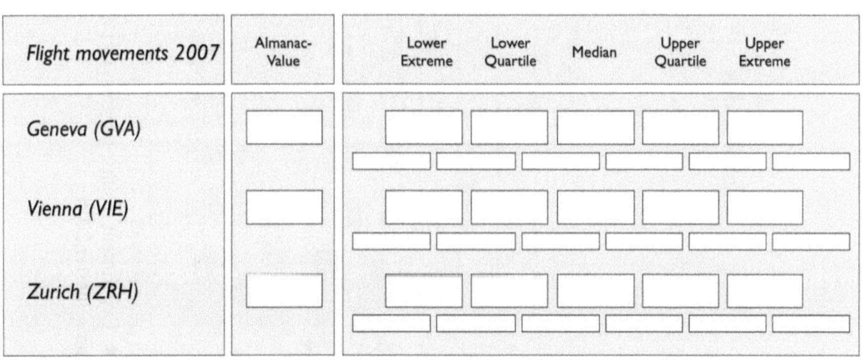

Fig. 3 Experiment question form (Signer)

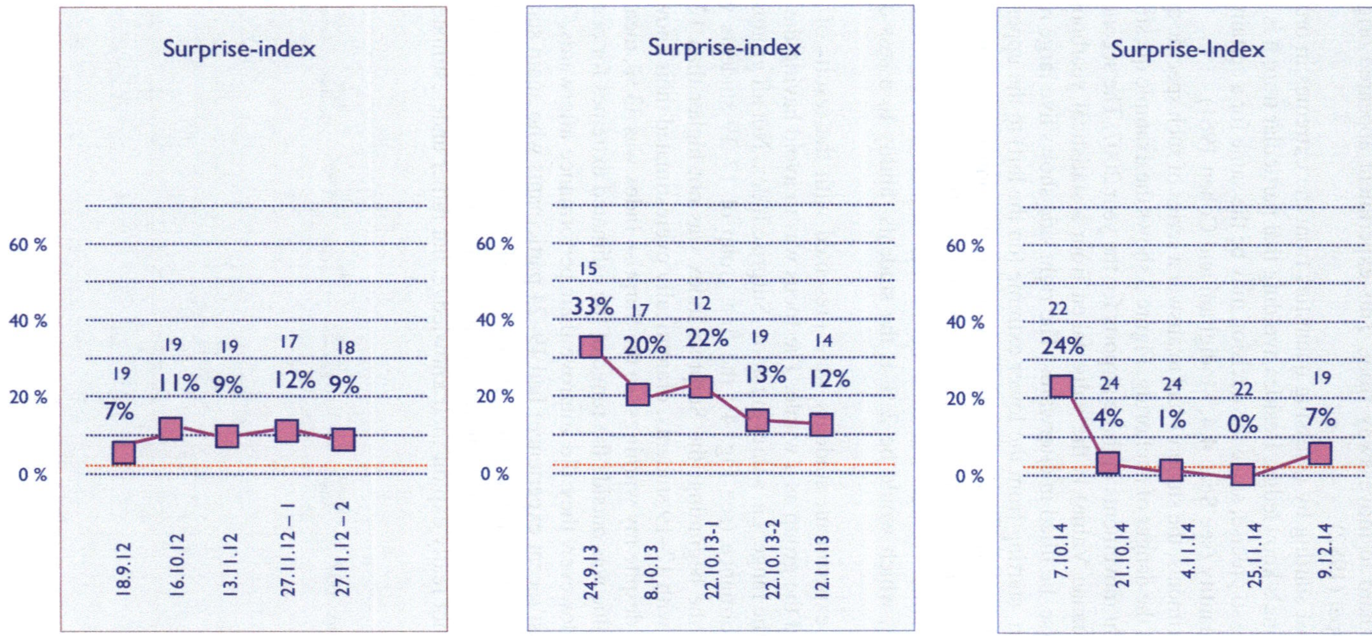

Fig. 4 Experiment results: progress of the Surprise Index of the student groups (Signer). From *left* to *right*: autumn semester 2012; autumn semester 2013; autumn semester 2014

All these estimations concern questions about the past, which can be checked in a statistical yearbook. Spatial planning, however, deals with future situations, which does not have a corresponding possibility of control. Adequate measures to avoid the trap of overestimating one's knowledge are discussed in Sect. 4.

3.4 Neglecting Possibilities with Low Probabilities

In spatial planning, the systematic consideration of circumstances is critical. It is about what might happen, i.e. about possibilities. Here, the question of reality is addressed. According to Bunge (1977), reality is partitioned into actuality and real possibility—even if the possibility does not materialise. The Stoic philosopher, Chrysippus, defines the possible as 'that which is not prevented from happening by anything, even if it does not happen.' Chronos, however, says, 'the possible is that which either is or will be true'. Here, the notions of an unrealized potentiality and missed opportunities find no room. Chrysippos' ontology is called possibilistic, whereas Chronos' ontology is actualistic. As spatial planners, we should favour the possibilistic approach, since one of our main tasks is to decide on possible measures and to take possible circumstances into account that may influence the results.

One common trap is the confusion between possibility and probability. Schelling writes in the foreword of the German translation of Wohlstetter's (1966) book about Pearl Harbour: 'When planning, we tend to confuse the unknown with the improbable. A possibility we do not take into consideration seriously seems to be strange. We consider strange things to be improbable, and improbable things must not taken into consideration seriously.' This trap leads immediately to the maxim 'Possibilities first, Probabilities later', meaning that one has to make a great effort to explore the possibilities, whether they are options, circumstances or results.

4 Maxims

This collection of maxims is presented in order to provide a framework of precautionary measures to avoid traps when reasoning in a macro level spatial context. According to Maurer (1995), 'Someone who knows about maxims and doesn't need them is better off than someone who needs them and doesn't know them'.

4.1 Total Evidence

As this contribution has indicated, it is of crucial importance not to overlook anything of importance; 'importance' being understood here as having the potential to influence or change a decision. This is easy to understand, but not always easy to attain. The next two maxims will help make this maxim operational.

4.2 Competition in Three Cycles

Clarification processes in spatial planning should be structured in three cycles with groups of experts working competitively. This allows the participants to look at the problem and its possible solutions several times and thus profit from what Hadamard calls 'the incubation period' (Simon 1966). The assumption behind the three-cycle approach, in contrast to the usual sequence of phases, is that several approaches with different thematic priorities are needed to get a broader range of possible solutions and to detect gaps and misunderstandings more easily, thus providing a platform for learning.

The Test Planning Method is based on this maxim. Three events are important in a test planning process, where all participants meet for a day: a workshop meeting, an intermediate presentation and a final presentation of the results (Scholl 2011). Each of these events represents a cycle, and at each event, several teams of experts can share and discuss their ideas and possible solutions in a competitive manner.

4.3 Possibilities First, Probabilities Later

The maxim about competition and three-cycles provides a good framework for avoiding the danger of neglecting possibilities with low probability. Since it is possible to bring bold ideas into discussion in the early stages of a clarification process, the pros and cons can be discussed thoroughly, even if some of the participants may consider this a waste of time. These ideas concern possible options, possible circumstances and possible results. Dealing with possible circumstances that may lead to the failure of an option is in this stage one of the most challenging tasks. This, in fact, is not a pessimistic attitude, since it makes it easier to be prepared in case an option fails. The Swiss writer Friedrich Dürrenmatt (1962) goes along with this idea: 'If you start out with a story, you must think it through to its conclusion. … A story has been thought out to its conclusion when it has taken its worst possible turn'. This is part of Dürrenmatt's *21 Points to the Physicists*, and the play *The Physicist* itself is considered a comedy…

4.4 Mixing Codes

Spatial planners are accustomed to using texts, numbers and pictures in their work. They draw plans, principle sketches and diagrams, they use statistical numbers about the past and estimated numbers for the future, and they use words to explain and comment. These three ways of communicating are called cultural techniques [*Kulturtechniken*]. In many cases, texts, numbers and pictures can be found at the same time, which is then called a 'whole text' [*Gesamttext* or *Supertext*]. Texts with diagrams are a common example (Signer 2012).

Figure 5 is an example of such a 'whole text' and shows how to make different aspects of a macro level space visible, such as the Alpine-Rhine Transport Corridor.

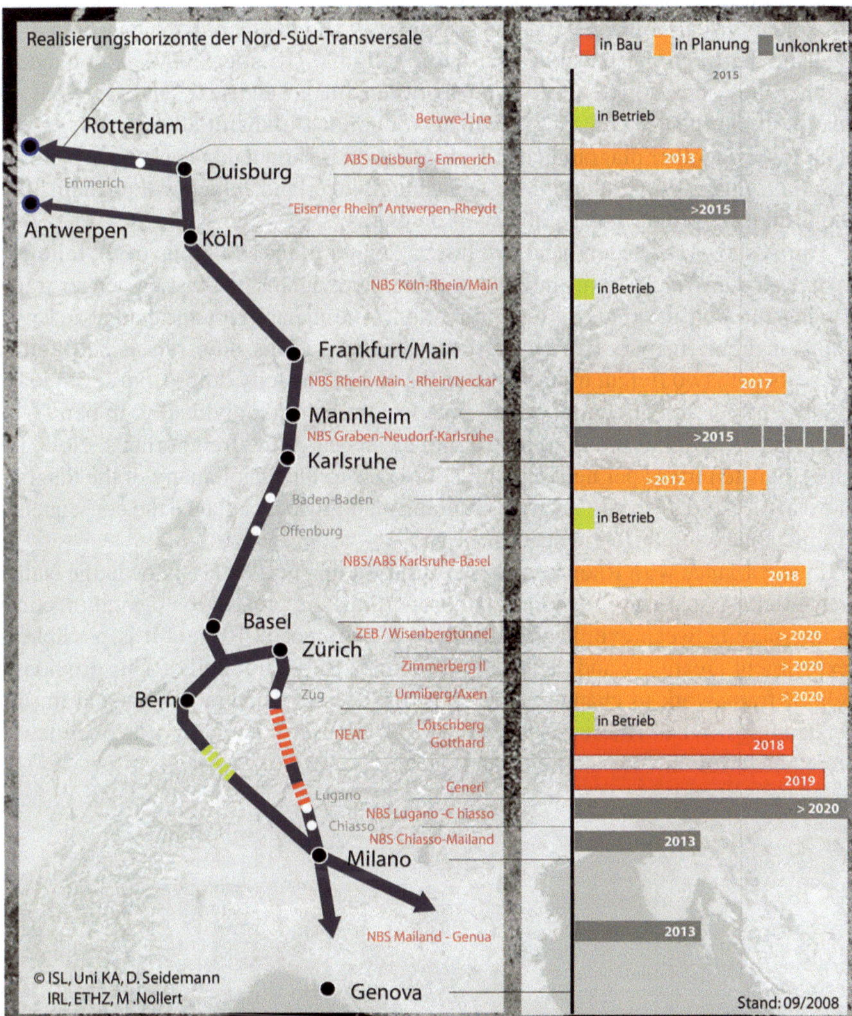

Fig. 5 A diagram with accompanying text showing an exemplary use of pictures, texts and numbers. It contains information about the location, time, and planning state of building projects on the north-south railway route from Rotterdam to Genoa as of 09/2008, now called the Alpine–Rhine Corridor. *Source*: University of Karlsruhe (Dirk Seidemann), ETH Zurich (Markus Nollert)

On the left, the transport corridor is drawn on a map with the three base tunnels in Switzerland highlighted (NEAT). On the right, the diagram shows the expected timeline of certain important sections of the corridor and whether they are under construction (red), in the planning phase (orange) or have an indefinite status (grey). This is of special interest because the information on the different sections shows the lack of synchronicity at a glance. It is what we call, along with Kant, a diagrammatic demonstration (Schmidt-Burkhardt 2009).

According to Krämer (2010), diagrams are 'a form of symbolic depiction . . . that create a hinge between thought and view . . . This interim, graphic sphere makes it possible to visualise the general in a sensual way and to embody the conceptual.' It mediates between the sensitive and the intelligible (Krämer 2009). The diagram ignores the 'competitive relationship between image and text' and develops its quality in interplay with its text elements (Schmidt-Burkhardt 2009).

In Fig. 6, another diagrammatic demonstration is shown that concerns a pictorial plan for trains in the Gotthard Base Tunnel in Switzerland, which will open in 2016, and is an example of the operativity of diagrams.

Part A shows on the left hand two fast passenger trains as a group (red), followed by five slower freight trains in a block (blue) and, again, two fast passenger trains passing through the tunnel in one direction. The capacity of the tunnel is seven trains an hour. In part B, two the fast passenger trains now run at half-hourly intervals with two freight trains between them. The capacity drops from seven to six trains per hour and direction, because the fast trains run individually. In part C, all of the trains travel at the same speed, i.e. the speed of the freight trains. Now, the capacity is ten trains per hour and direction. Due to the simultaneity of the diagram one can 'see', so to speak, the trains running, and grasp at a glance the consequence of timetables with different speeds.

In accordance with Rheinberger (see Bredekamp et al. 2008), knowledge is thus created in a visual way, 'relations and proportions are created via mental practices in interplay between eye, hand and intellect' (Krämer 2010) and it is possible to 'experiment creatively and pictorially' (Schmidt-Burkhardt 2009). This provides a helpful framework to change views and identify gaps, contradictions and misunderstandings. What's more, such 'whole texts' are useful for gaining an overview, a prerequisite for any decision of importance.

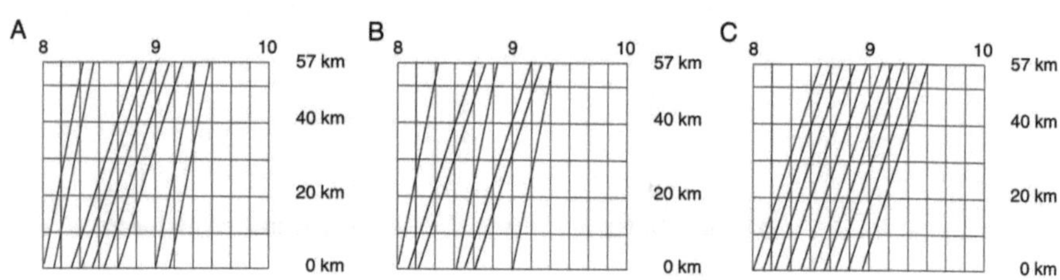

Fig. 6 Pictorial plan in one direction for trains in the Gotthard Base Tunnel showing time in hours. X-coordinates: example hours from 8 to 10. Y-coordinates: distance of the tunnel (0–57 km); the angle of the *lines* are indicators for speed: the steeper the line, the higher the speed

Going back to the example of the trap of misunderstandings in Sect. 3, if one has to express the probability of an event, it can be done with a phrase or with numbers. The phrase is risky because the audience can interpret it in different ways. The number is risky because it may reflect overconfidence, unless ranges or thresholds are given. There are situations, however, where the vagueness of a phrase may be useful; this is what Kissinger called 'constructive ambiguity' (see Berridge and James 2004). 'Constructive ambiguity avoids premature closure of options' (Jönsson and Aggestam 2007). However, when justifying a certain decision (the close of the cloud-tree metaphor), no vagueness is allowed. Nevertheless, clear arguments are needed, though without spurious accuracy. This leads us to the maxim of economical investigation.

*4.5 Economical Investigation [***Sparmaxime***]*

More than 50 years ago, the economists Modigliani and Cohen formulated their maxim (Modigliani and Cohen 1961): 'Don't devote resources to estimating particular aspects of the future if, no matter what you might find out (with due consideration to what you might conceivably find out), you would not be led to act differently from the way you would have acted without finding out.' Emphasis is on 'act differently', i.e. choose a different option.

This maxim is important for two reasons: First, its focus is on limiting the amount of resources (time, money, brains) spent on researching and analysing information on the background of a planned action, i.e. resources spent on information that will not affect the decision. This is something that surveyors have been considering for a long time: no important measuring campaign takes place without a sound theoretical plan. In other words, the quality of the conceivable results of the campaign must satisfy the needs of a specific decision—but not more!

The second reason for this maxim concerns the necessity for robust decisions, i.e. decisions that are not based on the spurious accuracy of the justifying arguments (remember the trap of anchoring). Good (1962) says: 'The art of being correct lies in making the weakest possible statements.'

5 Conclusions

Dealing with problems in a macro level spatial context requires a careful and thorough approach. It is important to be aware of an overall paradigm that can help with recognising and deciphering a series of possible difficulties (traps), such as hidden gaps, hidden misunderstandings, overestimating one's knowledge and neglecting possibilities with low probabilities. The paradigm includes a collection of helpful guidelines (maxims) for dealing with such large-scale and, at the same time, fragmented difficulties. They concern basic aspects, such as organising the clarification process as a competition carried out in several cycles, recommendations on simultaneously using different sign systems (codes) for communication purposes, recommendations on allowing bold ideas, despite their apparently low probabilities, and recommendations to limit the search for information, depending on the decision problem.

References

Althaus HP (2010) Kleines Lexikon deutscher Wörter jiddischer Herkunft, 3. Auflage. Beck, München, S 183

Behn RD, Vaupel JW (1982) Quick analysis for busy decision makers. Basic Books, New York, NY

Berridge GR, James A (2004) A dictionary of diplomacy. http://grberridge.diplomacy.edu/dict_comp_a_e.htm. Accessed 18 Feb 2009

Bredekamp H, Schneider B, Dünkel V (Hrsg) (2008) Das Technische Bild. Kompendium zu einer Stilgeschichte wissenschaftlicher Bilder. Akademie-Verlag, Berlin

Bunge MA (1977) Treatise on basic philosophy, Ontology: the furniture of the world, vol 3. Dordrecht

Bunge MA (2003) Dictionary of philosophy. New York, NY

Dürrenmatt F (1962) The physicists. (Including 21 Points to the Physicists)

Good IJ (ed) (1962) The scientist speculates. An anthology of partly-baked-ideas. Basic Books, New York, NY, p 212f

Jönsson C, Aggestam K (2007) Diplomacy and conflict resolution. Prepared for the NISA conference on 'Power, Vision and Order in World Politics', Odense, 23–25 May 2007

Krämer S (2009) Operative Bildlichkeit. In: Hessler M, Mersch D (ed.) (2009) Logik des Bildlichen. Zur Kritik der ikonischen Vernunft. Transcript, Bielefeld, pp. 94–122

Krämer S (2010) Zwischen Anschauung und Denken. Zur epistemologischen Bedeutung des Graphismus. In: Bromand J, Kreis G (ed.) (2010) Was sich nicht sagen lässt. Das Nicht-Begriffliche in Wissenschaft, Kunst und Religion, Akademie-Verlag, Berlin, pp. 173–192

Maurer J (1995) Maximen für Planer. Zürich (English version see: Scholl B (ed) HESP – higher education in spatial planning, positions and reflections. Zurich, 2012)

Modigliani F, Cohen KJ (1961) The role of anticipations and plans in economic behavior and their use in economic analysis and forecasting. University of Illinois, Urbana, IL

Schmidt-Burkhardt A (2009) Wissen als Bild. Zur diagrammatischen Kunstgeschichte. In: Hessler M, Mersch D (ed) (2009) Logik des Bildlichen. Zur Kritik der ikonischen Vernunft. Transcript, Bielefeld, pp. 163–187

Schönwandt W (2011) Probleme als Ausgangspunkt für die Auswahl und den Einsatz von Methoden. In: Akademie für Raumforschung und Landesplanung (ARL): Grundriss der Raumordnung und Raumentwicklung. Hannover

Scholl B (1995) Aktionsplanung. Zur Behandlung komplexer Schwerpunktaufgaben in der Raumplanung. vdf-Verlag, Zürich

Scholl B (2011) Maximen für Auswahl und Einsatz von Methoden. In: Akademie für Raumforschung und Landesplanung (ed.): Grundriss der Raumordnung und Raumentwicklung. VSB Verlagsservice, Braunschweig

Simon HA (1966) Models of discovery and other topics in the methods of science. In: Colodny R G (ed.) Mind and Cosmos. The Univerity of Pittsburgh Press, Pittsburgh, pp. 22–40

Signer R (1995) Argumentieren in der Raumplanung. Dissertation ETH Zürich, Zürich

Signer R (2011) Ein Klärungsprozess für komplexe Schwerpunktaufgaben in der Raumplanung. In: Akademie für Raumforschung und Landesplanung (ed.): Grundriss der Raumordnung und Raumentwicklung. VSB Verlagsservice Braunschweig

Signer R (2012) 'The Image Precedes the Idea' – Images in Spatial Planning. In: Internationales Doktorandenkolleg Forschungslabor Raum (ed) Research laboratory: space – the logbook. Jovis Verlag, Berlin

Signer R (2013) Planungstheorie und -methodik – Ansätze oder die Art des Denkens. Lecture within the framework of the Doctoral College 'Research Lab Space', TU Vienna, March 2013

Tversky A, Kahneman D (1974) Judgment under uncertainty: heuristics and biases. Science 185 (4157): 1124–1131, Reprint in: Kahneman (2011) Thinking fast and slow. Penguin Books, London

Wohlstetter R (1966) Pearl Harbor. Signale und Entscheidungen. Erlenbach (Original: Pearl Harbor. Warning and Decision, 1962)

Dr. **Rolf Signer** (*1948) first completed a diploma in cultural engineering at the Swiss Federal Institute of Technology Zurich (ETH). Subsequently, he completed post-graduate studies in spatial planning at the same university and was awarded a doctorate (Dr. Sc. Techn.). He works as a specialist planner in Switzerland, partner in an office of urban and regional planning in Zurich. Their projects encompass works on an urban and regional scale, both at home and abroad. As a member of the International Society of City and Regional Planners (ISOCARP), he was head of the national Swiss delegation from 2000 to 2006. Since 2009, he has been president of the Zurich Study Society for Construction and Traffic Issues (ZBV).

Exploring the Affordances of Collaborative Problem-Solving Technologies in the Development of European Corridors

Isabella M. Lami and L. Alberto Franco

Abstract Collaborative problem-solving (CPS) technologies are increasingly being used to support spatial planning processes. Despite these advances, published accounts of their use have largely failed to explain if and how uses of these technologies differ from other forms of spatial planning support. In this chapter, we adopt an affordance perspective to examine how the material features of CPS technologies can support stakeholder interactions. We illustrate our theorising by offering an empirical case vignette of the application of *InViTo*, a particular CPS technology designed and used to support the planning of Eurocorridor 24, which is now called the Rhine–Alpine Corridor. We conclude with a brief discussion of the implications of adopting an affordance lens for assessing the current use of CPS technologies in spatial planning processes. Directions for future research are also proposed.

1 Introduction

Any territorial transformation creates potential conflict, raised expectations, and unanticipated effects. Cities and territories have always been subject to change, but, compared to the past, there is now an increased difficulty in prefiguring future developments, as well as managing their effects. This greater complexity favours the adoption of new decision models that are able to both assess and design new transformation scenarios. Moreover, the transformation of the territory is almost always undertaken by a public administration agency, and public choices have a binding and often irreversible character for all citizens. So, it seems reasonable that

I.M. Lami (✉)
Interuniversity Department of Regional and Urban Studies and Planning, Politecnico di Torino, Viale Mattioli 39, 10125 Torino, Italy
e-mail: isabella.lami@polito.it

L.A. Franco
School of Business and Economics, Loughborough University, Leicestershire LE11 3TU, UK
e-mail: l.a.franco@lboro.ac.uk

H. Drewello, B. Scholl (eds.), *Integrated Spatial and Transport Infrastructure Development*, Contributions to Economics, DOI 10.1007/978-3-319-15708-5_4

any transformation not only has to be assessed in advance, but also to be expressed and communicated in a satisfyingly understandable way. This latter aspect, in particular, has an impact on the assessment activity by giving it an iterative and interactive character (Lami and Ferretti 2014).

The iterative and interactive nature of the assessment activity is also enabled by the current availability of data. Indeed, in the past, one of the difficulties in urban and territorial planning was the lack of data-measuring activities. Nowadays, the problem is the opposite: there is almost too much data on urban matters, such as people's activities, density, pollution, traffic, climate, environment and so on, with large amounts of information organised into extensive databases. Such an amount of data reflects the complexity of the reality they describe, but they are often difficult to read and need to be processed in order to extract useful information (Pensa et al. 2014). Several methods are available to mine the content of databases with a view to informing urban and territorial planning. Indeed, collaborative problem-solving (CPS) technologies, such as geo-visualisation tools, multi-criteria evaluation models, cartographies of controversies, are increasingly being used to support spatial planning processes (Geertman and Stillwell 2009; Malczeski 2006; Dykes et al. 2005).

Despite their widespread application, published accounts of the use of CPS technologies have largely failed to explain if and how uses of these technologies differ from other forms of spatial planning support. In this chapter, we argue that CPS technologies are important to spatial planning processes because they afford behaviours that were difficult or impossible to achieve in combination before these technologies entered the spatial planning field. We suggest that these 'affordances' can influence key aspects of multi-stakeholder spatial planning processes, such as negotiation and knowledge creation.

The remainder of this chapter is organised as follows. In Sect. 2, we outline the characteristics of spatial planning problems and the role that CPS technologies can play in supporting the spatial planning process. Section 3 introduces the concept of affordance, and explores its usefulness for the study of CPS technologies within the context of spatial planning processes. Section 4 describes the characteristics of *InVITo,* a particular CPS technology designed to support the planning of the Rhine–Alpine Corridor. In Sect. 5, we show, via an empirical vignette, how perceptions of affordance called forth by the *InVITo* technology affect stakeholder interactions in a pilot spatial planning workshop. Finally, Sect. 6 offers some considerations about the usefulness of the affordance concept for assessing the current use of CPS technologies within spatial planning processes. Section 6 also presents concluding remarks and directions for future research.

2 The Characteristics of Spatial Planning Problems and the Need for Problem-Solving Support

For planners and geographers, spatial planning problems concern the development of a territory, namely the implementation of public policies, programmes and projects, which are generally defined by the literature as the subject of territorial governance (Faludi 2012; Davoudi et al. 2008; Healey 1997). The key dimensions of territorial governance include (Tango-Espon 2013): coordination of actors and institutions; integration of policy sectors; mobilisation of stakeholder participation; ability to adapt to changing contexts; realisation of place-based/territorial specificities and impacts. Spatial planning is an analytical and cyclical process, and the development of a territory is considered to involve the improvement in the efficiency, equality and environmental quality of an area at urban, regional and cross-border levels.

By contrast, geo-visualisation experts characterise the nature of spatial planning problems from the perspective of spatial planning support (Andrienko et al. 2007, 2010; Malczewsky 2006; Boroushaki and Malczewsky 2010). Andrienko et al. (2007) define spatial planning support as 'computerised assistance to people in the development, evaluation, and selection of proper policies, plans, scenarios, projects, or interventions where the problems have a geographic or spatial component' (p. 840). The challenge for spatial planning support is to find the most efficient way to tackle geo-temporal character of spatial planning problems, combined with the heterogeneity of the information, in order to bridge the knowledge and communication gap between experts and public users. Three aspects characterise spatial planning problems from the perspective of spatial planning support (Andrienko et al. 2007):

1. *The complex nature of geographic and temporal spaces*. This means the need to take into account the metric properties, the topological relationships between objects in the space and above all the heterogeneity of the space, i.e. objects, phenomena and events occur in a physical space at different times. Moreover, dealing with a real space is quite different from an abstract one: the distances in a real territory are not the same as Euclidean distances on a plane, this implies that sometimes the theories are difficult to apply to an area with its orographic features.
2. *Multiplicity of actors*. Typically, spatial planning processes involve various stakeholders, e.g. citizens, developers, organisations, with different roles, interests and levels of knowledge of the spatial problem. Spatial planning problems thus need to be tackled through a collaborative multi-stakeholder process. This is consistent with the current trend to democratise planning and decision-making (Simao et al. 2008).
3. *Tacit criteria and knowledge*. As Malczewski (2006) notes, 'Spatial decision problems typically involve a large set of feasible alternatives and multiple, conflicting and incommensurate evaluation criteria' (p. 703). These criteria

relate to properties of spatial and/or temporal distribution that cannot be easily quantified to enable automatic processing, or they may be hard to articulate. Since they are unavoidable in decision processes (Newo and Wand 2005), it is necessary to develop spatial planning to stimulate the use of tacit criteria and knowledge to make the spatial planning process more effective and efficient.

4. *Nature of data to be handled*. There are specifics of time, dependencies between observations, issues about the scale and problems connected to data quality (Andrienko et al. 2010). Time has an inherent semantic structure, and two specific aspects of the dimensions of time have to be considered in spatial problems: the temporal primitives that make up the temporal dimension (time points or time intervals); and the structural organisations of the temporal dimension (ordered time, branching time and multiple perspectives). In spatial problems, it is often the case that spatial and temporal characteristics are correlated. These dependences are not only constraints, they can in fact also provide opportunities for data processing, such as interpolation and extrapolation (filling gaps in incomplete data), and integration of information using references to common locations and/or time units. The scale of spatial analysis, reflected in the units in which phenomena are measured and the measurements are aggregated, may significantly affect the results, for instance, patterns or relationships perceived at one scale may not be detected when analysed at another scale. Currently, there are no systematic methods to detect the better scale at which a process should be studied. Finally, spatio-temporal analysis has to consider the different aspects of uncertainty (MacEachren et al. 2004), for instance, (in) accuracy, (im)precision, (in)completeness, (in)consistency, and subjectivity, etc.

Current spatial planning support technologies, such as geographical information systems (GIS), are often incapable of coping with one or more of the characteristics of spatial planning problems described above (Boroushaki and Malczewski 2010; Andrienko et al. 2010; Masala 2014). Consequently, users of spatial planning support technologies are forced to simplify the problems in order to adapt them to the capabilities of these technologies. A number of spatial planning support technologies, known as collaborative problem solving (CPS) technologies, are increasingly being used to address the complexity of real-life spatial problems. These include, among others, geo-visualisation tools (Malczewski 2006; Andrienko et al. 2011; MacEachren et al. 2004) and multi-criteria decision analysis (MCDA) tools (Figueira et al. 2005; Roy and Slowinski 2013; Abastante et al. 2013). The most common spatial planning problem areas within which CPS technologies are used can be grouped into three types (Andrienko et al. 2007):

1. *Site selection*. This class of problems has become a classical case where the combination of visualisation tools (such as GIS) with MCDA works fairly well (Malczewski 1999, 2006; Ferretti 2012; Pensa et al. 2013). However, it should be noted that much of relevant information is often not taken into account in order to make a problem manageable;
2. *Operational planning*. A range of computational methods has been developed to deal with the complexity of real-world spatial planning problems. Methods such

as simulation and optimisation can consider a large number of variables, constraints, and stakeholder interests to generate and evaluate possible scenarios. However, new methods and tools, or new uses of existing methods and tools, are necessary to incorporate tacit knowledge and 'soft' evaluation criteria.

3. *Emergency management*. Using remote sensing to support disaster management activities has become a common approach, principally thanks to the large and timely availability of various kinds of remotely sensed data as well as geospatial information acquired in the field (Boccardo and Tonolo 2015). In emergency situations, analysts and decision-makers do not have time to consider all the possible variants of alternative options in detail, or to search for an optimal variant (the cost of an error may be human lives). Furthermore, decision making in emergency situations is a dynamic process in which key parameters change quickly. Therefore, tools must go beyond the display of results of a database query or of alternatives by way of computational analysis; they must support the entire problem solving process so that human expertise and technology capabilities are iteratively applied and mutually reinforcing. Maps and related visualisation tools have a fundamental role in this case (Andrienko et al. 2007).

While the literature on the use of the CPS technologies to address the complexity of real life spatial problems is increasing, the application of methods to measure their efficacy in facilitating the negotiation in spatial development processes are, in contrast, rare. The argument advanced in this paper is that the concept of 'affordance' could be particularly useful in this sense.

3 The Concept of Affordance

James Gibson (1986), the ecological psychologist, coined the term 'affordance' following his research on visual perception. His theory of affordances aims to explain how an individual perceives the behavioural possibilities of an object, noting that the same object can call for different possibilities for action. Thus, for example, a rock can be used as a weapon or as a paperweight. Similarly, an apple tree can be used as a shelter or as a source of food. Gibson claims that what we perceive is not what an object is but rather what kinds of uses it 'affords'. Obviously, we would still be able to discriminate an object's material features, e.g. colour, shape, if prompted to do so. However, the empirical evidence suggests that what we normally pay attention to is what the object affords us (Grezes and Decety 2002; Symes et al. 2007; Tucker and Ellis 2004). Beyond the individual, the theory of affordances is also taken to be relevant to social behaviour (Gaver 1996; Hutchby 2001). In this context, because an object can be perceived to afford multiple uses, it is possible that the same object can produce multiple behavioural outcomes.

Gibson's formulation has been further elaborated by numerous studies in psychology, sociology, and information management. Recent developments of the notion of affordance emphasise its relational character (Chemero 2003; Faraj and

Azad 2012; Hutchby 2001; Leonardi 2011). In this view, affordances are constituted in relationships between social actors and the materiality of the objects with which they interact. The term 'materiality' here refers to the features of an object, including physical or digital features (Leonardi 2012; Leonardi et al. 2012). In this formulation, materiality exists independent of social actors, but affordances do not. Because social actors come to the materiality of an object with diverse goals, they perceive that object as affording distinct possibilities for action. Furthermore, affordances of an object can change across different contexts of use even though its materiality does not. Similarly, people may perceive that an object offers no affordances for action, perceiving instead that it constrains their ability to carry out their goals.

We argue that to gain a deeper understanding of the use of CPS technologies in spatial planning processes, it is important to focus attention on how the perceived affordances affect stakeholders' behaviour. Take the models created by geo-visualisation tools (Geertman and Stillwell 2009; Klosterman 1997) to support spatial planning processes. These models are typically used by a group of stakeholders working in a workshop environment and can be edited 'on-the-spot' if needed. The high degree of editability enabled by geo-visualisation tools is possible because of computer support (Klosterman 2012) Perhaps more importantly, stakeholders will construct individual perceptions of whether this editability affords them the possibility to achieve their goals or not, and take action accordingly. Thus, affordance may have direct consequences for how stakeholders interact in the spatial planning process. To understand what these consequences are and the conditions under which they are likely to appear, we must first recognise that CPS technology makes certain actions possible and others impossible, or at least more difficult, to achieve. This insight has been used by organisation science scholars to show how variations in the level of affordances called forth by the same CPS technology can lead social actors to engage in similar or disparate dynamics (Franco 2013).

4 *Invito* as a CPS Technology

In this section, we will briefly outline the material features of *InViTo*, a particular CPS technology developed as a visual method to communicate spatial information. The main purpose of *Invito* was to improve the understanding of spatial data in spatial planning processes through the exploration of alternative transformation scenarios (Masala and Pensa 2014; Masala and Melis 2014). *Invito* offers a way to represent different typologies of geo-referenced data and combine them in order to visualise the 'hidden connections' (Dodge 2005) among these data. In contrast to the standard GIS tools commonly used, the parametric features of 3D modelling embedded in *Invito* allow for direct and immediate changes in shapes and colours of volumes, thus offering rapid interactive visualisations of geospatial information (Pensa et al. 2011).

Fig. 1 *Invito* workshop setting. *Source*: Author's photo, workshop 2013

In addition, *Invito* can be enhanced by integrating it with multi-criteria decision analysis tools, such as those based on the Analytic Network Process (ANP) (Saaty 2001; Saaty and Vargas 2006), which provide assistance in structuring and evaluating decision options. As a CPS technology, the enhanced *Invito* can thus offer support along two distinct dimensions of a spatial planning process: (1) the analytical dimension that involves the generation of possible solutions that best represents the preferences of those involved (Feick and Hall 1999, 2004; Jankowski and Nyerges 2001; Simão et al. 2009; Boroushaki and Malczewsky 2010); and (2) the deliberative and communicative dimension that involves the gradual building of consensual agreement on a coordinated action (Jankowski and Nyerges 2001; Rinner 2006).

Figure 1 shows an *Invito* workshop setting, and Table 1 summarises *Invito*'s key material features: tangibility, associability, and editability. In the next section, we present an empirical case vignette to help ground and illustrate how *Invito*'s materiality and its perceived affordances interact to shape stakeholder behaviour in a spatial planning workshop.

5 Empirical Case Vignette

In this section, we draw on the use of *Invito* in a pilot workshop that was part of the Interreg IVB NWE Project CODE24, an 8-month research project concerned with developing decision process support for the development of the Genoa–Rotterdam railway corridor (Abastante and Lami 2012; Lami et al. 2014). The workshop was designed to explore the Frankfurt–Mannheim section of the corridor with a group of

Table 1 Materiality of *Invito*

Material feature	Description	*Invito* illustrative features
Tangibility	Ability to make its content visible and concrete	• Multiple map (2D and/or 3D) and data views • Shapes can be shown in different views according to specific parameters
Associability	Ability to relate its contents based on shared attributes and transform input into output	• Linking (via laptop) • Each geographic element is described with its main characteristics and attributes in order to use data in model calculation • Clustering • Preference and ranking • Each element can be associated to a mathematical function, e.g. based on distance, that describe its behaviour; the function can be manually implemented or read from an external file • Mathematical functions and relationships among them define shapes and their behaviour
Editability	Ability to modify its contents immediately	• Concept editing and deleting (via laptop) • Style editing (via laptop) • Customised output graphs; output can be displayed in multidimensional maps in real-time in Google Earth according to input data and actor choices, i.e. quantitative maps, symbolic comparative maps, volume graph maps • Output changes in real time according to user choices following the rules of assessment method

relevant key stakeholders representing the areas under consideration. The group discussion focused on the areas where priorities and development strategies were not yet clear and thus needed to be established. The workshop was co-facilitated by the first author as part of a larger facilitation team (for further details, see Lami 2014), and held at the Value Lab of ETH Zurich. The room layout included three large vertical screens and two large horizontal screens.

Three scenarios concerning the development of the Frankfurt–Mannheim area were designed for the workshop by the researchers at the Institute of Spatial Planning, ETH Zurich. They were mainly focused on railway transport characteristics, as shown in Table 2. A multi-criteria evaluation model was built interactively and iteratively during the workshop, which included key parameters, such as benefits, costs, risks and opportunities.

The following interaction segment shows team members' discussions on different aspects of the evaluation model. Specifically, stakeholders were asked to assign

Table 2 Three alternative scenarios for the transformation of the Frankfurt–Mannheim area

Scenario 1: Development of a metropolitan region Frankfurt-Mannheim	Development of a metropolitan area between Frankfurt and Mannheim. Creation of a new high-speed line to directly connect the two cities (maximum speed = 250 km/h; total travel time = 20/25 min). This would relieve the Frankfurt rail hub, by reducing suburban traffic. No stop in Darmstadt; by-pass Mannheim. Renewal of brownfield areas in the Mannheim Region, primarily for residential and service use (available area = 1475 ha; potential new inhabitants = 180,000/250,000).	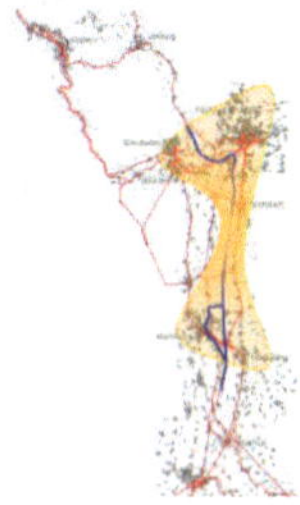
Scenario 2: Empowerment of the existing resources	Development of Mannheim independent from Frankfurt. More rational rail traffic management through the separation of traffic flows. A specialisation of existing lines in freight, regional and long-distance traffic and new signalling systems are required. Small changes in travel times. Hierarchical regional structure on three levels focusing on the regional transport (development is also concentrated on small nodes). By-pass Mannheim. Renewal of brownfield areas in the Mannheim Region, with focus on on-going activities: logistics, R&D, industry (available area = 1475 ha).	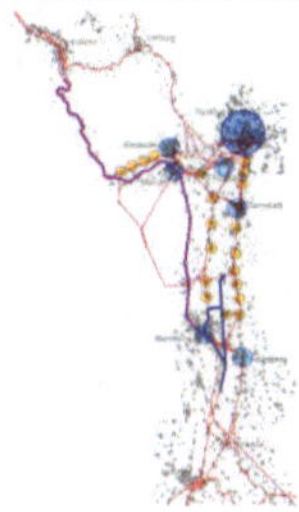
Scenario 3: Development of Mannheim logistic hub	Development of Mannheim as a logistics hub compared to the rest of Germany and Europe (considering the railway Corridors crossing Europe). Upgrading the railway junction in Mannheim. Improving connections with the River Port up-grading the Shunting Yard in Mannheim. By-pass Mannheim. Separation of traffic flows focusing on the development of the freight railway line Mainz-Mannheim. New signalling is required. Renewal of brownfield areas in the Mannheim Region, focusing on intermodal centres. Available area = 1475 ha.	

Source: Tosoni and Günther, ETH Zurich, IRL for Interreg IVB NEW project CODE24

weights to different aspects of the decision problem by making pairwise comparisons. Weights can vary between a value of 1 (equally important) to a value of 9 (extremely more important). The segment illustrates how stakeholders' behaviours and *Invito*'s material features (tangibility, associability, editability) become interwoven during the interaction.

F1 The second question is still related to the general criteria in order to choose the best scenario. With reference to the choice of the best alternative development scenario (looking at map), which one of these two aspects do you think is more important? The economic benefits or the environmental benefits? (Pointing at evaluation model) So, the economic benefits are 'increase in the level of attractiveness' and 'valorisation of the real estate market', while the environmental benefits are 'reduction in noise pollution' and 'reduction in pollutant emissions'. […]

ST1 Ok! I'd like to give you a comment. First, I started to go in the other direction, but I 'read' the aspects (pointing at evaluation model and maps) and I would go for a five for environmental aspects, because I think that maybe if we try to resolve our environmental problems, maybe we can have a better quality of life. This could enhance the economic benefits. So, let's say go for the environmental benefits first and you will also have economic benefits.

F1 So, you think that the environmental aspects are so important that…

F2 The environmental quality is so important that it can influence the real estate market, etc.

ST1 Yes. If we lose quality, we will lose everything.

F1 Yes. I see your point.

Key: F1=Facilitator 1; F2=Facilitator 2; ST1=Stakeholder 1. The use of an ellipsis in brackets […] following a statement indicates that comments by one or more participants have been edited out.

In the segment above, ST1 is making a comment about how he developed his views to the point of changing his opinion, by using the word 'read'. In this context, ST1 is effectively both reading data from maps and evaluation model, as well as following the discussion among the participants. F1, F2 and ST1 use *Invito's* tangibility and associability to highlight certain expected effects in the areas under discussion. It should be noted that ST1 changes his original position: he is not just reducing the weight assigned to the criterion that had received the best score earlier in the workshop, but also showing a preference for a competing criterion. *Invito* gave this stakeholder the opportunity to change his mind.

The discussion continues with other participants giving their preferences. The possibility of changes to participants' temporal perspectives is enabled by *Invito*'s immediate editing feature.

ST2 We can see here in this question (pointing at evaluation model) that it's important what time scale we have in mind. In the long term, I would say environment 3 (looking at maps).

F2 Actually, when we approach this complex decision (looking at evaluation model), we mean that benefits are positive aspects in the short term and the opportunities that we are not going to discuss today, but which are part of the decision model, refer to a longer time period.

F1	Because, we also have...a subnet for opportunities. One for benefits, one for costs for the short term and opportunities and risks, for the effects in the long term. However, this morning we don't have enough time to discuss everything.
ST2	So this is short term.
F1	This is in theory a shorter time period.
ST2	Ok. So, (points at model) I have switched to environment 3 for economic benefits.
F1	So, 3 economic, 5 environment (makes changes to evaluation model).

Key: F1=Facilitator 1; F2=Facilitator 2; ST2=Stakeholder 2. The use of an ellipsis in brackets [...] following a statement indicates that comments by one or more participants have been edited out.

It is worth noting that *Invito*'s associability feature produced unexpected results that caused stakeholders to engage in deep reflection about the main aspects of the decision. As one stakeholder put it:

> I think that in this solution, one can see that, according to the economic development, solution number 1 is a mainstream solution: everybody from our region says we need this high-speed traffic, we need this investment to have a contemporary infrastructure network. But, in terms of the overall aspects that we have discussed this morning, with all this differences of costs, about environmental aspects and urban planning or costs, for example, as brownfield development, the point of view changes a little bit and solution number 2 shows that the project is much more than pure economic investments, it is for the overall area, the overall territory and there is also further development that is much more complicated—in my opinion. And so, we have this more conservative and more preserving attitude and just to bring it into a more sustainable development was a step-by-step strategy. (Stakeholder 4)

Thus, the use of *Invito* may have provided those involved in the workshop with the possibility to approach the territorial problem in a less conventional way. An external observer, who had been invited to attend the pilot workshop, also noted this aspect. Despite his initial reservations, the external observer highlighted how the different affordances offered by the spatial maps and the evaluation model allowed an efficient comparison of the alternative scenarios, triggering a learning process in which all stakeholders participated.

> Coming from the Frankfurt region, this is a very new experience for me. When you see something new like this (looks at maps and model)... maybe, I'm a little bit critical, but what I see now is very interesting because the first positive point I've seen now is that there can be the possibility to have a comparison between different alternatives in public infrastructure and to compare this (points at map), to have this comparison in public discussion. This is very rare in Germany and we had this a little bit for the Stuttgart 21 project, but we didn't talk much about alternatives, in Stuttgart only, but not in the background area. And, we can open up a very new perspective on public infrastructure in terms of investments or the use of rare resources because we can't only spend money. This is one point. The second was very interesting and amazing for me to see the results of the discussion of the choice for alternative 2 (points at map and model). [...] What you have shown in your decision on the basic empowering objective is that we can have regional

> development as an alternative... use of this public money and this can be much more interesting when we compare the benefits and the costs at the end (look at maps). It opens a much wider perspective and different possibilities for using this money.
>
> (External Observer)

6 Discussion and Conclusion

Demands for CPS technologies that integrate visualisation tools with multi-criteria assessment procedures to support complex spatial planning processes are steadily increasing. This is particularly true for the case of major territorial transformation processes, such as the development of European rail corridors. The argument advanced in this chapter is that CPS technologies that meet these demands, such as *Invito*, differ from other forms of spatial planning support because of their perceived affordances, which have the potential to positively affect stakeholders' interactions in spatial planning processes. As briefly illustrated in the case study, to the extent that social actors have the ability to change their thinking and the ability to make changes to a CPS-supported model, their perceptions of affordances and constraints may produce behavioural changes that have important interactional consequences for those involved.

The perspective presented here has at least two implications for the research and practice of spatial planning processes. First, the affordance perspective can help us understand how stakeholders' thinking and the materiality of CPS technology interact to shape their behaviours. For the spatial planning community, adopting the language of affordance pushes planners to ask themselves many important questions about the impact of their interventions. For example, is the intended purpose of a particular spatial planning process to afford behaviours and work practices in particular ways and, if so, how to embody these intentions in spatial planning design? Did the intervention design subsequently have the effects intended? If so, why? If not, why not?

A second implication is to highlight that the effectiveness of any spatial planning process supported by a CPS technology is likely to be dependent on the extent to which that technology is perceived to afford or constrain stakeholders' ability to change their thinking during interaction. How can CPS technologies be perceived by stakeholders to afford, rather than constrain, opportunities for changing their thinking? Or, for 'changing their minds'? For the case of evaluation models, one aspect that may significantly affect how stakeholders perceive a model's affordances is the ability to sustain appropriate levels of equivocality and precision in the models created (Belton et al. 1997; Eden and Ackermann 2004, 2010; Franco 2013). A model that is too precise with little room for equivocality will not enable stakeholders to reconcile initial understandings and positions, which are key to changes in their thinking and knowledge.

Future work is needed to substantiate the generalisability of the argument we have presented here. A potentially useful direction for future study would be to implement research designs that compare the use of identical or similar CPS technologies in different spatial planning contexts, or that compare radically different CPS technologies in the same or similar contexts. An example of the former would be to compare a range of CPS technologies in small- versus large-group settings; the latter would involve comparing the use of 'hard' versus 'soft' CPS technologies within the same organisation. This type of research is likely to produce a classification of affordances (and constraints) that would help to better predict the nature and extent of the changes enabled by CPS technologies.

Another potentially useful avenue for future research relates to the relational character of the affordance concept. The notion of affordance always presupposes a perceiving stakeholder, and different stakeholders may be afforded different behaviours by the same CPS technology. Thus, future research could consider different types of stakeholders. For example, familiarity with a particular CPS technology or differences in cultural or professional background is likely to affect how stakeholders perceive the affordances (or constraints) called forth by the CPS technology.

To implement this research agenda would require that empirical accounts of CPS technology use in spatial planning processes include details of the social interaction context in which the technology is used. This means being able to capture and examine how stakeholder behaviours and the materiality of CPS technology become entangled during interaction and what the effects are.

References

Abastante F, Lami IM (2012) A complex analytic network process (ANP) network for analyzing Corridor24 alternative development strategies. In: CCCA' 2012, international conference on communications, computing and control applications, Marseilles, France, 6–8 Dec 2012

Abastante F, Bottero M, Greco S, Lami IM (2013) Dominance-based rough set approach and analytic network process for assessing urban transformation scenarios. Int J Multicrit Decis Mak 3(2/3):212–235

Abastante F, Guenther F, Lami IM, Masala E, Pensa S, Tosoni I (2014) Analytic network process, interactive maps and strategic assessment: the evaluation of Corridor24 alternative development strategies. In: Lami IM (ed) Analytical decision making methods for evaluating sustainable transport in european corridors. Springer International Publishing AG, Cham, pp 205–232

Andrienko G, Andrienko N, Jankowski P, Keim D, Kraak MJ, MacEachren AM et al (2007) Geovisual analytics for spatial decision support: setting the research agenda. Int J Geogr Inf Sci 21(8):839–857

Andrienko G, Andrienko N, Demsar U, Dransch D, Dykes J, Frabikant SI, Jern M, Kraak MJ, Schumann H, Tominski C (2010) Space, time and visual analytics. Int J Geogr Inf Sci 24 (10):1577–1600

Andrienko G, Andrienko N, Keim D, MacEachren A, Wrobel S (2011) Challenging problems of geospatial visual analytics. J Vis Lang Comput 22(4):251–256

Belton V, Pictet J (1997) A framework for group decision using a MCDA model: sharing, aggregating or comparing individual information? J Decis Syst 6(3):283–303

Belton S, Stewart TS (2002) Multiple criteria decision analysis. An integrated approach. Kluwer Academic Publishers, Norwell, MA
Belton V, Ackermann F, Shepherd I (1997) Integrated support from problem structuring through to alternative evaluation using COPE and VISA. J Multi-Criteria Decis Anal 6(3):115–130
Boccardo P, Tonolo FG (2015) Remote sensing role in emergency mapping for disaster response. Eng Geol Soc Territ 5:17–24
Boroushaki S, Malczewski J (2010) Measuring consensus for collaborative decision-making: a GIS-based approach. Comput Environ Urban Syst 34:322–332
Chemero A (2003) An outline of a theory of affordances. Ecol Psychol 15(2):181–195
Davoudi S, Evans E, Governa F, Santangelo M (2008) Territorial governance in the making. Approaches, methodologies, practices. Boletinn de la A.G.E.N. 46:351–355
Dodge M (2005) Information maps: tools for document exploration. CASA, working paper series 94
Dykes J, MacEachren AM, Kraak (eds) (2005) Exploring geovisualisation. Elsevier, Amsterdam
Eden C, Ackermann F (2004) Use of 'Soft OR' models by clients: what do they want from them? In: Pidd M (ed) Systems modelling: theory and practice. Wiley, Chichester, pp 146–163
Eden C, Ackermann F (2010) Decision making in groups: theory and practice. In: Nutt PC, Wilson DC (eds) Handbook of decision making. Wiley, Chichester, pp 231–272
Faludi A (2012) Multi-level (territorial) governance: three criticism. Plan Theory Pract 13(2):1–15
Faraj S, Azad B (2012) The materiality of technology: an affordance perspective. In: Leonardi PM, Nardi BA, Kallinikos J (eds) Materiality and organizing: social interaction in a technological world. Oxford University Press, Oxford, pp 237–258
Feick R, Hall GB (1999) Consensus building in a multiparticipant spatial decision support system. URISA J 11(2):17–23
Feick R, Hall GB (2004) A method for examining the spatial dimension of multicriteria weight sensitivity. Int J Geogr Inf Sci 18(8):815–840
Ferretti V (2012) Verso la valutazione integrata di scenari strategici in ambito spaziale. I modelli MC-SDSS. Celid, Torino
Figueira J, Greco S, Ehrgott M (2005) Multiple criteria decision analysis: state of the art surveys. Springer, Boston
Franco LA (2013) Rethinking soft OR interventions: models as boundary objects. Eur J Oper Res 231(3):720–733
Gaver WH (1996) Affordances for interaction: the social is material for design. Ecol Psychol 8 (2):111–129
Geertman SCM, Stillwell J (2009) Planning support systems: new methods and best practice (advances in spatial science). Springer Publishers, New York
Grezes J, Decety J (2002) Does visual perception of an object afford action? Evidence from a neuroimaging study. Neuropsychologica 40(2):212–222
Healey P (1997) Collaborative planning: shaping places in fragmented societies. Mac-Millan, Basingstoke
Hutchby I (2001) Technologies, texts and affordances. Sociology 35(2):441–456
Jankowski P, Nyerges T (2001) Geographic information system for group decision—making. Towards a participatory, geographic information science. Taylor and Francis, New York
Klosterman RE (1997) Planning support systems: a new perspective on computer-aided planning. J Plan Educ Res 17:45–54
Klosterman RE (2012) Simple and complex models. Environ Plann B Plann Des 39(1):1–6
Lami IM (ed) (2014) Analytical decision making methods for evaluating sustainable transport in European corridors. Springer International Publishing AG, Cham
Lami IM, Ferretti V (2014) Spatial decisional processes: evaluation tools and new challenges. In: Masala E, Melis G (eds) Interactive visualisation tool for brownfield redevelopment. A European experience. Celid, Torino

Lami IM, Abastante F, Bottero M, Masala E, Pensa S (2014) Integrating multi-criteria evaluation and data visualization as a problem structuring approach to support territorial transformation projects. In: Special issue on problem structuring research and practice, Euro J Decis Process (EJDP) (forthcoming)

Leonardi PM (2011) When flexible routines meet flexible technologies: affordance, constraint, and the imbrication of human and material agencies. MIS Q 35(1):147–167

Leonardi PM (2012) Materiality, sociomateriality, and socio-technical systems: what do these terms mean? How are they related? Do we need them? In: Leonardi PM, Nardi BA, Kallinikos J (eds) Materiality and organizing: social interaction in a technological world. Oxford University Press, Oxford, pp 25–48

Leonardi PM, Nardi A, Kallinikos J (eds) (2012) Materiality and organizing: social interaction in a technological world. Oxford University Press, Oxford

MacEachren AM, Gahegan M, Pike W, Brewer I, Cai G, Lengerich E, Hardisty F (2004) Geovisualization for knowledge construction and decision-support. Comput Graph Appl 24 (1):13–17

Malczewski J (1999) GIS and multicriteria decision analysis. Wiley, New York

Malczewski J (2006) IS-based multicriteria decision analysis: a survey of the literature. Int J Geogr Inf Sci 20(7):703–726

Masala E (2014) Visualisation as a support to spatial decision processes: some considerations on the concepts behind the construction of a strategy image. In: Masala E, Melis G (eds) Interactive visualisation tool for brownfield redevelopment. A European experience. Celid, Torino, pp 81–94

Nevo D, Wand Y (2005) Organizational memory information systems: a transactive memory approach. Decis Support Syst 39:549–562

Pensa S, Masala E, Marietta C (2011) The effects of decision-making on urban form: A tool for supporting planning processes. In: Pinto NN, Tenedorio JA, Santos M, Deus R (eds) Proceedings of the 7th international conference on virtual cities and territories, Lisbon, 11–13 Oct 2011. Department of Civil Engineering of the University of Coimbra and e-GEO, Research Centre in Geography and Regional Planning of the Faculty of Social Sciences and Humanities of the Nova University of Lisbon, Coimbra, pp 41–44

Pensa S, Masala E, Lami IM (2013) Supporting planning processes by the use of dynamic visualisation. In: Geertman S, Toppen F, Stillwell J (eds) Planning support systems for sustainable urban development. Springer, Berlin, pp 451–467

Pensa S, Masala E, Lami IM, Rosa A (2014) Seeing is knowing: data exploration as a support to planning. In: Proceedings of the institution of civil engineers—Civil Engineering Special Issue 167, May 3–8

Rinner C (2006) Argumentation mapping in collaborative decision making. In: Balram S, Dragicevic S (eds) Collaborative geographic information systems. Idea Group Publishing, Hershey, PA, pp 85–102

Roy B, Slowinski R (2013) Question guiding the choice of a multicriteria decision aiding method. EURO J Decis Process 1:69–97

Saaty TL (2001) The analytic network process. RWS Publications, Pittsburgh

Saaty TL, Vargas LG (2006) Decision-making with the analytic network process. Springer Science, New York

Simao A, Densham PJ, Haklay M (2009) Web-based GIS for collaborative planning and public participation: an application to the strategic planning of wind farm sites. J Environ Manage 90:2027–2040

Symes E, Ellis R, Tucker M (2007) Visual object affordances: object orientation. Acta Psychol (Amst) 124(2):238–255

Tango—Espon (2013) Territorial approaches for new governance. Final Report

Tucker M, Ellis R (2004) Action priming by briefly presented objects. Acta Psychol (Amst) 116 (2):185–203

Isabella M. Lami is Associate Professor of Planning Evaluation and Project Appraisal at Politecnico di Torino, Italy. She has expertise in multicriteria analysis methods and spatial decision processes within the context of territorial transformations, particularly in connection with large transport infrastructure. She was coordinator of several European projects. Professor Lami has published books and papers on both national and international scientific journals, and received many awards for her work.

L. Alberto Franco is Professor of Management Sciences at the School of Business and Economics, Loughborough University. His main research interests focus on the practice of operational research, particularly that which involves the use of visual decision support technologies in a group setting. As a management scientist, Professor Franco has led several projects to support strategy making, value-focused thinking, scenario planning, and multi-stakeholder collaboration in a variety of sectors including: construction, transport, health, defence, and local government. His latest work has appeared in the *British Journal of Management*, *European Journal of Operational Research*, *OMEGA*, and *Group Decision and Negotiation*.

Corridor Development from a Regional Perspective: The Case of the Frankfurt/Rhine-Main Region

Birgit Simon and Peter Endemann

Abstract The metropolitan region Frankfurt/Rhine-Main is located in the heart of Europe and in the middle of the Rotterdam-Genoa corridor, also known as the Rhine-Alpine corridor, which is part of the EU transport core network. This central location, along with excellent transport (airport, highways, rail) and telecommunication infrastructure, attracts a highly skilled workforce and a large number of businesses from abroad. Substantial freight volumes are transhipped in the region, moving in all directions, and rail infrastructure plays a key part in this transport network. Rail is the backbone of the region's development. It is key to regional and long-distance passenger and rail freight transport and must be promoted to prevent further increases in the volume of road transport as the main competitor of rail transport.

To achieve this, appropriate capacity increases in rail infrastructure are needed, and negative impacts on built-up areas, caused by rail noise, have to be prevented. Furthermore, integration into the trans-European corridor Rotterdam-Genoa is necessary. The area between Frankfurt and Mannheim is one of the key sections of the Rhine-Alpine corridor.

This paper outlines what form a bottom-up strategy to develop this segment of the corridor should take. It also shows that integration into the overall Rotterdam-Genoa corridor and coordination with other partners is necessary to handle both benefits and risks. Here, a comprehensive strategy is needed to reconcile both aspects. Furthermore, the paper emphasizes that the regional voice be heard by the competent European and national institutions.

List of Abbreviations

BMVI	Bundesministerium für Verkehr und digitale Infrastruktur (German Ministry for Transport and Digital Infrastructure); formerly BMVBS (Bundesministerium für Verkehr, Bau und Stadtentwicklung)
DB	Deutsche Bahn (German railway company)
ERTMS	European rail traffic management system

B. Simon • P. Endemann (✉)
Regionalverband FrankfurtRheinMain, Poststr. 16, 60329 Frankfurt am Main, Germany
e-mail: endemann@region-frankfurt.de

H. Drewello, B. Scholl (eds.), *Integrated Spatial and Transport Infrastructure Development*, Contributions to Economics, DOI 10.1007/978-3-319-15708-5_5

EC	Eurocity-train; European conventional long-distance interurban train
EU	European Union
GDP	Gross domestic product
HSR	High-speed rail
HST	High-speed train
IC	Intercity-train; national conventional long-distance interurban train
ICE	Intercity-express (German high-speed train)
TGV	Train à Grande Vitesse (French HST)
TEN-T	Trans-European transport network

1 Aim of the Contribution

This paper describes a bottom-up strategy capable of ensuring the development of a key segment of the corridor. It further outlines why integration into the Rotterdam-Genoa corridor strategy and coordination with other partners are necessary, given that the related benefits and risks are not limited to a sole regional or local entity. Moreover, the paper identifies the need for a comprehensive strategy reconciling both aspects. As a result of this discussion, it should then become clear that the regional voice must be heard by the competent European and national institutions.

2 The European Framework for the Rotterdam-Genoa Corridor

Europe will face major challenges in the future, given the expected decrease in its total population and changes in its structure and economic framework. These developments will have a number of consequences for transport infrastructure and have already led to a revision of the Trans-European Transport Network (TEN-T) guidelines. One key revision has been the definition of a priority core network and focusing of Trans-European infrastructure development on major axes, with the Rhine-Alpine corridor (Rotterdam-Genoa) being one of them.

In our view, refocusing the TEN-T activities was necessary for the following reasons:

- Some areas will face a larger population decrease than others. This has implications for future employment and housing allocation.
- A shortage of financial resources makes it impossible to implement every TEN-T measure considered to date.
- The European Commission's White Paper of 2011 (European Commission 2011) refers to the need for a 60 % reduction in the transport sector's CO_2 emissions by 2050. Road traffic is required to make a substantial contribution to

this reduction—this is essential not only in environmental terms but also from an economic point of view.

- Given the limited financial resources and circumstances outlined above, the future European transport policy must focus on a priority network of key rail corridors to connect the large and further expanding metropolitan regions. As a result, investments in remote and demographically declining areas will be reduced.
- Another aspect is the strive for a more energy-efficient use of infrastructure. The core network[1] contains several bottlenecks and gaps which need to be addressed in order to avoid negative impacts.
- The core network itself needs to prove its energy efficiency, and consequently land-based movements of goods and passengers need to be reviewed.
- It is important to stress the need to decouple economic growth from transport demand. The EU White Paper published in 2001 regards it as a high-priority objective to foster economic growth without an increase in transport. It goes on to suggest that it is important "to replace existing transport system taxes with more effective instruments for integrating infrastructure costs and external costs (European Commission 2001, p. 72)." The 2011 White Paper adopts the objectives of higher efficiency but argues: "Overall, transport infrastructure investments have a positive impact on economic growth, create wealth and jobs, and enhance trade, geographical accessibility and the mobility of people. It has to be planned in a way that maximises positive impact on economic growth and minimises negative impact on the environment (European Commission 2001, p. 4)."

However, this approach has never been concretised and applied to the (inter-) regional level. How much mobility is needed to achieve the reduction goals by 2050? To what extent are fast links between major cities a priority when regions in between lack good accessibility?

3 Rotterdam-Genoa Corridor: Relevance—Impact—Potential

The Rhine-Alpine corridor (Rotterdam-Genoa) is one of nine multi-modal corridors making up the newly defined TEN-T core network.[2] It links major economic centres along the river Rhine and on both sides of the Alps with the North Sea in the Netherlands and Belgium and the Mediterranean Sea in Italy. It is one of the major north-south freight axes (Fig. 1).

---

[1] http://ec.europa.eu/transport/infrastructure/tentec/tentec-portal/site/en/maps.html, accessed 20 June 2014.

[2] http://ec.europa.eu/transport/infrastructure/tentec/tentec-portal/site/en/maps.html, accessed 20 June 2014.

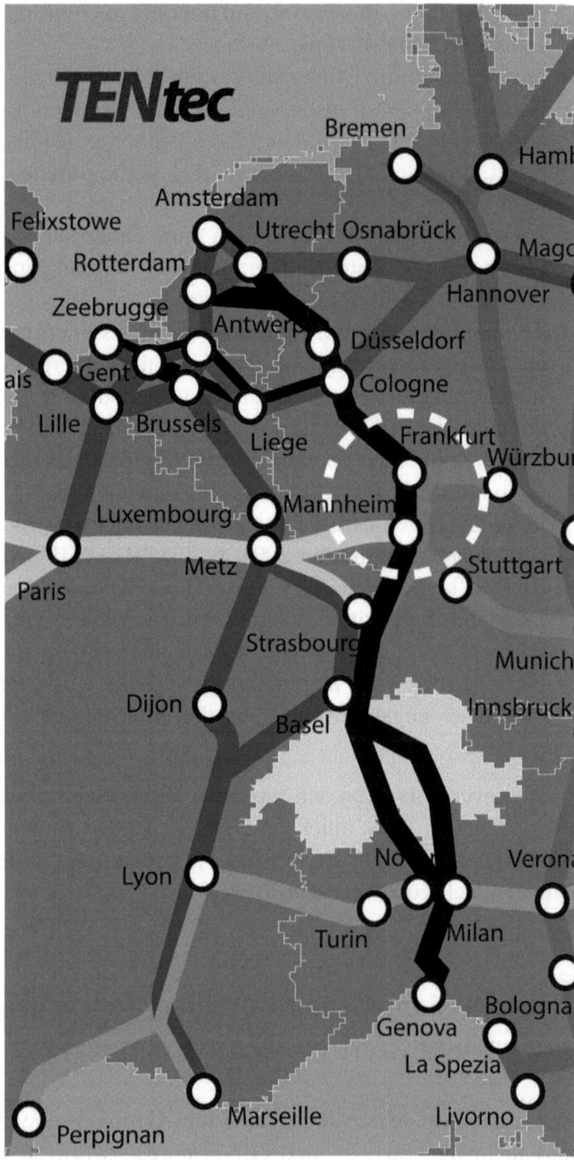

Fig. 1 Geographical location of the Frankfurt/Rhine-Main (Frankfurt) and Rhine-Neckar (Mannheim) regions on the Trans-European corridor Rotterdam-Genoa. *Source*: EU Commission, TEN-T core network corridors (Regulation (EU) no. 1316/2013—O.J.L 348-20/12/2013

While there are considerable freight flows along the entire corridor, there are no end-to-end transport movements, i.e. from one sea to the other (Fig. 2). The port of Rotterdam has a dominant position as a gateway for goods shipped to/from key origins/destinations along the corridor, as a study commissioned by Regionalverband FrankfurtRheinMain reveals (ProgTrans 2013). It also shows that there is an imbalance between inbound and outbound flows in this wide catchment area: while nearly 4 million tonnes from the hinterland are transhipped in

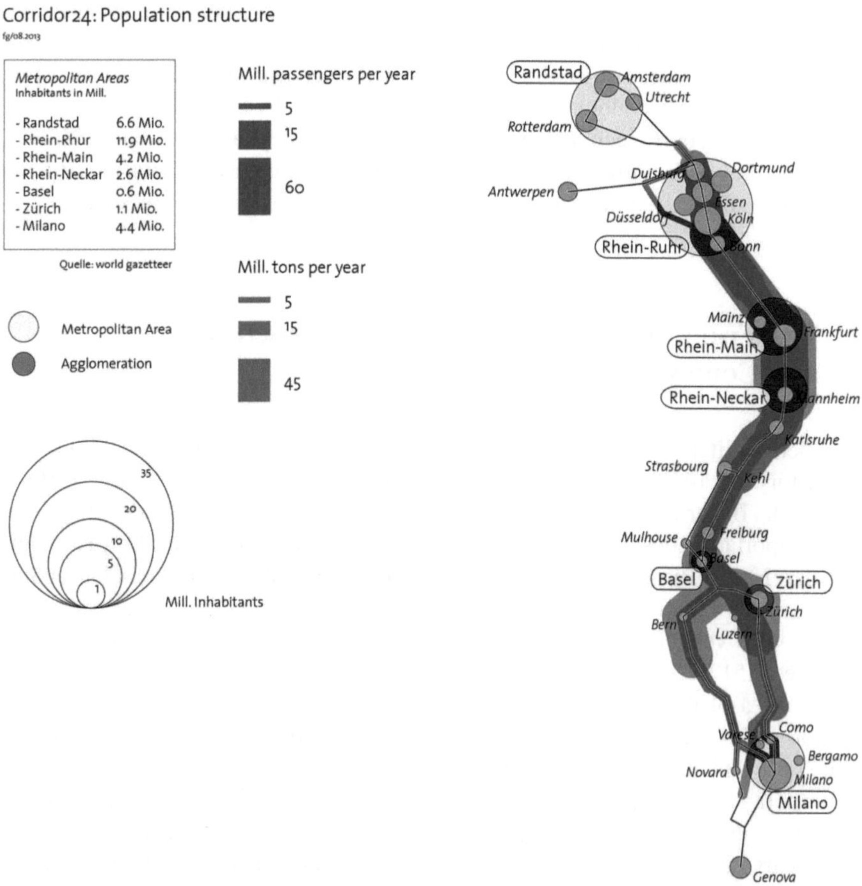

Fig. 2 Rhine-Alpine corridor in the CODE24 context: population—passengers—freight. *Source*: F. Günther, ETH Zurich IRL, prepared for CODE24

Rotterdam, 14 million tonnes arrive in Rotterdam from abroad and, for the most part, continue along the corridor. Of these 14 million tonnes transported from Rotterdam to the hinterland by rail, 71 % go to Germany, 10 % to Italy, 3 % to Switzerland, and 10 % stay in the Netherlands (ProgTrans 2013).

The general aim should be to reduce long-distance seaport-to-hinterland shipments, whether by road or rail, and achieve a better use of transport infrastructure (including ports). In line with this, the opening of the Alpine base tunnels in Switzerland in 2007 and 2017 offers the potential to shift more freight from road to rail and make better use of freight and passenger rail infrastructure. The new tunnels will help to improve the accessibility of the Northern Italian ports. They will serve as the southern gateways to the corridor, providing suitable transhipment facilities for goods from Asia or Africa going to Switzerland or southern Germany that are currently transhipped at the North Sea ports, i.e. they are shipped over longer distances by rail or road (ProgTrans 2013).

A better landside distribution of goods will decrease through-traffic in some sections along the corridor suffering from high noise and vibration levels caused by freight rail transport, such as Upper Middle Rhine Valley, an important UNESCO world natural and cultural heritage site. If these challenges are not tackled, the acceptance of rail freight as a sustainable means of transport is jeopardized.

4 The Frankfurt/Rhine-Main Region: Structure and Context

The metropolitan region is a large, polycentric area with 5.5 million inhabitants and a territory of 14,755 km^2 which includes parts of the federal states of Hesse (Hessen), Bavaria (Bayern) and Rhineland-Palatinate (Rheinland-Pfalz). The most important cities are Frankfurt am Main, Offenbach am Main, Bad Homburg vor der Höhe, Rüsselsheim, Wiesbaden, Mainz, Darmstadt and Aschaffenburg (Fig. 3). The region has a total of 2.1 million jobs and exhibits a car ownership rate of 672 vehicles per 1000 inhabitants. This is above the German national average of 517 vehicles per 1000 inhabitants (BMVBS 2011).

The area of the Regionalverband FrankfurtRheinMain has a population of 2.2 million and is the core of the metropolitan region. This region under the responsibility of this planning association is defined in the Act on the Frankfurt/Rhine-Main Metropolitan Region that entered into force on 1 April 2011. Compared to the metropolitan region, the territory of the planning association or agglomeration has fewer cars per habitant (659 vehicles/1000 inhabitants) and 1.3 million jobs. It is also worth mentioning that Frankfurt International Airport covers a sizeable area and is one of the region's biggest employers with over 78,000 jobs (Fraport 2014).

The Regionalverband FrankfurtRheinMain is the regional platform for the strategic alignment of regional associations. It is responsible for controlling and coordinating regional development and land use planning for its 75 member municipalities. The regional preparatory land use plan (Regionaler Flächennutzungsplan), which became effective in 2011, defines inter alia the allocation of land use until 2020 which includes open spaces and transport infrastructure. A European perspective is key to the Regionalverband's commitment in its role as a member of European networks such as METREX, Purple or the Airport Regions Conference or as part of various European projects. Participation in such transnational activities is indispensable, given the international character of the region, whose transport connections extend far beyond its institutional boundaries. This also offers the opportunity to share knowledge on many issues common to regions across the continent. More significantly, involvement in CODE24 and the idea of developing a common strategy for the Rotterdam-Genoa axis is groundbreaking for

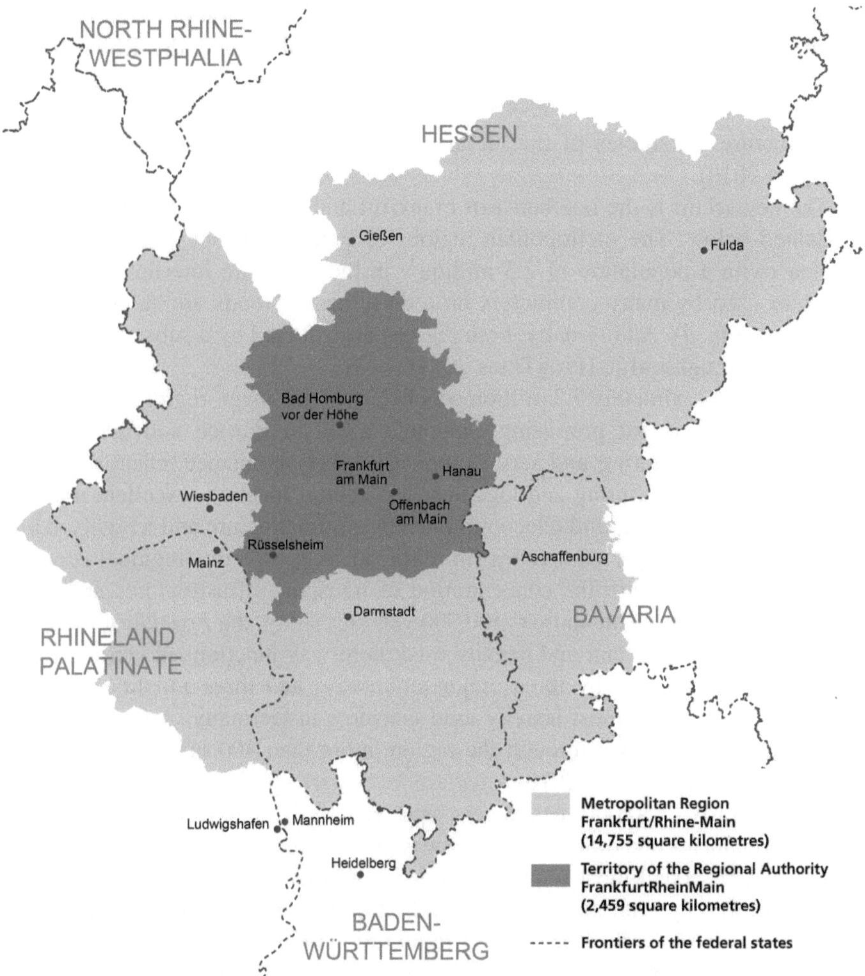

Fig. 3 Geographical location of the metropolitan region and Regionalverband FrankfurtRheinMain. *Source*: Regionalverband FrankfurtRheinMain

transnational coordination and calls for bottom-up coordination. This is further outlined hereafter.

5 The Need for Interregional Cooperation

The Frankfurt/Rhine-Main metropolitan region is located in the middle of the Rotterdam-Genoa corridor and has been identified as a key section of what is now called the Rhine-Alpine Corridor of the TEN-T core network. Its central geographical location on this corridor, together with its interrelationships with neighbouring regions such as the Middle-Rhine Valley and the Rhine-Neckar region and significant flows of passengers and goods, make it an area of strategic importance for international coordination with regard to managing heavily used infrastructure, as studies of the German Transport Ministry suggest (BVU and Intraplan 2010).

The key sector is the one between Frankfurt and Mannheim, as will be further explained below. The metropolitan regions of Frankfurt/Rhine-Main and Rhine-Neckar (with a population of 2.3 million[3]) in the south are interlinked by infrastructure used by many commuters on a daily basis. Goods are shipped in both directions (Fig. 4). Additionally, both regions are affected by a substantial amount of freight through-traffic (ProgTrans 2013).

In total, approximately 7.3 million people live in this mega-region. Both regions are amongst the most promising economic areas in Europe and are home to important manufacturing and service industries such as finance, chemistry, pharmacy, automobile, printing and logistics. The central location, excellent transport (airport, highways, rail) and telecommunications infrastructure and a highly skilled workforce attract many companies from abroad. A prosperous regional economy goes hand in hand with the concentration of transport infrastructures and traffic flows in the region. With approx. 300,000 vehicles a day, the *Frankfurter Kreuz* is one of the most important and heavily used motorway junctions in Germany. The two regions are linked by three major motorways and three rail lines (Fig. 4), making up one of the most heavily used corridors in Germany. A total of almost 700 trains per day travel through the region, more than 400 of which are freight trains (BVU and Intraplan 2010).

With approx. 450,000 passengers and visitors,[4] Frankfurt Central Station (*Hauptbahnhof*) is one of the largest railway stations in Europe. Mannheim Central Station is the second most important node in the two regions with 100,000 daily users, followed by Frankfurt Airport Rail Station (*Fernbahnhof*) with over 66,000 users.[5] In a 2013 worldwide survey, Frankfurt Airport ranked 12th in terms of passengers with approximately 58 million passengers and ninth for freight transport with over 2 million tonnes of cargo.[6] The ports of Ludwigshafen, Mannheim and Frankfurt are in the top 10 of German inland ports (Statistisches Bundesamt (2014).

The current situation for the section between Frankfurt and Mannheim, as one of the key parts of the corridor, is shown below:

---

[3] http://www.m-r-n.com/start/forschen-und-studieren/rhein-neckar-in-zahlen.html, accessed 11 July 2014.

[4] According to new figures found at: http://dmm.travel/news/artikel/lesen/2014/10/taeglich-450000-reisende-am-frankfurter-hauptbahnhof-63731/, accessed 9 February 2015.

[5] Source: www.bahnhof.de, accessed 20 June 2014.

[6] http://www.panynj.gov/airports/pdf-traffic/ATR2013.pdf, accessed 18 June 2014.

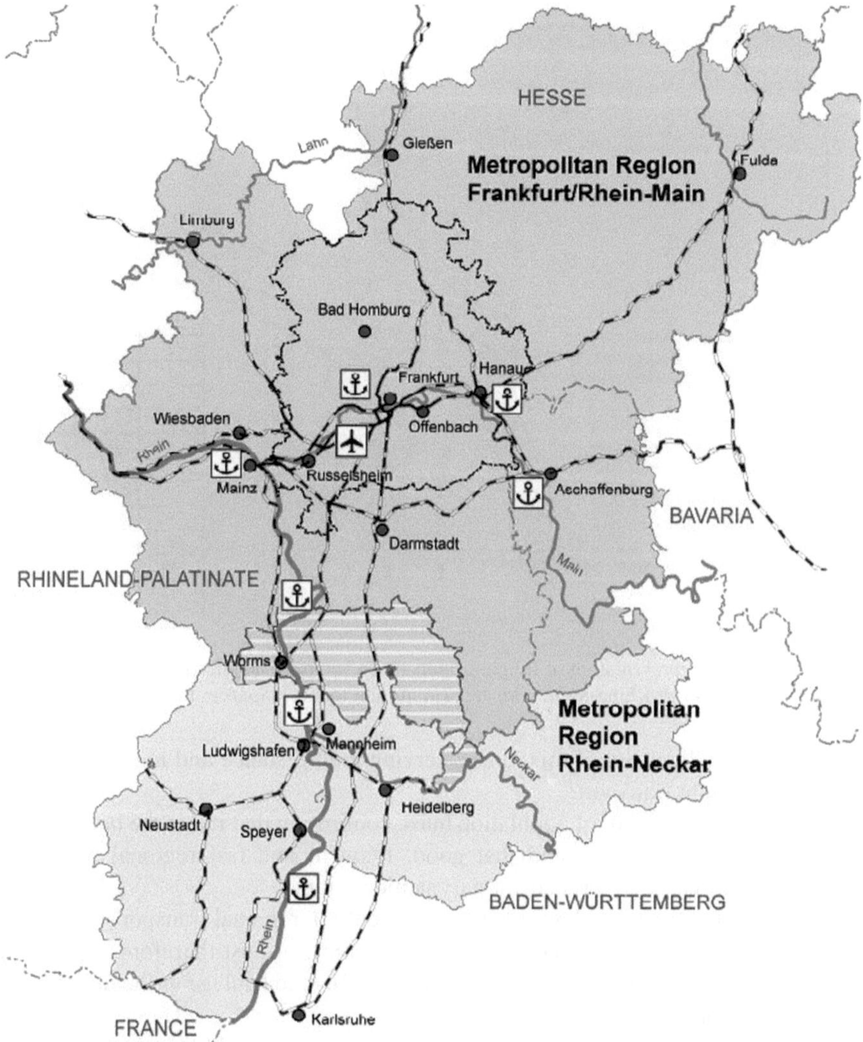

Fig. 4 Structural and functional links between Frankfurt/Rhine-Main and Rhine-Neckar metropolitan regions. *Source*: Regionalverband FrankfurtRheinMain

- Considerable freight volumes are transhipped in both regions, with many nationwide and European connections.
- Figure 5 visualises the multi-directional freight flows passing through the Frankfurt node and neighbouring areas. The flows along the Rhine-Alpine corridor amount to approx. 16 million tonnes p.a. Other major segments are the eastbound section with 14.9 million tonnes and the north-south axis with 6 million tonnes (ProgTrans 2013).

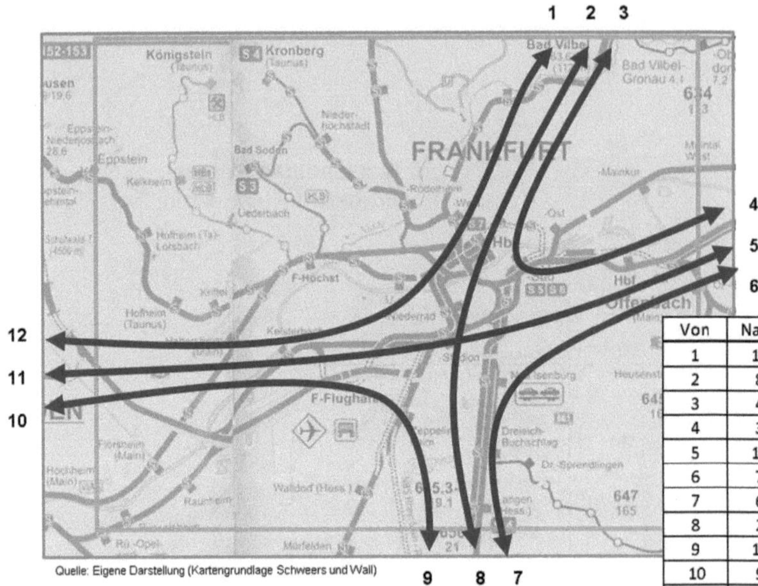

Von	Nach	Mio. t
1	12	0,2
2	8	3,5
3	4	0,0
4	3	0,0
5	11	5,7
6	7	2,8
7	6	1,7
8	2	2,5
9	10	4,8
10	9	12,0
11	5	9,2
12	1	0,2

Fig. 5 Schematic representation of freight origins ("von" = from) and destinations ("nach" = to) transiting the Frankfurt/Rhine-Main core area (in million tonnes). *Source*: ProgTrans 2013

- The existing lines are of mixed use, serving long-distance and regional passenger and freight transport.
- All lines serve important population hubs, confirming that rail is the backbone of economic development, and that good, frequent and fast regional passenger transport services are crucial for a dynamic region.
- The rail infrastructure plays a key role not only in regional transport but also in long-distance passenger and rail freight transport. It must therefore be actively promoted in order to prevent a further increase in road haulage as the rail sector's strongest competitor.
- Inefficient rail freight services lower accessibility and might threaten the quality of public transport as an alternative to the car and as the backbone of development.
- Noise protection is not sufficiently guaranteed and the quality of life could suffer.
- There are already capacity problems on some segments which will exacerbate by 2025, according to a German Transport Ministry forecast for 2025 (BVU and Intraplan 2010).
- The earlier proposal for an exclusive high-speed passenger rail line between Frankfurt and Mannheim will not solve the capacity problems.

– As a consequence, noise mitigation is not possible and regional passenger transport services cannot be adequately provided, as there will still be operational constraints.

This is why the German Transport Ministry intends to identify other and more sustainable solutions and has commissioned a feasibility study for the Cologne-Karlsruhe corridor under the name of *Mittelrheinstudie* (Middle Rhine Valley Study).

6 The CODE24 Process and Stakeholder Involvement

As part of the Rotterdam-Genoa corridor analysis, the following criteria were applied to identify areas where sustainable development is jeopardized by major risks. These risks are related to an increase in the volume of freight transport and include noise pollution and the displacement of regional transport, both of which have the potential to decrease public transport options. These infrastructure constraints also have a negative impact on the performance of the regional economy. The area between Frankfurt and Mannheim has been identified as one of four key sections along the whole Rhine-Alpine corridor which need thorough planning and coordination beyond the regional level (Table 1).

To arrive at a common strategy, all major stakeholders need to be involved. Within CODE24 this has been achieved with a series of regional workshops held by Verband Region Rhein-Neckar (Rhine-Neckar Regional Association) and Regionalverband FrankfurtRheinMain and with substantial input from the project partners, Zurich University of Technology (ETH Zurich) and the Higher Institute on Territorial Systems for Innovation in Turin (SiTI Turin).

In 2010, as a first step at regional level, roundtable discussions were held in both regions. Each meeting attracted 30–40 participants from federal state ministries, regions, cities, districts, logistics and network operators, forwarders, chambers of commerce, research institutes and specialist authorities, who discussed the situation in each region. A detailed report on both workshops can be found on the CODE24 website at www.code-24.eu.

Table 1 Selection criteria for corridor assessment

Railway system	Logistic hubs	Built-up area/landscape
• Insufficient capacity • Upgrading regional system (foreseen or desirable) • Infrastructure upgrading • New lines • Junctions with limited capacity	• Providing insufficient development potential • Providing development potential	• Threatened by noise • Urban areas: weak demand and high potential • Urban areas: high demand and no potential • Urban areas: high demand and high potential

Source: ETH Zurich and SiTI Turin (2012)

In the Frankfurt/Rhine-Main region, the participants stressed the need to remove bottlenecks in the rail network, which have so far limited regional transport development and reinforced the negative impact of rail freight. These issues make a modal shift from road to rail more difficult. It was also argued that striking the right balance between growth opportunities and the preservation of quality of life and natural resources should be a key priority in any strategy to optimise transport. High accessibility was acknowledged to be necessary and acceptable as long as it does not impede the high standard and advantages of a location, which is often the result of its good accessibility in the first place (Endemann et al. 2010). This statement was made by Bernd Hartz from Regierungspräsidium Darmstadt, the institution responsible for the regional plan in Southern Hesse. As an approach to steering land use, Thomas Busch from the Rhein-Main-Verkehrsverbund (RMV), the association organising transport services in the Rhine-Main region, emphasised the need to maintain rail infrastructure. He went on to highlight the importance of relieving heavily used rail infrastructure by separating passenger and freight services, without impairing access for residential and commercial areas along the lines. To this end, new and upgraded rail infrastructure was deemed necessary. The participants agreed and stressed that negative impacts on built-up areas caused by rail noise must be prevented. Furthermore, integration into the Trans-European axis Rotterdam-Genoa is necessary (Endemann et al. 2010).

The regional workshop in the Rhine-Neckar metropolitan region revealed similar concerns about the lack of sufficient financial resources to remove all bottlenecks. The attendants argued that a coordinated strategy involving all stakeholders is necessary in developing major infrastructure, which includes the section between Frankfurt and Mannheim and the provision of logistics-related areas (Verband Region Rhein-Neckar, ETH Zurich 2010) (Fig. 6).

Based on the two workshops, an assessment workshop moderated by ETH Zurich and SiTI Turin was held in Zurich in March 2012. Experts from both regions representing relevant stakeholders, such as the Chamber of Commerce, DB Netz (German rail network operator), environmental and passenger interest groups and regional and municipal planners, were present. Experts from the two project partners Regionalverband FrankfurtRheinMain and Verband Region Rhein-Neckar participated, too. The experts were asked to evaluate three scenarios that can be categorized as follows (ETH Zurich and SiTI Turin 2012):

- High-speed: Development of a dedicated high-speed passenger rail line between Frankfurt and Mannheim, reducing travel time between the two cities and thus bringing them closer to each other.
- Net-Regio: The existing transport as well as residential and commercial areas will be strengthened, as all lines meet the requirement of rail-oriented development. Additional rail infrastructure is not planned, but rail nodes might be partially upgraded and improved.
- Euro-hub: This freight-focused scenario foresees the development of a dedicated freight rail line between both regions and a stronger concentration of logistics facilities in the Rhine-Neckar region.

Fig. 6 Schematic diagram of the two agglomerations of Frankfurt and Mannheim illustrating the two saturated lines (*dashed lines*) and a potential new line (*double continuous line*). *Source*: F. Günther, ETH Zurich IRL, prepared for CODE24

A detailed description of the scenarios is provided in the input paper (ETH Zurich and SiTI Turin 2012) or on the CODE24 website.[7]

As a main outcome, the experts preferred neither a dedicated passenger nor a dedicated freight solution for the two regions. The Net-Regio scenario was considered a good starting point for a further strategy refinement, as it takes into account growing built-up areas and the need to keep them well served by both regional and long-distance rail services. Adopting this approach strengthens interregional cohesion with rail as the backbone of successful regional development. However, capacity constraints need to be further addressed.

Based on this outcome, a follow-up roundtable held in Mannheim in April 2013 discussed an integrated scenario and came to the following conclusions:

– There is a need for mixed use and development on all three existing lines.

[7] http://www.code-24.eu/activities/collaborative-assessment/, accessed 9 July 2014.

- Rail freight on existing infrastructure should be diverted to the greatest possible extent to new infrastructure at night between Frankfurt and Mannheim.
- No incremental implementation of partial upgrades, but a dedicated new line that is capable of resolving capacity problems and serves both freight and passenger transport.
- Potential links between old and new lines remain to be clarified.

No mega-hubs for logistics are needed, but it is necessary to safeguard the concentration of activities on existing sites.

The regional debate in Frankfurt/Rhine-Main showed that the German Transport Ministry, as the body responsible for financing and defining the new rail line between Frankfurt and Mannheim, needs to be involved at this stage of the process. Prior to this involvement, the various institutions in the Frankfurt/Rhine-Main region must be consulted in order to reach a consensus on undisputable aspects. This is why Regionalverband FrankfurtRheinMain held a high-level political meeting in Frankfurt am Main in January 2014. Attendants included official representatives such as mayors, directors and managers from federal state ministries, regions, cities, counties, chambers of commerce and interest groups in the areas of environmental protection, civic design and passenger concerns.

The participants agreed on the following objectives to be pursued with the development of the new Frankfurt-Mannheim rail line. These objectives were stated in a letter sent to the German Minister of Transport and Digital Infrastructure (BMVI).

- **Safeguarding the performance of regional passenger transport**
  Rail is the backbone of development and strengthens the polycentric structure of and cohesion between regions. This requires a fully operational regional passenger rail network with frequent service and short transfer times at regional nodes such as Frankfurt, Darmstadt, Mainz, Wiesbaden, Mannheim, Ludwigshafen and Heidelberg. An integrated timed-transfer timetable is key to such a strategy.
- **Better long-distance rail services for both regions and integrated timed-transfer with regional passenger services**
  Existing standards in the long-distance segment (i.e. ICE, IC/EC, TGV trains) need to be improved by increasing the number of long-distance services between Frankfurt and Mannheim and reducing travel times between both cities. The long-distance passenger train connections (incl. HSR) to other transfer nodes of regional importance, such as Mainz, Wiesbaden, Frankfurt, Darmstadt and Heidelberg, must not be neglected. To enhance regional accessibility, integration with regional passenger transport is necessary.
- **Noise protection for built-up areas close to new and existing rail lines**
  Active noise measures (brakes, rolling stock, rail infrastructure) have to be implemented. This must also apply to built-up areas along existing rail infrastructure which have no legal right to any noise abatement measures, despite the steady increase in rail freight transport.

- **A European perspective is key to supporting and enhancing the performance of both regions.**

  Both regions benefit from the movement of goods and international trade. But with the rail freight volume predicted to increase in the future, better connections from the Mediterranean ports to the Italian hinterland and up to Southern Germany could contribute to making more efficient use of rail infrastructure capacity, particularly with the opening of the two Swiss Alpine base tunnels. The implementation of measures to enhance the accessibility of the Mediterranean ports and the associated rail infrastructure need to be promoted, although the population's quality of life must not suffer as a consequence.
- **The Frankfurt/Rhine-Main and Rhine-Neckar metropolitan regions need to be involved in relevant studies.**

  The German Transport Minister must ensure that both regions are involved at an early stage in the current Cologne-Karlsruhe corridor study (*Mittelrheinstudie*) and other relevant studies, such as the German National Transport Plan (*Bundesverkehrswegeplan*) and the related processes. German Transport Ministry agreed to invite stakeholders from both regions as well as members of the national and federal state parliaments to a presentation of the outcome of the aforementioned corridor study in July 2014 and in March 2015. This was followed by an informal participation process enabling public and private stakeholders as well as individuals to comment on the results of this study.

7 Conclusions

The two regions and municipalities along the Rhine-Alpine corridor benefit from their geographical location, as it gives them better accessibility and connectivity. The Rhine-Alpine corridor has a long history, providing good connectivity and enabling exchanges over a long period of time. The increase in freight volumes and passenger movements over the years has helped the regions and municipalities to grow and achieve a certain standard of living.

However, one of the regional CODE24 workshops highlighted that negative impacts must not be allowed to impair the quality of life and the preservation of natural resources. While transport infrastructure contributes to higher GDP and overall wealth, it also increases the attractiveness of an area, thereby creating a higher demand for mobility.

This is why new infrastructure investments and measures along the Rhine-Alpine corridor must be carefully assessed and priority should be given to measures to improve operational performance of the corridor, such as the European Rail Traffic Management System (ERTMS). We therefore conclude that before considering new investments, better use of existing infrastructure along with improved organisational and infrastructure management (especially in rail nodes such as Frankfurt am Main) may be able to meet the considerable demand along the corridor. This approach also offers the advantage of more moderate land use and better integration into the existing rail network than infrastructure upgrades or new infrastructures. For the Frankfurt/Rhine-Main and Rhine-Neckar regions, a new dedicated rail line should increase the operational performance of both freight and long-distance passenger trains, justifying the investment and demonstrating its benefits. The mixed use also allows greater flexibility in the future, supporting the rail line as a long-term investment.

Above and beyond this, the new rail line needs to be integrated into the existing rail network. Infrastructure links between the new and the existing lines would allow more flexibility, for example better distribution of trains, better accessibility for local freight trains, more options for future train routes and flexibility in terms of traffic diversion in the event of disruptions on any of the other lines. As the regions and municipalities are affected by both top-down and bottom-up impacts, the development and management of infrastructure need to be approached from these two perspectives. The current approach pursued by the German Transport Ministry and the development process for the nine Trans-European core network corridors envisaged by the European Commission should acknowledge that a purely top-down approach is no longer adequate to ensure acceptance among the population affected. The activities initiated by Regionalverband FrankfurtRheinMain and Verband Region Rhein-Neckar have the intention of achieving a broad consensus and acceptance for interregional transport infrastructure at all levels of society. This outcome would also be of great benefit to the stakeholders responsible at EU and national level.

References

Bundesministerium für Verkehr, Bau und Stadtentwicklung (BMVBS) (2011) Verkehr in Zahlen 2011/2012. Berlin

BVU und Intraplan (2010) Überprüfung des Bedarfsplans für die Bundesschienenwege, study commissioned by the German Federal Transport Ministry. Freiburg/Berlin

Endemann P, Günther F, Tosoni I, Wempe J, Kleinwächter E (2010) Perspektiven des Korridors 24 an Rhein und Main, Ergebnisse—Erster Regionaler Workshop für Südhessen/FrankfurtRheinMain. Frankfurt am Main/Zürich. http://www.region-frankfurt.de/media/custom/2005_650_1.PDF. Accessed 4 July 2014

ETH Zurich and SitI Turin (2012) CODE24 Collaborative assessment—input paper. Unpublished document

European Commission (2001) The European transport policy for 2010: time to decide. Luxembourg. http://ec.europa.eu/transport/themes/strategies/doc/2001_white_paper/lb_com_2001_0370_en.pdf. Accessed 18 June 2014

European Commission (2011) Roadmap to a single European transport area—towards a competitive and resource efficient transport system. COM(2011) 144, Brussels. http://eurlex.europa.eu/LexUriServ/LexUriServ.do?uri=COM:2011:0144:FIN:EN:PDF. Accessed 18 June 2014

Fraport AG (2014) Zahlen, Daten, Fakten 2014 zum Flughafen, Frankfurt am Main. www.fraport.de/zahlendatenfakten. Accessed 11 July 2014

ProgTrans (2013) Auswertungen der Verkehrsverflechtungen entlang des Korridors Rotterdam - Genua. Study commissioned by Regionalverband FrankfurtRheinMain. Basel

Statistisches Bundesamt (2014) Güterverkehrsstatistik der Binnenschifffahrt, Fachserie 8, Reihe 4, Wiesbaden. https://www.destatis.de/DE/Publikationen/Thematisch/TransportVerkehr/Schifffahrt/BinnenschifffahrtJ.html. Accessed 11 July 2014

Verband Region Rhein-Neckar and ETH Zurich (2010) Minutes of the regional workshop on 14 September 2010. http://www.code-24.eu/fileadmin/code24/pdf/Documents_Public_Events/Reg%20Workshop%20CODE24_Protokoll_Mannheim_de.pdf. Accessed 4 July 2014

Birgit Simon is first deputy director of Regionalverband FrankfurtRheinMain (Authority FrankfurtRheinMain) since 2012 and member of the environment committee of the Hessian Association of Cities. Since 2015 she is Vice Chair of the newly founded European Grouping of Territorial Cooperation "Interregional Alliance for the Rhine-Alpine Corridor EGTC". Her work focuses on mobility with an emphasis on cycling, rail freight, the expansion of the local public transport network infrastructure as well as municipal affairs and integrated development. Prior to this, she was mayor and councilor of the City of Offenbach, here responsible for the department of transport, environmental and social affairs.

Peter Endemann has a background in Geography, Spatial Planning and Urban Design. Since 2003 he is employed at Regionalverband FrankfurtRheinMain (Regional Authority FrankfurtRheinMain). His activities comprise travel behaviour analysis and interaction with land-use, transport infrastructure development, rail freight and long-distance passenger rail. Within CODE24 he was co-responsible for the development of the Online Rail Freight Exchange (ORFE) and the action on high speed rail integration.

Challenges for Corridors: Future Perspectives on European Corridor Development

Patrick Witte and Tejo Spit

Abstract In many European policy documents, the integration of transport, spatial and economic dimensions in corridor development is advocated as a means to cope with challenges that occur within corridors. For many years, however, knowledge about corridors was only developed from a sector perspective and with a more or less technocratic approach—despite a growing call for a more integrated analysis of corridor issues. To address these challenges efficiently, the integration argument needs specific knowledge on the different problem areas. Therefore, the question remains as to what extent an integrated analysis would be beneficial in resolving persistent corridor issues, such as the existence of bottlenecks. This contribution aims to highlight the outcome of two related research projects in which the possible added value of an integrated perspective on corridor development is addressed. The evidence presented reflects the current state-of-the-art knowledge regarding integrated corridor development and the possible future challenges for European corridors that may result from this. Whether the added value of the integration argument in corridor development can be found in practice will be researched as well as whether this leads to restating the importance of corridors for present-day European policy objectives.

Note: This chapter is an adaptation of research carried out in the context of a doctoral thesis (Witte 2014, pp. 19–37 and 127–144) and related research papers. This chapter will summarise the main argument and present the most important findings of the entire research. Overlap between this chapter and the thesis is intentional.

P. Witte (✉) • T. Spit
Department of Human Geography and Planning, Utrecht University, Heidelberglaan 2, 3508 TC Utrecht, The Netherlands
e-mail: p.a.witte@uu.nl

H. Drewello, B. Scholl (eds.), *Integrated Spatial and Transport Infrastructure Development*, Contributions to Economics, DOI 10.1007/978-3-319-15708-5_6

1 Introduction

The inception of the Connecting Europe Facility by the European Commission (2011) has triggered the evolution of yet another definition of corridor routes and programmes. This time, nine transnational corridors have been defined, which all together are called Europe's Core Network Corridors (European Commission 2013). For years, such corridor programmes have been driven to a large extent by policy ambitions regarding the integration of the different scales and dimensions involved in corridor development. The topic of interest in this chapter is the question of to what extent an integrated perspective on corridor development can provide added value for European policy-makers in their current and future governance strategies regarding European corridor development.

For over a decade now, European corridors have been receiving ongoing attention from policy-makers and academics alike. First, this can be observed from the considerable number of European policy programmes with regard to corridor routes, programmes and definitions (European Commission 1999, 2008). Second, the attention to European corridors can be observed from the academic literature. The attempts at introducing corridor development into the academic debate as a promising spatial concept to integrate different kinds of objectives resulting from different sectors and scales of policy-making is of relevance in this respect. This is reflected in, among others, the work of Priemus and Zonneveld (2003), Albrechts and Coppens (2003), Chapman et al. (2003), Romein et al. (2003), Schönharting et al. (2003) and De Vries and Priemus (2003) on the governance of corridors. According to these authors, corridors and corridor development can be regarded as potentially important concepts for spatial policy-making on different levels of scale, enabling policy-makers to deal more easily with the challenging and complex spatial reality presented to Europe's urban regions nowadays.

However, despite this abundant attention, the corridor concept thus far seems not to have been sufficiently or correctly analysed in policy, practice or academia (Witte 2014). First, many policy programmes still have a limited scope in the sense that they merely take into account one-dimensional, transport-oriented issues related to logistics and transport operations. Second, many of the issues in the practice of corridor development have not been solved yet, but remain highly relevant, especially the persistence of bottlenecks along corridors, despite the fact that most of the issues have been known for over a decade—since the initial introduction of the corridor concept. Finally, fragmentation in the academic debate can be observed in the availability of knowledge, institutions and governance structures that seem fit to efficiently address present day issues in European corridors.

As a result, knowledge on corridors has been developed in a sector-based manner for many years, despite a growing call for a more integrated analysis of corridor issues. In other words, a discrepancy is observed between the call in policy-making, as well as in the academic debate, for an integrated approach towards the development of European transport corridors, and the often isolated, local and sector-based practices of corridor development. This disparity raises the question of to what extent striving for an integrated approach to corridor development is the right way forward, in policy debate as well as in the academic debate on European transport corridors. This chapter addresses this question.

The chapter is organised as follows. Section 2 elaborates on the definition and issues of corridors and corridor development in Europe and the integration debate is brought to the fore. Next, Sect. 3 provides arguments stemming from the economic dimension that refute the benefits of integration for corridor development. In contrast, Sect. 4 does the exact opposite by highlighting the added value of an integrated perspective on corridors as seen from the transport dimension. Section 5 balances the pros and cons regarding the integration argument in corridor development. Lastly, Sect. 6 discusses the potential of corridors as a useful concept in planning and sketches some prospects for future corridor studies on the European transport network.

2 Corridor Development in Europe: Definitions, Issues and Debates

It has been over a decade since the *Journal of Transport Geography* published a special issue on the governance of corridors (Priemus and Zonneveld 2003). This section will take the findings from that special issue as a starting point to outline the definition and issues of corridors and corridor development in Europe. The integration debate mentioned above will also be introduced in greater detail. This sets the scene for discussing the pros and cons of the integration argument in corridor development in the remainder of this chapter.

2.1 What are Corridors?

Essentially, at least in theory, corridors can be viewed as narrow bundles of infrastructure that connect two or more urban regions dispersed over a certain physical space (Fig. 1). These bundles usually exist in three modes: motorways, railway links and inland navigation or short sea connections. One can also include ICT infrastructure such as power lines, cables and oil pipes to arrive at a broader definition of a corridor. In general, however, the concept of corridors concerns connections that use at least one (or more) of the three previously mentioned modes (road, rail and inland waterways) and include both passenger and freight transport (Priemus and Zonneveld 2003). For years, however, the corridor concept has been studied from a broader point of view, which is exemplified by greater attention to the various spatial scales at which corridors operate, and the various (sector) dimensions that seem to be integrated into corridor development (Chapman

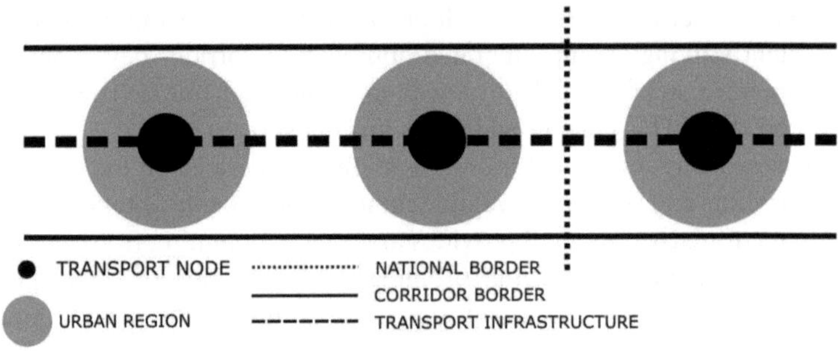

Fig. 1 Transport corridor conceptualisation (Witte 2014)

et al. 2003; Priemus and Zonneveld 2003; Romein et al. 2003; Schönharting et al. 2003; De Vries and Priemus 2003).

It must be emphasised that corridors occur at many spatial scales, often at the same time, ranging from tramway corridors in urban regions to high-speed intercity rail corridors and freight corridors at the global level (Pain 2011). Other examples of corridors are 'string-of-pearls' corridors for channelling smart growth at the local to regional levels and corridors from production areas to port areas. Moreover, corridors can be perceived as a structuring concept for infrastructure development (Bruinsma et al. 1997) and urban development plans (Banister et al. 1995), as well as a network structure in freight and passenger transportation (Hesse and Rodrigue 2004; Woxenius 2007), but also as a policy concept in the European cohesion discourse (Peters 2003; Dühr et al. 2007) or as a vehicle to trigger economic development (European Commission 1999, 2011). According to Rodrigue (2004), corridors can be viewed as the place where transport, economic and demographic processes are linearly articulated. In summary, what all approaches have in common is that the corridor concept strives to integrate policies on infrastructure, urbanisation and economic development (Priemus and Zonneveld 2003). The crucial factor is the multi-dimensional and multi-scalar nature of present-day corridors. In this sense, the concept refers to corridors not only as infrastructure axes, but also as economic development and urbanisation axes (Priemus 2001; Pain 2011).

On the basis of the foregoing argument, it can be stated that a definition of corridors should not only be concerned with the different scopes (freight and passenger) and modes (road, rail and inland waterway) involved in corridor development, and that the different scales (local, regional and trans-national) and dimensions (transport, spatial, institutional and economic) are also of relevance (Fig. 2). This line of reasoning is in line with the corridor conceptualisation by Chapman et al. (2003). Corridors are thus perceived to incorporate multi-modal infrastructure connections that serve both freight and passenger transportation, operate on multiple scales and impact multiple dimensions. In other words, present-day corridor development is concerned with a complex interrelatedness between transport capacity, economic benefits and spatial structures. This chapter is especially

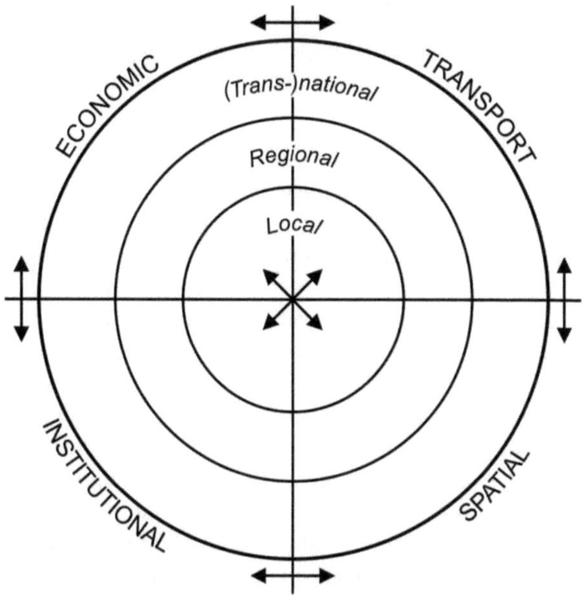

Fig. 2 Conceptual framework for integrated corridor development (Witte 2014)

interested in the variety of scales and dimensions involved in corridor development, because knowledge is lacking on many of these scales and dimensions. Moreover, these are the levels to which many of the present-day issues in corridor development can be related. In particular, the focus of this chapter will be on the economic and transport dimension of corridor development.

2.2 *What are (Still) the Issues?*

Numerous issues can be found in literature and practice regarding corridor development in Europe, mostly related to difficulties in achieving the wished-for successful transnational spatial governance in European corridors. For example, a common remark is on the lack of institutional involvement in the management of corridors. Although there is no great support among stakeholders for a governmental authority for complete corridors, the need to coordinate central government policies with local land use and transport policy at the corridor level is felt (Chapman et al. 2003). Chapman et al. (2003) also point to a strategic conceptual choice to be made between developing corridors in general, and developing at dense, nodal points. In addition, the key issues for corridors are

- Poor transnational connectivity
- Conflicts between long-distance and short-distance traffic
- The inability to manage infrastructure congestion
- Competitive pressures

- Inequalities between regions
- The environmental impact of increasing demands for transport and development
- Development patterns increasing the need to travel
- Institutional discontinuities
- Lack of coordination in decision-making

As can be observed, the majority of issues in corridor development are related to either the multi-scalar or the multi-dimensional nature of corridors. Zooming in on the multi-scalar, multi-dimensional nature of corridors, Albrechts and Coppens (2003) argue that corridors have become trapped between the global and the local scale. In this way, European policy for efficient transportation and communication systems intertwines with local policy aimed at quality of life and the environment. This is related to the argument presented by Bertolini and Spit (1998) on node development and Scholl (2012) on corridor development: while the direct costs of node development are likely to remain at the lowest spatial level, its benefits tend to spread over a wider area. Therefore, the aims of economic development and transport improvement on an interregional or national level must be accompanied by aims of environmental protection and social integration on a local to regional scale. To this end, advance is favoured in governance structures able to support the integration of different kinds of objectives resulting from different sectors and scales of policy-making (Priemus and Zonneveld 2004).

2.3 *Integrated Corridor Development in Europe?*

What is becoming evident from this brief overview of the corridor concept and its major issues is that a call for a more holistic approach to corridor development is desired to adequately address the wide variety of issues. It should be noted, however, that the call for integration is not at all new and stems from the traditional debate in spatial planning on the self-evident efficiency of sector-based planning versus the sector-transcendent benefits of integrated planning (e.g. Spit 1998). When this debate is related to European corridor development, the spatial impact of transport infrastructures and the positioning of corridors within these spatial and transport dimensions are of interest. In other words, the spatial dimensions of the growing transport sector and the implications of corridor development for European policy strategies are relevant. From this perspective, it is remarkable to note that either way, planning has thus far largely failed to produce a systematic approach to deal with corridor issues (Witte et al. 2012, 2013a; Witte and Spit 2014).

On basis of the characteristics mentioned above, corridors can be seen as integrating both multiple dimensions (i.e. transport, spatial, institutional and economic) and multiple spatial scales (i.e. local, regional and trans-national). As many of the present-day issues in corridor development can be related to this multi-scalar, multi-dimensional nature of corridors, it has been put forward that a more holistic approach to corridor development is desired to adequately address the variety of issues. However, given the sector-based practices and the fragmented nature of the available knowledge, it can easily be understood that up to now there has been little research available that has evaluated the added value of an integrated perspective on corridor development in a satisfactory way to solve the remaining corridor issues. As the empirical support for corridors and integration is limited, analyses of the potentials of and challenges for corridor development at different scales and across different dimensions is desired (Witte 2014). This chapter reports on the findings of two problem areas of integrated corridor development that are of interest in this respect: the (limited) potential for integrated corridor development stemming from the economic dimension (Witte et al. 2013b) and the challenges for integrated corridor development stemming from the transport dimension (Witte et al. 2012, 2014).

3 Why Integration Doesn't Work: Evidence from the Economic Dimension

This section addresses the supposed added value of integrated corridor development in achieving patterns of structural economic growth in corridor regions. Witte et al. (2013b) have focused their attention on the question of whether corridors have a special function in regional economic growth due to agglomeration advantages, and whether corridors can consequently be seen as a useful planning instrument to help connect urban regions into large-scale development zones across Europe. The starting point was the often-heard assumption in policy documents that corridor development contributes positively to regional economic growth (e.g. European Commission 1999, 2011), in contrast to the notion that the impact of corridors on regional economic development lacks substantial empirical support (Bruinsma et al. 1997; Louter et al. 1999; Van Oort and Raspe 2005).

Corridors have been absent in the agglomeration debate, even though corridors link larger urban agglomerations and may facilitate larger markets and knowledge spill-overs (McCann and Shefer 2004; Frenken et al. 2007; Thissen et al. 2013). Witte et al. (2013b) have contributed to this hypothesis by means of an empirical analysis of the economic potential of corridors, and the added value of the corridor concept for explanations of regional economic growth in terms of positive externalities and spill-overs. In this chapter, their main findings will be highlighted. The main problem to be addressed is whether corridors can operate as independent economic clusters (Bathelt 2005), thus showing functional (specialisation- or diversity-based) clustering, as opposed to merely reflecting co-located agglomeration advantages of connected large urban regions (Louter 1999).

The foremost conclusion to be drawn, based on the empirical material, is that there is little empirical support for a corridor effect on productivity and employment growth externalities. In other words, the results seriously question the provability of the added value of corridors for growth and agglomeration. However, general relationships between agglomeration economies and regional economic growth have been found that are in line with accepted insights from new economic geography theorising (compare Frenken et al. 2007; Dogaru et al. 2011).

At least five problems related to the supposed contribution of corridors (i.e. independent clustering effect) to achieving regional economic growth have been found. First, corridor regions cannot be distinguished from non-corridor regions in terms of spatial-economic determinants of productivity growth without incorporating the urban dimension in the same analysis. Second, whereas diverging specialisation effects between core and peripheral regions were observed, corridors are not the driving force of this effect. Third, non-corridor regions are more conducive to employment growth than corridor regions. Fourth, employment growth is especially dependent on urban contexts, and corridors appear to hamper this relationship more than they foster it. Finally, both the urban dimension and the European core-periphery dimension dominate the corridor dimension in determining the decisive coefficients of much of the modelling (Witte et al. 2013b).

Thus, it can be concluded that the outcome shows significant spatial heterogeneity when applying varying concepts of space to the relationship between agglomeration economies and growth differentials in Europe. Remarkably, there is little support for the special function of corridors in economic growth due to agglomeration advantages; although the magnitude and direction of agglomeration effects generally are as expected, the findings are either not systematically stronger inside corridors than outside them, or are not altogether a result of a genuine corridor effect. The limited corridor effect was already confirmed on a local to regional scale by Bruinsma et al. (1997), Louter et al. (1999) and Van Oort and Raspe (2005), but now research shows that it also has empirical validity on the European regional scale. Still, the scale problem is a recurring issue in measuring agglomeration effects (Frenken et al. 2007) and deserves further elaboration. Other specific measurement issues such as the cut-off points of certain spatial regimes (of corridors and the size of cities) and robustness analyses of time- and sector-varying dynamics could be considered in future research. And of course, the causality problem also remains unaddressed.

The findings have important implications for European policy objectives regarding corridor development and regional economic growth, because the variety that has been shown in the research is little recognised in EU policy. Since a Europe in which regions develop at different rates has been observed, the remark of Puga (2002), that governments have no clear indication of which way to push when seeking efficiency, still holds. The results of Witte et al. (2013b) show a highly varied picture of corridor effects with tight conditions: what is beneficial in some corridors and urban regions is not necessarily beneficial in other regions, even when the same conditions apply. In other words, the type of agglomeration economies in combination with the structure of the economy is important for the prospects of structural economic growth in regions. This confirms the recently suggested need for a place-based approach in regional development policy in Europe (complementary to a generic, people-based approach) that takes into account these regional differences and requirements, so that each region has its own specific approach to economic development (Barca et al. 2012).

In conclusion, the empirical support for the economic potential of corridors and the resulting positive externalities are not strong, and the added value of the corridor concept in explaining the spatial heterogeneity of structural growth patterns is not proven. However, whereas the empirical evidence presented in this section is in contrast to the positive stance towards the relatedness between space, economy and transport in relation to corridors, as was initially put forward in Witte et al. (2013a), thus refuting the integration argument in corridor development, this does not imply that the corridor concept therefore is irrelevant. This provides the outlook for the next section on the possible benefits of the transport dimension to integrated corridor development in Europe.

4 Why Integration Works: Evidence from the Transport Dimension

In contrast to Sect. 3, this section highlights evidence that supports the integration argument in corridor development by zooming in on findings from the transport dimension of integrated corridor development (Sect. 2.3). Witte et al. (2014) have addressed—following the implications of Witte et al. (2012)—the heightened need for empirical support regarding the supposed multi-dimensional nature of issues in global freight transportation and corridor development. Within Europe, the port and inland navigation network can be seen as the backbone on which these global freight transportation issues take physical shape. Especially when considering the 'weakest link' principle, the functioning of inland ports is important for the overall efficiency of corridors. For years, however, the port system development literature (e.g. Hesse and Rodrigue 2004; Rodrigue 2004; Notteboom and Rodrigue 2005) has shown a strong focus on the maritime context within a network-based perspective (Outside–In). In contrast, it is argued that inland ports are growing in complexity and importance, and that the same port system development literature should also be sensitive to the independent role and structure of inland ports in transportation networks and corridors (Inside–Out). Thus, the attention should focus more on the challenges that possibly exist within the context of inland ports and the ways in which these challenges are influencing the independent role of inland ports and the shaping of inland ports' governance strategies.

The results of Witte et al. (2014), which are based on case-study evidence from the Dutch section of the Rotterdam–Genoa corridor, have both theoretical and practical implications, which will be discussed in greater detail later in this chapter. With regard to the theoretical implications, two major consequences arising from the growth in cargo volume and the expansion of distribution facilities, both in the seaports and in the hinterland, can be observed and are reflected in the theoretical models (e.g. Notteboom and Rodrigue 2005; Wilmsmeier et al. 2011; Monios and

Wilmsmeier 2012). These are the increasing importance of inland ports as cornerstones of inland accessibility (i.e. port regionalisation) and the increasing extent to which seaport areas are facing port–city challenges (Wiegmans and Louw 2011; Daamen and Vries 2013). However, on basis of the 'weakest link' principle and the directional development debate (Outside–In/Inside–Out), Witte et al. (2014) show that two problems emerge. First, in the port regionalisation concept, insufficient attention is paid to the independent role of inland ports (i.e. Inside–Out). Second, there is hardly any consideration of the possibility that port–city challenges may also arise between inland ports and cities within transnational corridors. Witte et al. (2014) have thus contributed a next step in port system development to the ongoing discussion in the literature, i.e. the emergence of inland port–city challenges.

With regard to the practical implications, an empirical analysis of the development strategies inland ports has been performed (Witte et al. 2014), using an institutional methodological approach that is in accordance with the recent 'institutionalist turn' observable in port literature (e.g. Daamen and Vries 2013). This research approach opens up the possibility to shed more light on the exact nature of the challenges emerging in the context of inland ports (Inside–Out). The results can also be seen as an empirical follow-up to the analytical framework (Fig. 3) that was presented in Witte et al. (2012), in which the framework was tested systematically, using both deductive and inductive types of analysis. The most important conclusions to be drawn are related to the multi-dimensional nature of the inland port–city challenges on one hand, and the ways in which these challenges are shaping inland ports' governance strategies on the other.

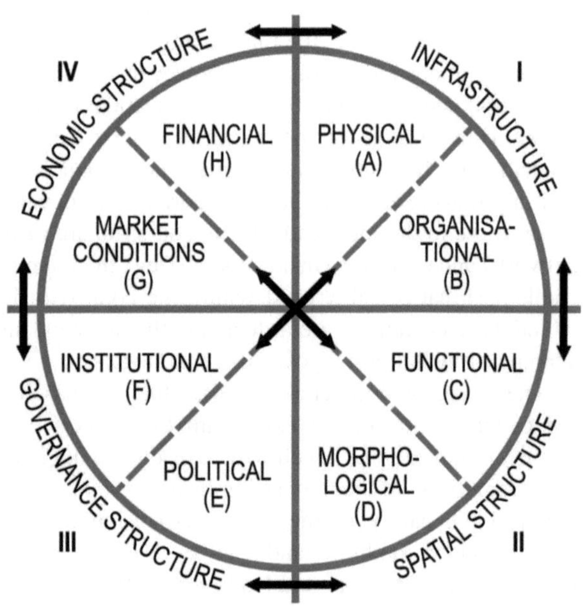

Fig. 3 Analytical framework for bottlenecks in the European transport network (Witte 2014)

One general conclusion was that all the dimensions of the analytical framework (Fig. 3) have been found in practice. This highlights that the framework can be considered to have an added value in identifying the multi-dimensional nature of inland port–city challenges and the ways in which these are related to one another. Next, it is shown that new challenges arise when these dimensions tend to overlap (in particular, industrial and distribution functions versus residential, leisure and nature functions). Although the challenges between inland ports and cities that have been identified take a specific form, a commonality has been found in the imbalance regarding the supra-regional benefits and local to regional negative externalities of inland ports. This probably results from difficulties in the trade-off between land-use functions in making plans (e.g. the conflicting functions of water) or the problematic relationship between infrastructure and spatial structure. A better consideration of the governance dimension of the analytical framework might help ease such conflicts between transport and land use.

This drives us to another conclusion, for several governance strategies have been observed that inland ports use in dealing with the emergence of inland port–city challenges. It has been found that a pro-active and positive stance towards zoning contributes to an efficient accommodation of mutually exclusive dimensions of inland port development. An interesting finding in this respect is the importance of institutions and the dominance that the institutional dimension can have over other dimensions. This can either be positive, thus contributing to the efficiency of inland ports, or negative, thus hampering the further development of inland ports. This finding is in line with the stated importance of institutional forces and the implications for the prioritisation of certain issues or dimensions, as was mentioned in Witte et al. (2012). In other words, the willingness or reluctance of actors and institutions to interfere in inland port development might either stimulate or hamper the overall efficiency of inland ports and transport corridors. The implications of these findings for the integration debate in corridor development will be discussed in the next section.

5 Evidence Pro and Con the Integration Argument

In this chapter, the question has been put forward of whether an integrated conceptualisation of corridor development has added value for European policy-makers in their current and future governance practices regarding corridors and corridor development. This section will examine whether the added value of the integration argument can be proven and whether this leads to a restating of the importance of corridors for present-day European policy objectives. In particular, by reviewing evidence from two problem areas of integrated corridor development, i.e. the economic dimension and the transport dimension. The remainder of this section outlines the arguments pro and con integration. This can be considered a stepping stone towards the reflections posed in the final section of the chapter.

A first notion regarding this chapter's contribution to the integration debate is that the findings in principle are uniform, but they are also contradictory in the sense that some evidence that supports the integration argument is provided (Sect. 4), and other evidence is provided that refutes the integration argument (Sect. 3). In other words, it is shown that integrated corridor development is inconsistent within itself, which affects the problem-solving capacity of the corridor concept fundamentally. Thus, when this is related to the integrated versus sector-based discussion (Sect. 2), a definite answer can never be provided, for it depends strongly on the issues at stake, the context and the conditions. The opportunities of an integrated corridor approach (or authority) always combat the efficiency of local and sector-based solutions for corridor issues. However, some indications can be given regarding the dimensions and scales in which integrated corridor development is or is not of added value for corridor governance practices.

The arguments that plead for the integration argument in corridor development can be extracted from the transport dimension (Sect. 4). First, from Witte et al. (2012) it was concluded that many policy documents are insensitive towards the multi-dimensional nature of bottlenecks, while the theoretical and empirical evidence presented in that paper clearly pointed at the inevitable interrelatedness of the multiple dimensions of bottlenecks. Moreover, Witte et al. (2014) provided additional empirical support for this analytical framework of cumulative bottlenecks by showcasing the multi-dimensional nature of challenges existing between inland ports and cities along corridors. This has proven to support the supposed added value of the multi-dimensional framework for the integration argument in corridor development. Also, it should be noted that these publications share a commonality in the sense that the relevance of the integration argument is emerging especially from the traditional, transport-oriented starting point of both papers. In other words, integration 'works' when the initial situation is dominated by a sector transport-oriented perspective (compare Witte and Spit 2014).

The other side of the integration debate is formed by evidence that pleads against the existence of any added value in integrated corridor development (Sect. 3). The findings presented in Witte et al. (2013b), in particular, have made clear that the initial support for the integration argument in Sect. 2 should be refuted, since the empirical evidence consistently invalidated the importance of corridors for growth and agglomeration. A clustering effect in corridor regions is observable, but the corridor itself cannot be considered exclusively responsible for this. Thus, viewed from the economic dimension, the added value of the integration argument is not proven.

This implies that the added value of the integration argument in corridor development is at least not uncontested, which reinforces the conclusion from Witte et al. (2013a) that the sector-based practices of corridor development in Europe are not really surprising. On the contrary, they are very realistic, especially in the light of the recent economic downturn. The overall conclusion concerning the integration argument in corridor development is that the contribution of an integrated approach to promote a more efficient corridor development is certainly not self-evident—but it is not irrelevant either. The findings are not conclusive in supporting either side of the integration argument. Universally speaking, this chapter has contributed to differentiating the debate regarding corridors and integration. It has shown that, although the empirical evidence in some cases contradicts the integration argument in corridor development (Sect. 3), corridors can still be seen as useful linkages between regions sharing a commonality in their respective issues (Sect. 4).

6 Future Perspectives on European Corridor Development

The debate on an integrated perspective versus a more sector-based perspective in European corridor development has been at the heart of this chapter. In this final section, some reflections will be made regarding the contribution of the integration argument to the positioning of corridors in European policy and its implications for transnational corridor development for multi-level governance strategies. It can be concluded that the contribution of the integration argument to the positioning of corridors within European policy must be considered limited. In terms of economic development (Sect. 3), the added value of integration is hard to prove. However, integration does pose promising common ground for regions sharing a commonality in their respective transportation issues (Sect. 4). For instance, the inland ports in the Rotterdam–Genoa range might find a commonality in dealing with trans-border water management issues along the Rhine River. This is a topic for future study.

Nevertheless, one could question whether the corridor concept is a proper reflection of the spatial reality in Europe and whether corridors as a spatial phenomenon can be legitimised (Witte et al. 2013a). What could be really interesting in this respect is a consideration of the 'why' question. The findings have not yet provided enough insight into the reasons *why* the corridor concept has difficulties in becoming a mainstream and accepted spatial concept (compare Pain 2011). Is it because of conceptual ambiguity with regard to the multi-dimensional nature of corridors? Is it because of the transnational scale at which corridors usually operate? Or does it have to do with institutional fragmentation? These questions could be the starting point for future research into the contributions of the corridor concept to achieving European policy objectives. In particular, the findings might contribute to the formulation of new corridor studies under the umbrella of the Connecting Europe Facility and the tender process, which is ongoing for studies on the Core Network Corridors, following the revision of the TEN-T guidelines.

A more practical consideration is what implications transnational corridor development can have for multi-level governance strategies (e.g. in a European Grouping of Territorial Cooperation (EGTC)). What does the possible added value of transnational corridors imply for policy-making and in practice on the national, regional and local levels? Witte (2014) has brought the relationship of positive and negative externalities with the importance of public–private constellations in this respect to the forefront. In Sect. 4, it was argued that many challenges in corridor development in the end reflect a structural imbalance regarding the supra-regional benefits on one side and local to regional negative externalities on the other. This is an important outcome that should be considered in policy-making for corridors on different spatial scales.

Another, but also interesting, matter for future research concerns the 'break-even point' where positive externalities turn into diseconomies. With respect to the public–private constellations that are of importance in corridors, future research could focus on the contribution of corridors in avoiding the unplanned extension of urban areas, on the question of whether corridors can manage without public interference, or even whether the governance of corridors is needed at all (compare Schönharting et al. 2003; Chapman et al. 2003). These questions could lead to an intriguing new way to view European corridor development from a joint and interdisciplinary planning perspective.

References

Albrechts L, Coppens T (2003) Megacorridors: striking a balance between the space of flows and the space of places. J Transp Geogr 11(3):215–224

Banister D, Capello R, Nijkamp P (1995) European transport and communications networks: policy evolution and change. Wiley, Chichester

Barca F, McCann P, Rodriguez-Pose A (2012) The case for regional development intervention: place-based versus place-neutral approaches. J Reg Sci 52(1):134–152

Bathelt H (2005) Geographies of production: growth regimes in spatial perspective (II): knowledge creation and growth in clusters. Prog Hum Geogr 29(2):204–216

Bertolini L, Spit T (1998) Cities on rails: the redevelopment of railway station areas. Spon Press, London

Bruinsma F, Rienstra S, Rietveld P (1997) Economic impact of the construction of a transport corridor: a multilevel and multi-approach case study for the construction of the A1 highway in the Netherlands. Reg Stud 31(4):391–402

Chapman D, Pratt D, Larkham P, Dickins I (2003) Concepts and definitions of corridors: evidence from England's Midlands. J Transp Geogr 11(3):179–191

Daamen T, Vries I (2013) Governing the European port–city interface: institutional impact on spatial projects between city and port. J Transp Geogr 27(1):4–13

Dogaru T, van Oort F, Thissen M (2011) Agglomeration economies in European regions: perspectives for objective 1 regions. Tijdschr Econ Soc Geogr 102(4):486–494

Dühr S, Stead D, Zonneveld W (2007) The Europeanization of spatial planning through territorial cooperation. Plan Pract Res 22(3):291–307

European Commission (1999) European spatial development perspective: towards balanced and sustainable development of the territory of the European Union. Office for Official Publications of the European Community, Luxembourg

European Commission (2008) ERTMS Corridor route comparison. http://ec.europa.eu/transport/modes/rail/doc/2008_ertms_corridor_route_comparison.pdf. Accessed 16 May 2014

European Commission (2011) Connecting Europe: the new EU core transport network. MEMO/11/706, Brussels, 19 October 2011

European Commission (2013) Core network corridors on the TEN-T. http://ec.europa.eu/transport/infrastructure/tentec/tentec-portal/site/en/maps.html. Accessed May 16 2014

Frenken K, van Oort F, Verburg T (2007) Related variety, unrelated variety and regional economic growth. Reg Stud 45(5):685–697

Hesse M, Rodrigue J (2004) The transport geography of logistics and freight distribution. J Transp Geogr 12(3):171–184
Louter P (1999) Economische netwerken: determinanten van de ruimtelijk-economische dynamiek. INRO/TNO, Inro/Ecn 1999/25
Louter P, Puylaert H, Raspe O (1999) Megacorridors en stadsgewesten: analyse van ontwikkelingen in de bedrijvigheid. TNO Inro, Afdeling Planning/Afdeling Economie, Delft
McCann P, Shefer D (2004) Location, agglomeration and infrastructure. Pap Reg Sci 83 (1):177–196
Monios J, Wilmsmeier G (2012) Giving a direction to port regionalisation. Transp Res A 46 (10):1551–1561
Notteboom T, Rodrigue J (2005) Port regionalization: towards a new phase in port development. Marit Policy Manag 32(3):297–313
van Oort F, Raspe O (2005) Kennisassen en Kenniscorridors. Over de structurerende werking van infrastructuur in de kenniseconomie. Ruimtelijk Planbureau, Den Haag
Pain K (2011) New worlds for old? Twenty-first-century gateways and corridors: reflections on a European spatial perspective. Int J Urban Reg Res 35(6):1154–1174
Peters D (2003) Cohesion, polycentricity, missing links and bottlenecks: conflicting spatial storylines for pan-European transport investments. Eur Plan Stud 11(3):317–339
Priemus H (2001) Corridors in the Netherlands: apple of discord in spatial planning. Tijdschr Econ Soc Geogr 92(1):100–107
Priemus H, Zonneveld W (2003) What are corridors and what are the issues? Introduction to special issue: the governance of corridors. J Transp Geogr 11(3):167–177
Priemus H, Zonneveld W (2004) Regional and transnational spatial planning: problems today, perspectives for the future. Eur Plan Stud 12(3):289–297
Puga D (2002) European regional policies in light of recent location theories. J Econ Geogr 2 (4):373–406
Rodrigue J (2004) Freight, gateways and mega-urban regions: the logistical integration of the Bostwash corridor. Tijdschr Econ Soc Geogr 95(2):147–161
Romein A, Trip J, de Vries J (2003) The multi-scalar complexity of infrastructure planning: evidence from the Dutch–Flemish megacorridor. J Transp Geogr 11(3):205–213
Scholl B (2012) SAPONI: spaces and projects of national importance. ETH Zürich, Zürich
Schönharting J, Schmidt A, Frank A, Bremer S (2003) Towards the multimodal transport of people and freight: interconnective networks in the RheinRuhr Metropolis. J Transp Geogr 11 (3):193–203
Spit T (1998) Ruimtelijke ordening: integraliteit van beleid als probleem. Bestuurswetenschappen 52(6):289–296
Thissen M, van Oort F, Diodato D, Ruijs A (2013) Regional competitiveness and smart specialization in Europe. Edward Elgar, Cheltenham
de Vries J, Priemus H (2003) Megacorridors in north-west Europe: issues for transnational spatial governance. J Transp Geogr 11(3):225–233
Wiegmans B, Louw E (2011) Changing port–city relations at Amsterdam: a new phase at the interface? J Transp Geogr 19(4):575–583
Wilmsmeier G, Monios J, Lambert B (2011) The directional development of intermodal freight corridors in relation to inland terminals. J Transp Geogr 19(6):1379–1386
Witte P (2014) The Corridor Chronicles: integrated perspectives on European transport corridor development. Eburon Academic Publishers, Delft
Witte P, Spit T (2014) Sectoral drawbacks in transport: towards a new analytical framework on European transport corridors. In: Lami I (ed) Analytical decision-making methods for evaluating sustainable transport in European corridors. SxI – Springer for Innovation, vol 11. doi:10.1007/978-3-319-04786-7_4
Witte P, Wiegmans B, van Oort F, Spit T (2012) Chokepoints in corridors: perspectives on bottlenecks in the European transport network. Res Transp Bus Manag 5:57–66

Witte P, van Oort F, Wiegmans B, Spit T (2013a) Capitalising on spatiality in European transport corridors. Tijdschr Econ Soc Geogr 104(4):510–517
Witte P, van Oort F, Wiegmans B, Spit T (2013b) European corridors as carriers of dynamic agglomeration externalities? Eur Plan Stud 22(11):2326–2350
Witte P, Wiegmans B, van Oort F, Spit T (2014) Governing inland ports: a multi-dimensional approach to addressing inland port–city challenges in European transport corridors. J Transp Geogr 36:42–52
Woxenius J (2007) Generic framework for transport network designs: applications and treatment in intermodal freight transport literature. Transp Rev 27(6):733–749

Patrick Witte (1988) is a junior assistant professor in the Department of Human Geography and Planning at Utrecht University. In 2014 he obtained his PhD degree. His present field of work in teaching and research is focused on infrastructure planning and integrated area development.

Tejo Spit (1955) is a full professor in Spatial Planning in the Department of Human Geography and Planning at Utrecht University since 2002.

His teaching and research focusses on sustainability issues, land policy, planning methodology, infrastructure planning and administrative aspects of spatial planning. Until now he has published more than 200 papers, books, book chapters and research reports.

Part II
Regional Accessibility: An Important Locational Factor

Bottlenecks and Regional Economic Impact: Simulations with the CODE24 Transport Model

Hansjörg Drewello, Marcel Huschebeck, and Norbert Schick

Abstract Bottlenecks in transport infrastructure will change the behavior of logistics companies because of increasing costs of transportation. If a bottleneck in a railway network occurs, railway freight transport will be shifted to roadways. Railway congestion causes, for example, a decrease in reliability and a considerable increase in transport costs, and thus road transport becomes less expensive. The CODE24 Transport Model supports decision-making about choice of transport modes. All available information potentially affecting those decisions, such as regions, transport networks, terminals and logistic services, costs as well as transport or monetary flows, are integrated into the model. The model tries to find the shortest or most efficient route, e.g. with regard to transport costs.

In this article, economic effects along the railway corridor Rotterdam-Genoa were analyzed in a simulation of three different types of bottlenecks on railways by using the CODE24 Transport Model. The simulation shows a shift of freight transport from rail to road or from rail to barge, depending on different transport distances. The change of intermodal behavior increases transport costs, which can be calculated within the model. The simulation allows a better insight into the regional and overall costs generated by transport bottlenecks. The analysis indicates that an increase in transport costs for one transport mode, due to a bottleneck, has an impact on all the regions along the corridor and even more so on specific regions outside the corridor.

H. Drewello (✉)
University of Applied Sciences Kehl, Kinzigallee 1, 77694 Kehl am Rhein, Germany
e-mail: drewello@hs-kehl.de

M. Huschebeck • N. Schick
PTV Group, Haid-und-Neu-Straße 15, 76131 Karlsruhe, Germany
e-mail: marcel.huschebeck@ptvgroup.com; norbert.schick@ptvgroup.com

H. Drewello, B. Scholl (eds.), *Integrated Spatial and Transport Infrastructure Development*, Contributions to Economics, DOI 10.1007/978-3-319-15708-5_7

1 Introduction

Even if freight transport is still growing rapidly in Europe, building new rail infrastructure to further support modal shift is becoming more and more difficult due to public budget restrictions or opposition from civil society on environmental issues and route alignment. Frequently, transport associations or other economic stakeholders use the bottleneck argument to press for new tracks. From a local or national point of view, investments are first sought to expand infrastructure where bottlenecks are located. In the case of Germany, Deutsche Bahn denies the existence of bottlenecks along the German part of the corridor Rotterdam-Genoa. They maintain that the infrastructure manager DB Netz is able to satisfy all demands for new freight or passenger routes. This different view of reality probably results from a different concept of bottleneck.

Bottlenecks in transport infrastructure will change the behavior of logistics companies because of increasing costs of transportation. They will affect some regions more than others. If a bottleneck in a railway network occurs, railway freight transport will shift to road or barge because congestion causes, for example, a decrease in reliability and a considerable increase in transport costs, and transport by truck or by barge becomes less expensive. The CODE24 Transport Model supports decision-making about choice of transport modes. All available information potentially affecting those decisions, such as regions, transport networks, terminals and logistic services as well as transport-mode specific cost structures, are integrated into the model. The model tries to find the most cost-efficient route.

In this article, effects of bottlenecks along the railway corridor Rotterdam-Genoa are simulated in a shift of container freight transport from rail to road or from rail to barge. Three different simulations have been developed. The change of intermodal behavior increases transport costs, which can be calculated within the model. The simulation allows a better insight into the regional and overall costs generated by transport bottlenecks (costs of non-doing).

2 Bottlenecks in Transport Markets

2.1 What is a Bottleneck in Freight Transportation?

In public debate it is sometimes unclear what is meant by a bottleneck in transport infrastructure. In many of these political statements a lack of capacity is specified, usually combined with the projection of increasing freight transport.[1] Holzhey determined capacity bottlenecks by calculating the potential maximum of freight

[1] e.g. DB Schenker, Rail freight companies present their requirements for the most important European Corridor: Rotterdam-Genoa, Press release, 19 Dec. 2011; Port of Rotterdam, Newsflash 2012, February 2012, p. 4.

train runs per day for a corridor and comparing it with future needs (Holzhey 2010, p. 17, 2011, pp. 4–6).

As early as 1996 Rothengatter stated that technical capacity is not a sufficient measure to identify major deficiencies in railway networks. Accordingly, insufficient service levels of railway companies were more important than technical bottlenecks at that time (Rothengatter 1996, p. 1). Cipolina and Ghiara distinguish four different categories of bottlenecks in freight transportation: infrastructural, organisational, technical and bureaucratic (Cipolina and Ghiara 2011, p. 150).

One important aim of the CODE24 project was to better understand bottlenecks in logistics and their effects on the freight transport corridor Rotterdam-Genoa. In order to do so, an international and interdisciplinary expert group within the project (planners, architects, engineers, logisticians and economists) developed a common definition in 2011 in the course of a structured brainstorming process (metaplan):

> Bottlenecks always represent additional costs to logistic services by hindering them. They can be observed on a politico-legal, organisational or physical level. Such a bottleneck can be national or transborder. (Drewello and Günther 2012, p. 3)

On the politico-legal level, a bottleneck occurs if restrictions resulting from political decisions or legal frameworks hinder infrastructure planning and logistic processes. This can be e.g. regulation of competition, standards, regulation of noise protection, price and access regulation, taxes, nighttime bans and so on.

A bottleneck also emerges from inefficiencies of operation inside the logistics sector on the supply and/or on the demand side (organisational level). A common example of inefficiencies of operation is a lack of information e.g. concerning terminal services. The market structure can also lead to inefficiencies. Strong competition could hamper advantageous cooperation as well as a market-leader position of one supplier (Ahrens et al. 2007, p. 3).

On the physical level, bottlenecks emerge when demand for freight transport exceeds the available infrastructure capacity. Railway infrastructure includes tracks, junctions, signalling systems, terminals, tunnel heights, etc.

What are the consequences of emerging bottlenecks in transport infrastructure? Let us assume, for example, growing demand for freight transport runs on a railway corridor. If the price for train runs remains unchanged, this may lead to excess demand. A bottleneck appears where demand exceeds supply (capacity). Regulatory responses to the excess demand in the short run, which exclude new construction of infrastructure, could be to do nothing but accept the bottleneck or to increase the price for slots in order to create new market equilibrium. There are other possible substitutes for railway freight transport, specifically road or barge. Both regulatory options will therefore lead to a substitution effect. Freight which cannot be transported by rail will be transported especially on roads. This effect is contrary to the manifested transport policy of the European Union and Switzerland.

2.2 The "Rheintalbahn" as a Bottleneck

Bottlenecks with a physical characteristic emerge when demand for freight transport exceeds available infrastructure capacity. Many different kinds of information are necessary to compute the capacity of a railway line. Much of this data in Germany is only available to Deutsche Bahn, and that company has declared it to be a business secret. Nevertheless, transport capacity can be calculated for railway sections. But much effort is required.

One factor that is of major importance for determining the capacity of a railway line is the number of tracks available for travelling in a particular direction. Other key aspects influencing capacity are the distance between the block signals, buffer times, the speed difference between slow and fast trains, overtaking opportunities along the railway line and the sequence in which the various categories of trains follow one another (Drewello and Gütle 2013, 6 ff.).

In February 2013, the Universities of Kehl and Offenburg organised a train count on the Rheintalbahn near the town of Lahr. The analysis consisted of the calculation of a mean capacity for freight train runs per day for the two railway tracks near Lahr (Drewello and Gütle 2013, pp. 14–24) based on the blocking time model originally developed by Happel (1959). This information was then used as a baseline for the real use of capacity, observed and counted day and night for 2 weeks using a high-tech infrared camera.

The analysis shows that utilisation of the capacity of the Rheintalbahn on the section between Offenburg and Lahr is subject to fluctuations during the observation period. Capacity utilisation is at its highest on the weekdays of Tuesday to Friday. The data collected in February 2013 show that at this time of the year, at least at the weekend and more especially on Sunday, there is still capacity available for the transport of freight. On the weekdays of Thursday to Friday, this capacity reserve is, however, only marginal on the basis of the assumptions underlying the study (Fig. 1).

It is on Wednesday that the highest capacity utilisation is to be noted. The computed capacity limit is exceeded during the time segment between 8:00 a.m. and 4:00 p.m. in both directions (116 % from Offenburg southwards towards Freiburg). This can result in decreasing punctuality during this period. Capacity utilisation is slightly lower during the day's other two time segments, but only in one direction the capacity limit is not exceeded (see Fig. 2). The explanation for that during the morning time segment from midnight to 8.00 a.m. is to be found especially in the lower number of passenger trains. During the evening time segment from 4:00 p.m. to midnight, the number of freight trains is lower southbound. In the other direction, the number of trains still exceeds the capacity limit.

The statements by DB Netz experts also indicate that the observation period occurred during a phase of below-average capacity utilisation. It follows from this that there is a further reduction in capacity reserves in the months May, June, October and November with above-average capacity utilisation (Drewello et al. 2013, p. 50).

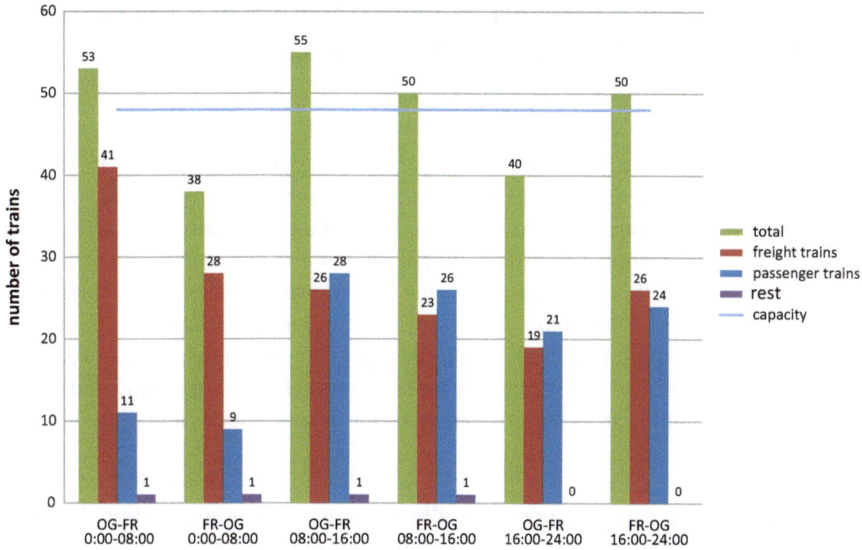

Fig. 1 Mean train numbers on Wednesday (Drewello et al. 2013, p. 51)

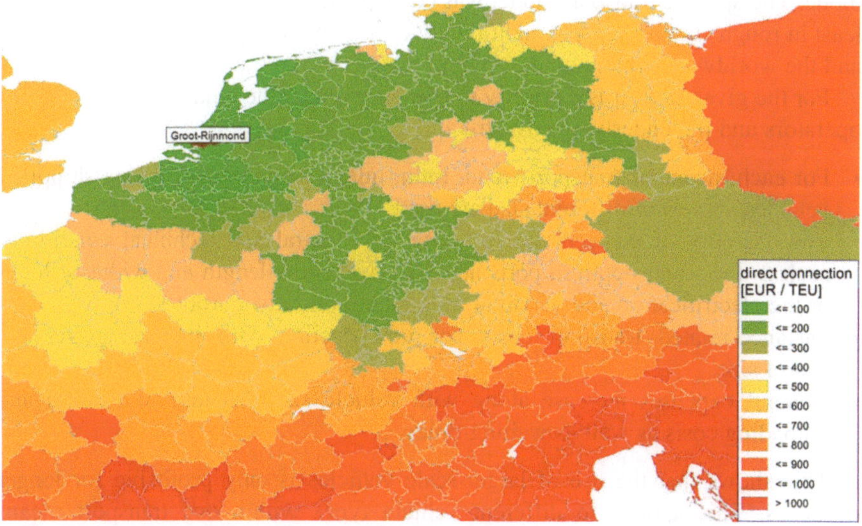

Fig. 2 Corridor and neighbouring regions in the CODE24 model. *Source*: Own illustration

A further effect to be expected is that most of the growth in freight traffic is therefore going to take place on the road. This effect can already be observed in the Upper Rhine Valley (Drewello and Gütle 2013, pp. 42–44).

3 The CODE24 Transport Model: Structure of a Traffic Model for the Calculation of Transport Costs and Accessibility

The CODE24 corridor is a central transport axis for freight transport in Europe. A major part of the imports and exports from and to European countries is transshipped at the seaports of Amsterdam, Rotterdam and Antwerp. The Ligurian ports, particularly Genoa, form the southern endpoint of the axis.

The seaports in this case do not frame the transport flows by forming their starting points and endpoints. They are the major transshipment points for global freight transport, especially container transport. The transport flows running through the CODE24 corridor can only be understood by looking at the underlying cost structure resulting from the combination of transport modes and transshipment possibilities.

A central issue in CODE24 is the location quality, i.e. the accessibility of regions by different means of transport. For the shipping and transportation industry, accessibility is expressed in terms of overall transport costs.

At the core of the CODE24 Transport Model is a cost modelling system derived from a bottom-up calculation approach based on generalized costs per transport mode. The specific transport model was conceptualized as a global model that—at least in rough simplification—additionally covers the transport route on the oceans and the worldwide origin and destination regions of the trade flows.

For the given task, it is a priority to correctly model the tenders of the transport operators and the underlying cost structures. This implies:

- For each means of transport (road, train, inland waterway, deep-sea shipping) vehicles/vessels of different size are used.
- The usability of the vehicles is limited by natural and technical conditions, such as the water depth in ports and waterways, the length and width of locks or the electrification and maximal train length on railway routes.
- Distance-related transport costs per unit decline with increasing vehicle/vessel size.
- General expenses increase along with vehicle size, because of loading and unloading costs as well as waiting times.

This shows that the use of larger-scale vehicles is only profitable for longer distances. This applies to different vehicle types within one transport means as well as to different vehicles from different transport means.

The cost structures illustrate that the operational transport costs play a less important role than in passenger transport. Of much more importance are the handling costs and time-related costs. Time costs originate from the deterioration of goods, the tie-up of capital and the resulting effects on subsequent production steps. These time costs require that a faster vehicle is used for higher-quality or perishable products. The transport costs for a faster vehicle are higher, but from the perspective of a goods owner, faster transport leads to decreased time costs and therefore, this is a reasonable action.

The relatively high proportion of time-dependent costs in the total cost structure means that on many shipping routes longer main-haul legs are chosen in order to avoid an additional transshipment process.

3.1 Development

The CODE24 Transport Model is an integrated intermodal transport network that in the first stage focusses on standard containers as transferable loading units across the transport modes. The networks for road, train track and waterway were created as three separate network models by using the transport planning software PTV VISUM. The road and train networks are limited to Europe. These networks were generated from existing network data and reduced to a reasonable network density. The waterway network contains the inland waterways and deep-sea routes connecting major ports overseas, such as those in Asia, Africa, North and South America and Australia. The inland waterways had to be updated with information on the maximum length of the vessels, the width of the vessels, maximum unloading depth and overhead clearance limitations.

The routes of the three networks are respectively opened or closed for the different vehicle types. This has also been done for deep-sea shipping. There, the biggest container ships are only accepted at a few ports. Eight vehicle types were identified for deep-sea shipping, 5 for inland waterway shipping, 12 for train and 4 for truck transport. Additionally, there are train- and truck-carrying ferries, which are part of the train or truck network. The network models calculate the vehicle-specific transport costs, which arise from distance-dependent costs (e.g. fuel costs) and time-dependent costs (e.g. driver costs, depreciation costs), taking into consideration the different velocities on the routes. For example, in the waterway network locks are depicted with a time supplement of 20–30 min. In road transport compulsory breaks for the drivers after 4 h are considered a time and cost supplement. For the total picture, the distance-speed-related calculation (instead of just a distance-related calculation) has advantages for regions with well-developed, dense and fast networks in all directions.

The zoning of the model is relatively accurate for the countries within the CODE24 corridor on a NUTS 3 level. In the neighboring European countries, a NUTS 2 level or a NUTS 1 level was realized. In total the model contains 963 transport districts.

A central part of the model is the terminal modelling. At these terminals the handling of different vehicles is possible. Handling can occur within one transport means and between different vehicle types, or between vehicle types which are assigned to different transport means. Tri-modal terminals are therefore included in all three networks and are connected to the terminals through the particular network. Railway yards are typical mono-modal terminals, at which a change between a short-distance freight train and a long-distance freight train can be depicted. In total, the model includes 1200 terminals which are depicted as additional and special terminal districts.

The cheapest route between all the transport districts is determined in terms of the assumption about the specific time costs (€/h) for the transported goods. For that, multiple sequential calculation steps are needed:

- For each vehicle/vessel type, the transport costs and the time needed between all transport districts (normal and terminal districts) without handling are determined.
- For each shipping route, the vehicle with the lowest total costs is determined.
- The time calculation assumes that different services are available with different frequencies for each transport system. Costs are considered as additional waiting time parameters relative to truck operations (truck frequency = 0)
- An assessment is made as to whether a handling connection makes cheaper transport possible. For that reason, all theoretically possible combinations with up to six route sections (five handling procedures, accordingly) are tested. In the case of the handling procedure, one must take into consideration the additional handling costs, handling times and waiting times.

3.2 Results

As a result of the calculations, a number is assigned to every shipping route (about 0.8 million shipping routes; theoretically 963×963 shipping routes; shipping routes that had neither origin nor destination in Europe where not considered) corresponding to the number of handling processes consisting of: transshipment terminals, the vehicles of each route section, the transport costs, the time required, the time costs and the total costs. Additionally, a main transport carrier (transport carrier whose vehicles manage the biggest route section) is determined (Fig. 3).

With this information, the transport districts within the research corridor can be rated in terms of their accessibility. A reasonable criterion is the spatial extent of the areas where the truck, the train or the ship is the respective cheapest main transport carrier. In general, good accessibility is given wherever cheap transport modes like inland navigation and rail are employed for the main haul.

After the assessment of some districts within the corridor, it becomes clear that the connection quality on the north-south axis of the corridor is significantly better than on the east-west axis. With regard to the Rhine waterway this is logical, but this fact also applies to the railway, whose network is better developed in the north-south direction between Scandinavia and Southern Italy than in the east-west direction. For that reason, in some of the transport districts in the east and west, though relatively far apart, truck transport is designated as the cheapest alternative by the model.

Fig. 3 Accessibility of Rotterdam: cost per TEU to/from Rotterdam. *Source*: Own illustration

An analysis with data from the CODE24 model shows very clearly the higher regional accessibility of the Rotterdam-Genoa corridor, especially to the north of the Swiss Alps, which one could also assume (Drewello 2014, p. 110). The results fit well with considerations of geographical economics, which explains the high regional GDP per capita in the corridor ("blue banana") with good transport infrastructure since Roman times (Brunet 1989).

It was also quite evident during the evaluation that the Rhine-Main-Danube Waterway, the Moselle, the Saar and the Neckar do not play a significant role in container transport from the perspective of the CODE24 corridor. This is because of the slow speed of transport, which is due to the multiple locks and loops in the waterways. On the other hand, this is not applicable to the Dutch and Northern German channel network, especially the Midland Canal.

On a final note, it should be taken into consideration that the shipping routes should not be regarded as equivalents, since the geographical units in the districts represent quite different population figures, economic performance and economic structures. Since these parameters cannot be included in one common average-costs scheme, more and in-depth structural indicators which determine regional accessibility are needed.

4 Simulation of the CODE24 Transport Model

The approach to modelling bottlenecks is to modify the cost structure on single links within the transport network. It is assumed that in the case of bottlenecks, substitution effects within the intermodal network structure will take place. Therefore, a bottleneck simulation is expected to provide the following adaptations:

- Shift to another mode, in the event that alternative access to inland waterways or road is possible. This might lead to additional transshipment processes which increase the overall transport costs;
- Shift to another railway route. The overall costs for making a detour are less than the increased costs resulting from the bottleneck;
- The initial transport route is retained while higher transport costs are accepted due to a lack of alternatives.

We defined three scenarios for bottleneck modelling based on an increase in transport costs for railway services. For the scenario set-up, one must take into account the fact that the cost structure in goods transport is characterized by high fixed costs resulting from the time-dependent costs. Therefore, in order to visualize the impact of bottlenecks on the overall transport operation cost variations as compared to the basis scenario, a significant increase in transport costs was chosen. A significant increase in distance and the time-dependent costs of railway operations are the basis for the three scenarios. At the same time, the value of the time parameter was increased, being equivalent to additional waiting times. Bottleneck impact is assessed by using a before-and-after comparison of the total transport costs from one district to all of the other 962 districts within the CODE24 Transport Model.

Within scenario 1, all rail corridors parallel to the Rhine valley between Cologne and Zurich were considered as bottlenecks. The distance and the time-related costs parameter were increased by a factor of 2

Within scenario 2, the corridor between Karlsruhe and Basel was considered as a bottleneck. Along this corridor, a significant increase, a factor of 5, was assumed for distance and time-related costs.

Scenario 3 again addressed the corridor Karlsruhe to Basel. However, a tenfold increase in time-and distance-related costs was modelled. As an additional effect, this scenario included an assumed bottleneck for the "Gäubahn" between Stuttgart and Konstanz as well as for the "Schwarzwaldbahn" between Offenburg and Konstanz, thereby increasing costs by a factor of 5.

The figures below show the impact on regional accessibility based on the different bottleneck scenarios (Figs. 4, 5 and 6):

By means of the variation of transport costs for selected links in the railway network along the CODE24 corridor, all shipping routes within the CODE24 Transport Model were recalculated. Although limited to specific parts of the corridor, this increase in transport costs for one transport mode has an impact on most regions, even on regions outside the corridor. Obviously, the impact for

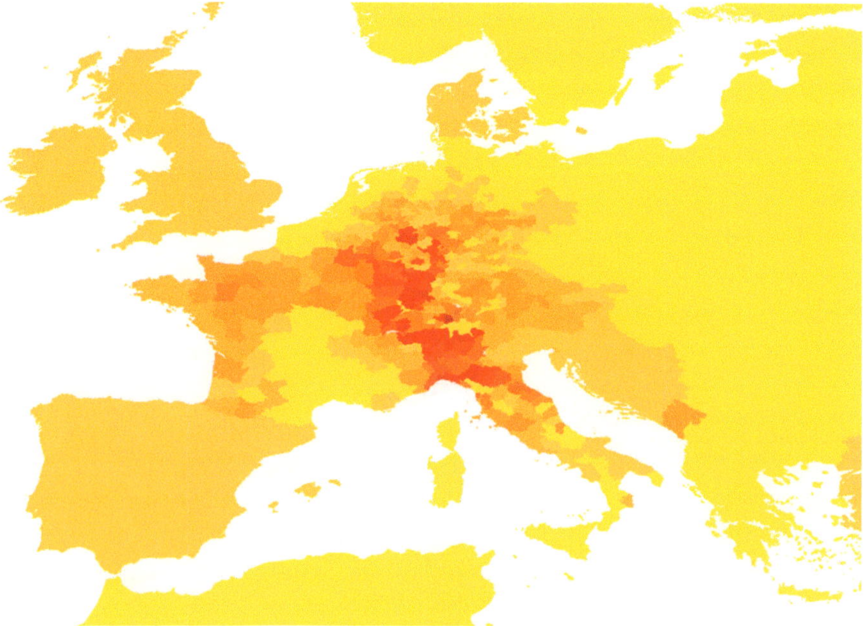

Fig. 4 Increased transport costs per region (*yellow*: 0 %, *red*: 10 %), Scenario 1: Cologne-Zurich, factor 2. *Source*: Own illustration

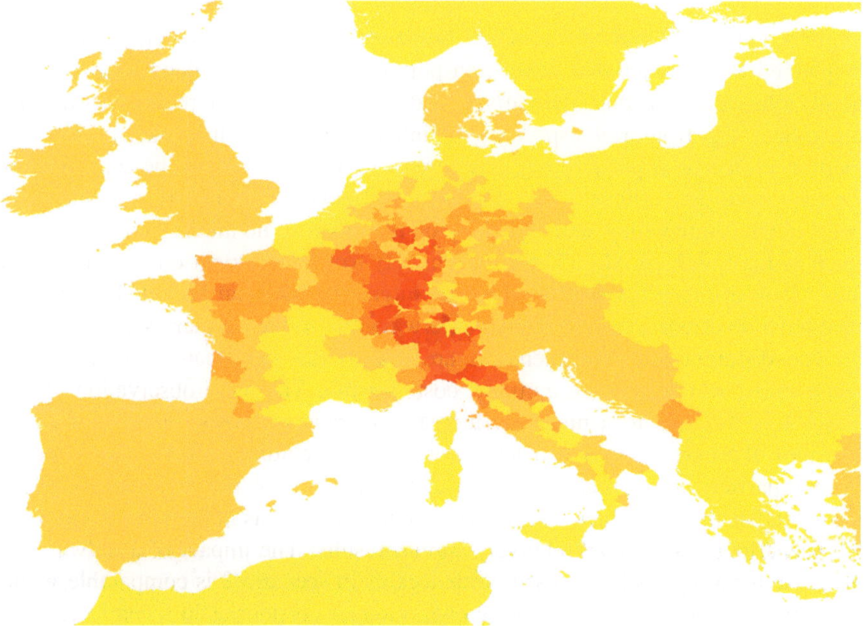

Fig. 5 Increased transport costs per region (*yellow*: 0 %, *red*: 10 %), Scenario 2: Karlsruhe-Basel, factor 10. *Source*: Own illustration

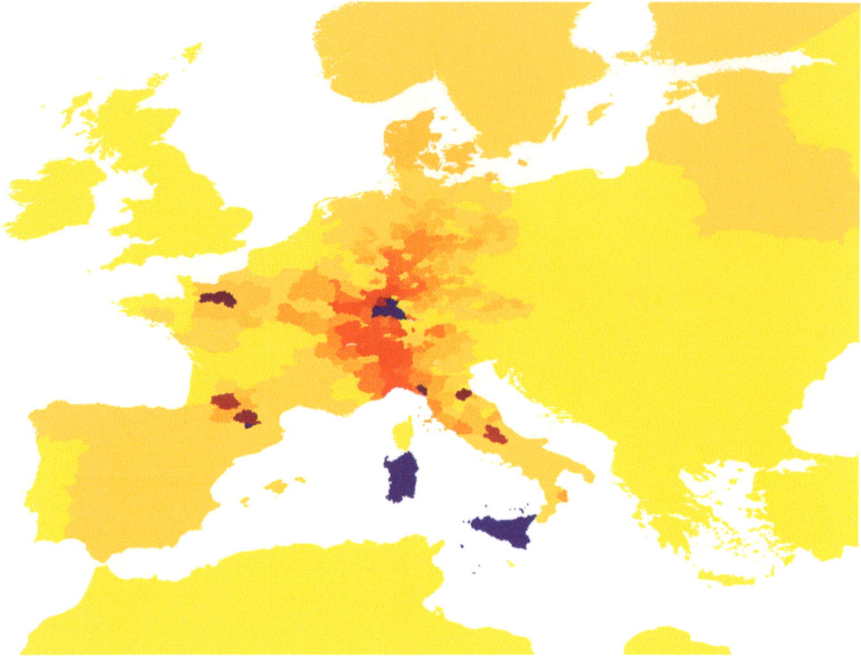

Fig. 6 Increased transport costs per region (*yellow*: 0 %, *red*: 10 %, *blue* 30 %). *Source*: Own illustration

regions located within the assumed bottleneck corridor is lower. This can be explained by the fact that other transport modes such as inland waterways or roads remain as transport alternatives, and that due to cost increases a shift in the main-haul leg from rail especially to inland waterways has taken place. The bottleneck has an especially negative impact on regions that can be considered as extensions of the bottleneck corridor, such as regions located to the east and to the west of the Rhine valley, and especially on regions with no inland waterway access such as the Southern Alps. These regions rely to a large extent on railways as the main long-distance transportation mode. In general, this can be stated for regions that have no direct access to inland waterways and to the northern seaports.

The differences in the scenarios are due to the significance of the chosen parameters, which increase transport costs for rail. The main observation of scenario 1 is that due to a moderate increase in transport costs, but over a longer corridor, the magnitude of impact is lower. More regions, however, are affected. In comparison, a significant increase in transport costs, but regionally focused, show higher impact values with fewer regions affected, as in scenario 2 and 3.

Scenario 1 and 2 show relatively similar results: The impact of a railway cost increase for a regionally focused bottleneck as in scenario 2 is comparable with a scenario of a moderate cost increase over a longer corridor as in scenario 1.

Scenario 3 addresses a scenario in which railway transport between Karlsruhe and Basel is practically blocked and all other regions are forced to accept additional transport costs, detours, additional transshipments and other transport modes on alternative routes, leading to higher costs for these regions.

A massive disruption, capacity limitations or cost increases for railway transport along the Rhine valley especially affect transport to and from Switzerland negatively. The necessity of additional transshipments in Basel to inland navigation or the use of alternative routes via France, Italy or Bavaria will lead to considerable extra costs.

5 Conclusions

Transport economics assumes that impediments on routes will increase with the number of users. In the case of alternatives, this leads to a split of transport flows. Within passenger and public transport modelling, this methodology provides realistic calculations of impediments and route assignments. Within the CODE24 Transport Model, this methodology has been applied to goods container transport, based on transport costs as the main impediment parameter. The calculated split of transport supply on the intermodal transport modes and routes is an impact of bottlenecks.

The modeled bottlenecks might not be considered as realistic in their magnitude. However, they show impressively the impact on regional accessibility and development. The analysis indicates that an increase in transport costs for one transport mode, due to a bottleneck, has a negative impact on most regions along the corridor and even more so on specific regions outside the corridor. Since we are analysing bottlenecks in the corridor, such negative impacts have to be considered and can be regarded as the "costs of non-doing."

A current example: The recent devaluation of the Swiss Franc is considered as a bottleneck on all Swiss railway routes, as it increases operational and infrastructure costs by 20 %. Since alternatives on inland waterways do not exist and road transport costs will also increase due to the HVC (Heavy Vehicle Charge), the conclusions from the CODE24 bottleneck modelling can be applied accordingly.

Within the CODE24 bottleneck analysis, two different analyses have been carried out: an empirical analysis on the capacity on a specific rail leg of the corridor Rotterdam-Genoa and an infrastructure-based modelling of regional transport costs (see Chapter "Comparative Analysis of Accessibility for Freight Transport in Corridor Regions: Results of Two Case Studies"). Transport demand issues suggesting interdependences with the supply side of the transport model have not been taken into account at this stage. In addition, the economies of scale that might provide an explanation for the concentration of transport flows on specific routes are not addressed in detail within the modelling approach. The empirical results as well as the modelling results suggest that an extension of the modelling approach into these research fields might complete the picture towards an optimal development and usage of CODE24 infrastructure.

References

Ahrens G-A, Baum H, Beckmann KJ, Brilon W, Eisenkopf A, Fricke H, Göpfert I, Hirschhausen C, Knieps G, Oeter S, Radermacher F-J, Rothengatter W, Schindler V, Schlag B, Siegmann J, Stölzle W (2007) Verkehrspolitische Handlungsfelder für eine effiziente Logistik, on behalf of Wissenschaftlicher Beirat beim Bundesministerium für Verkehr. Bau und Stadtentwicklung, Berlin

Brunet R (1989) Les Villes Européennes, Rapport pour la DATAR. Reclus, Montpellier

Cipolina S, Ghiara H (2011) Market situation and context analysis. Project report MoS24, Genoa

Drewello H (2014) Transport policy and regional development: the economic impact of regional accessibility on economic sectors. In: Lami IM (ed) Analytical decision making methods for evaluating sustainable transport in European Corridors. Springer International, Cham, pp 103–120

Drewello H, Günther H (2012) Bottlenecks in railway infrastructure—do they really exist? The Corridor Rotterdam-Genoa. Paper presented on the European transport conference, Crowne Plaza Hotel, Glasgow, October 2012

Drewello H, Gütle S (2013) The need for investment on the 'Rheintalbahn' in the Upper Rhine Valley. Results of a railway capacity analysis near the town of Lahr. 3rd Code24 report of action 9, August 2013, Kehl

Drewello H, Dittrich I, Gütle S (2013) Kapazitätsbelastung der Rheintalbahn—Zugzahlmessung mit Infrarottechnik. Internationales Verkehrswesen 4:48–51

Happel O (1959) Sperrzeitentreppe als Grundlage für die Fahrplankonstruktion. Eisenbahntechnische Rundschau 8(2):79–90

Holzhey W (2010) Schienennetz 2025/2030. Ausbaukonzeption für einen leistungsfähigen Schienengüterverkehr in Deutschland, ed. Umweltbundesamt, Dessau-Rosslau

Holzhey W (2011) Rail network 2025/2030. Expansion concept for an efficient rail freight service in Germany, Summary, Dessau-Rosslau

Rothengatter W (1996) Bottlenecks in European transport infrastructure. Paper presented on the European transport conference 1996, Brunel University, Uxbridge

Since 2009 **Hansjörg Drewello** has been a professor of economics, first at the University of Applied Sciences Göttingen and now at the University of Applied Sciences Kehl. He is the German director of the European Competence and Research Centre Cluster Management. His teaching and research deal with regional economics, in particular infrastructure development and the regional economic impact of cluster initiatives in a cross-border context. Until 2009 Hansjörg Drewello was the director of the Department of International Commerce, Transportation, Commerce and Tourism at the Chamber of Commerce and Industry Karlsruhe and a member of its management board. He was the lead partner of the study segment "Freight transport and logistics" of CODE24.

Marcel Huschebeck received his Diploma in National Economics at the University of Freiburg, Germany. He has gained more than 3 years' experience in the field transport, logistics and transport policy as officer for the German association Bundesverband Güterkraftverkehr Entsorgung und Logistik (BGL). Since 1999 he is involved in research and consultancy projects in the field of freight transport and logistics IT systems at PTV Planung Transport Verkehr AG. Today he has the position of a Manager responsible for the Logistics Systems group within the Research & Innovation unit at PTV. As project co-ordinator he has managed several EU research projects: IRIS, IDIOMA, AIMS, BESTUFS, BESTUFS II, PROMIT, BESTFACT, MODULUSHCA.

Norbert Schick received his master degree in civil engineering with emphasis on transportation planning at the University Karlsruhe, Germany. Since 1994 he is working with PTV as project manager and senior consultant in the fields of demand models (Prt, Put, Freight), network modelling, integrated transportation planning (PrT, PuT), urban and spatial planning as well as economic feasibility studies. He has been involved among others in the Interreg Project Corridor Development Rotterdam-Genoa (CODE24), the Transport Corridor Europe-Caucasus-Asia (TRACECA) and the Traffic Impact study Barwa city (Doha).

Comparative Analysis of Accessibility for Freight Transport in Corridor Regions: Results of Two Case Studies

Alessandro Africani, Roberta Delpiano, Hansjörg Drewello, Antonello Fontanili, Marcel Huschebeck, and Daniela Taake

Abstract Regional accessibility is the main 'product' of a transport system in a region. It determines the locational advantage of an area relative to other areas. To measure regional accessibility, various quantitative indicators were developed in the past. The idea of most of those indicators is that the accessibility of a region is directly proportional to the attractiveness or size of the region, and indirectly proportional to distance, travel time or cost. As those indicators are based exclusively on quantitative data, qualitative aspects as the quality of the infrastructure and the quality of the supply of logistic services are neglected.

Therefore, a logistics accessibility analysis is performed for regions along the corridor. A methodology is presented which allows the examination and evaluation of qualitative aspects of regional accessibility concerning freight transport. An important module of this methodology is an expert assessment which addresses terminal and infrastructure operators as well as important forwarders. The aim is to identify regional issues as well as to reach an overview about a region's logistics accessibility that allows a comparison with other regions in the corridor.

Finally the results of the exemplary application of this accessibility analysis in the corridor regions 'Technology Region Karlsruhe' and 'Liguria' are presented.

1 Introduction

Regional accessibility is an important location factor, which enhance the region and, therefore, promote economic growth. It is the virtual aim of transport policy, on European as well as on national level, to strengthen regional accessibility. But

A. Africani • R. Delpiano • A. Fontanili
Uniontrasporti scarl, Via E.Oldofredi, 23, 20124 Milan, Italy
e-mail: africani@uniontrasporti.it; delpiano@uniontrasporti.it; fontanili@uniontrasporti.it

H. Drewello
University of Applied Sciences Kehl, Kinzigalle 1, 77694 Kehl am Rhein, Germany

M. Huschebeck (✉) • D. Taake
PTV Group, Haid-und-Neu-Straße 15, 76131 Karlsruhe, Germany

H. Drewello, B. Scholl (eds.), *Integrated Spatial and Transport Infrastructure Development*, Contributions to Economics, DOI 10.1007/978-3-319-15708-5_8

available resources to achieve this are limited. Therefore, new investment and enlargement in transport infrastructure should be based on efficiency considerations. To measure regional accessibility, various indicators were developed in the past. The idea of those indicators is that the accessibility of a region is directly proportional to the attractiveness or size of the region, and indirectly proportional to distance, travel time or cost. But this kind of analysis does not allow a detailed insight into reasons of high or low regional accessibility.

This article presents a new methodology developed in Code24 to analyze regional accessibility in more details. With the help of a valuation method a bundle of regional indicators is evaluated by logistic experts. The method allows on one hand an in-depth overview about regional strength and weakness concerning accessibility. On the other hand it allows a comparison between regions and their accessibility situation. Finally two case studies in the Technology Region Karlsruhe and the region of Liguria illustrate the application of this method.

2 Theoretical Considerations: Accessibility and Transport Costs as Locational Factors

Locational factors are a concept of geographical economics. They significantly influence the choice of companies, where to produce, of households where to settle down or of governments, where to install public administrations and public agencies. The theory of industrial location tries to find principles and factors, which influence systematically the decision of industries, concerning their location (Riley 1973; Dicken and Loyd 1990). Weber describes locational factors as spatial differences, which lead to spatial advantages in costs (Weber 1909, p. 16). In a country or a region it will be possible to produce industrial products with lower costs in some places than in other places.

Accessibility is an important locational factor for transport intensive industries. But it is as well important in the decision making of private households where to live. Rodrigue, Comtois and Slack define accessibility as "the measure of the capacity of a location to be reached by, or to reach different locations. Therefore, the capacity and the structure of transport infrastructure are key elements in the determination of accessibility" (Rodrigue et al. 2006, p. 28). Good accessibility of regions improves their competitive position. Mobility and accessibility are key prerequisites for economic development of all regions of the EU. Another definition is that "accessibility indicators describe the location of an area with respect to opportunities, activities or assets existing in other areas and in the area itself, where 'area' may be a region, a city or a corridor" (Spiekermann and Wegener 2006).

Accessibility is the main product of the transport system. It determines the locational advantage of an area relative to other areas. With accessibility indicators, one can measure the benefits for households and firms, which profit from the existence and use of transport infrastructure.

To measure regional accessibility, various indicators were developed in the past. The basic principle of these indicators is relatively simple. It refers to Newton's law of universal gravitation, which states that any two bodies in the universe attract each other with a force that is directly proportional to the product of their masses and inversely proportional to the square of the distance between them. It was adapted by Stewart for use in geographical economics (Stewart 1947). The idea of those indicators is that the accessibility of a region is directly proportional to the attractiveness or size of the region, and indirectly proportional to distance, travel time or cost. Hence, regional accessibility is a function of regional attractiveness and transport cost

$$A_i = \sum_j g(W_j) \cdot f(c_{ij}) \tag{1}$$

where A_i is the accessibility of a region i, W_j the attractiveness or size of a region j, and c_{ij} the general cost to reach region j from region i. $g(W_j)$ is a function of regional attractiveness, e.g. regional GDP, population or employment rate, and $f(c_{ij})$ of space resistance, e.g. transport cost or time between regions (Wegener et al. 2001): the more attractions in region j and the more easily reachable from region i, the higher the accessibility of region i.

Various types of accessibility indicators can be created by specifying the functions g(Wj) and f(cij) differently (linear, or non-linear). Wegener et al. describe functions of travel time or weighted travel time, of daily accessibility, potential accessibility, multimodal or intermodal accessibility (Wegener et al. 2001).

Space resistance is measured basically in terms of transport cost, travel time or quality of transport. From an economic point of view, travel time and quality of transport can be included in travel cost, too. Long freight transport time increases transport costs because capital (vehicles and goods) works less efficiently than in shorter transport times. It is the same for passenger transport, where longer transport time creates higher costs because people cannot use the time alternatively. If quality of transport represents punctuality, a bad quality (declining punctuality) in the transport system creates higher costs because of increased uncertainty. The Code24 model contains all this information for regions along the Rotterdam-Genoa corridor (see Chapter "Bottlenecks and Regional Economic Impact: Simulations with the CODE24 Transport Model") and includes it in transport cost.

When a firm relocates, transport issues are not in general a first-order consideration, especially if these costs are only a small part of total production costs. However, these costs may be important in several sectors if the ratio of transportation costs to production costs is high or if accessibility to customers may influence the performance of a company (Banister and Berechman 2000).

3 CODE24 Accessibility Scheme for Freight Transport

CODE24 accessibility approach addresses the development of a quantitative and qualitative concept for defining and measuring accessibility for regional freight transport. Research on accessibility is dominated by approaches for passenger transport while approaches on measuring accessibility for goods transport are scarce. Accessibility described as the window of opportunities, activities and resources of one regions compared to other regions opens a variety of possible indicators. Obviously, there are a number of possible indicators to measure accessibility. Measuring the length of motorways available in a region is a simple approach, while measuring the impedance to access one region from the other in terms of transport costs is an example for a more complex indicator. However, with regard to possible conclusions on the impact on regional development and competitiveness these indicators show limitations in their validity.

Assessing the impact of the regional development and competitiveness a comprehensive analysis is needed applying and transferring the approach of competitiveness analysis of Porter to a regional assessment concept (Porter 1993). Thus, competitive branches benefit from their region by means of available factor resources, demand driven requirements, the existence of similar and supporting branches as well as from investment and competitiveness oriented framework conditions. With regard to infrastructure the advantages of regional opportunities can only be fully maximized when it is used efficiently and economically.

Obviously, goods transport accessibility is closely linked to the removal of bottlenecks within infrastructure by means of targeted investments. Studies (Bundesamt für Bauwesen und Raumordnung (2001)) showed difficulties in measuring impact of single infrastructure measures on spatial development. Especially, the link between infrastructure measures and the impact on improving efficiency, competitiveness and hence the welfare of single regions is quite difficult to make. CODE 24 focused on the analysis of bottlenecks along the corridor Rotterdam to Genoa and to assess their impact on the regional development and competitiveness. The accessibility analysis within CODE24 aims to develop a common evaluation concept to assess and compare regional accessibility from a goods transport and logistics perspective.

Within CODE24 a combined indicator approach has been developed. This includes the set up of a transport model to calculate regional transport costs taking into account the different transport modes calculating intermodal transport costs on NUTS 3 level as measure for the impedance of regional accessibility. The outcome of the modelling approach is extended by a dedicated analysis on the strength and weaknesses of a region with regards to parameters on the regional infrastructure and service quality as well as on the regional attractiveness (Fig. 1).

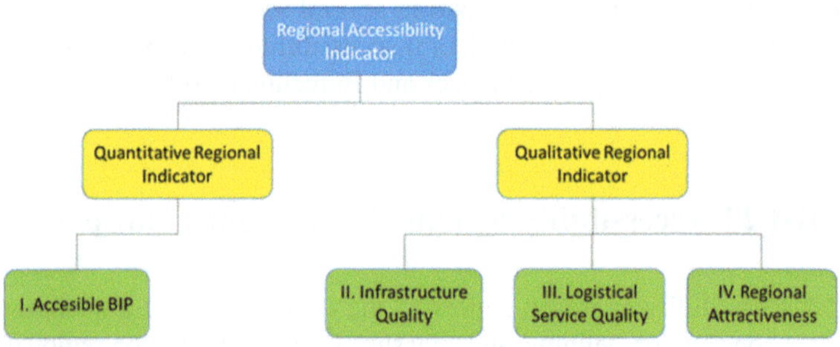

Fig. 1 Structure of accessibility analysis. *Source*: Own illustration

The measured parameters will be transferred into value neutral indicators on quantitative and qualitative level representing the CODE24 accessibility scheme.

4 Freight Transport Accessibility

CODE24 accessibility concept applies a twofold approach analyzing transport costs of single regions on NUTS3 level using modelled cost parameters. In addition an analysis on parameters describing the strength the weaknesses of infrastructure and equipment, service quality and the quality of the regions as a location is included. Quality issues should especially take into account the structural changes in transport business developing from mass production and transportations towards high value goods and individualized production and goods transport.

Qualitative accessibility indicators thereby complete quantitative indicators allowing identifying bottlenecks as perceived by local operators providing a comprehensive and differentiated picture on the regional opportunities to connect and interact with other regions. CODE24 accessibility concept addresses four indicators:

- Transport costs; using a standard cost calculation approach to analyse asset costs from one region to any other region. As indicator average transport costs or costs to access a certain market size (costs/BIP) can be chosen.
- Quality of infrastructure; assessing the capability and applicability of infrastructure located in the region. Parameters on the capacity of existing infrastructure such as limitations in the existing infrastructure as perceived by transport operators are of relevance
- Quality of service; assessing the availability of service providers in the region as well as the impact of good and poor service quality on the supply chain and production process are taken into account
- Regional attractiveness; assessing the quality of the region as location for industry and transport. Developing towards a specialized service economy implies changes in transport demand and services. Furthermore, the indicator should also address structural changes in the way that on regional level economic growth is not necessarily linked with a corresponding growth in goods transport. Relevant indicators are linked to the regional image, recreation or work life balance.

According to these considerations specific indicators are developed. An important module to assess the indicators of regional quality of infrastructure and logistics services is to include regional experts. Regional experts can for example be responsible staff from infrastructure operators, forwarders or shipping companies.

5 Methodology and Model Applied

The set up of the indicator model includes two working lines a quantitative analysis applying a PTV-VISUM transport model for cost calculation and a qualitative component based on a multi criteria approach to assess additional indicators. The outcome of the indicator model approach described a regional accessibility profile.

To ensure a structured and comparing analysis of quantitative and qualitative accessibility parameters across regions a specific Multi-Criteria-Analysis (MCA) methodology has been developed. Main benefits of a MCA is that it allows monetary and non-monetary evaluation of specific criteria and provides the necessary flexibility to come to comparable results even if there are differences in the underlying data basis. The CODE24 approach is to base the accessibility assessment on expert evaluation supported by additional surveys, such as questionnaires and interviews. Using a common scale to be applied for each indicator allows for a comparison even if the process or the data basis are not following the same evaluation structure.

For assessing the indicators a consistent indicator scheme has been developed using standard evaluation criteria. Assessment is made on the level of a numeric scoring using a scale from −2 to +2, the scoring scale represents the following scheme to be applied by all evaluators uniquely (Table 1).

For each indicator a description has been given as well as a reference scheme for evaluation (Table 2).

With the help of this specific MCA a qualitative assessment of the single indicators can be carried out by the CODE24 partner independently and also the approach allows a combination with the outcome of the quantitative assessment.

Table 1 Scoring scale for assessment

−2	−1	0	1	2
Very poor	Poor	Conform	Good	Very good

Table 2 Reference scheme for evaluation

Indicator		Assessment scale		
Indicator name	Indicator description	Weight	Scale	Assessment criteria
Connection to supra regional logistics nodes	Assessing how a region is linked to supra regional logistics hubs based on the infrastructure and infrastructure facilities, inter-modal modes (air, sea, rail) as well as the frequency of services	1.3	−2	Insufficient connection quality
			−1	
			0	Sufficient connection quality
			1	
			2	Good connection quality with potential to be used by additional capacity

6 Quantitative Evaluation

For the quantitative assessment the indicator accessible BIP was chosen. The assessment is a result using the average out of three assessment classes. Three classes for transport costs have been selected along the corridor. The quantitative indicators are assessed on the ration of the possible market that can be reached with 250, 450 and 650 euros transport costs (Fig. 2).

The transport model used is based on 804 relevant regions including EU and non EU regions. In total the market included in the transport model has a size of 14.8 billion euros. Along the corridor 24 between Rotterdam and Genoa 236 origin

Fig. 2 Number of possible markets per region that can be reached with 250 euros transport cost. *Source*: Own illustration

Table 3 Average accessible GDP

Region	Average accessible GDP (million euros) (a) . . .up to 250 €	(b) . . .up to 450 €	(c) . . .up to 650 €
TechnologieRegion Karlsruhe (€)	1,097,766	4,052,772	8,677,235
Corridor regions (€)	981,730	3,602,864	7,796,883
Deviation (%)	12	12	11
Assessment of the indicator	1	1	1
Total indicator value	1		

regions are taken into account. The results show that from the 236 Code 24 Regions towards the 804 destination regions the following clustering of transport costs can be made.

The assessment of the quantitative is made on the basis of the individual region values compared to the average corridor values. According to the difference in percent a scoring is made (Table 3).

This example for TRK region show a good accessibility compared to other regions. The existing infrastructure suggests good conditions for industry and goods transport to reach large markets. However, there are difficulties in deriving specific conclusions on the strength and weakness of the TRK region based on this indicator. In order to complete this assessment further details are needed based on a qualitative analysis providing additional significance to the results.

7 Qualitative Accessibility in the Technology Region of Karlsruhe and the Region of Liguria

The aim of this study is first to apply the methodology and second to assess the accessibility of the Technology Region Karlsruhe (TRK) in Germany and the Italian region Liguria.

Quantitative analyses that were executed on these two regions with the Code24 model show that the TRK generally benefits from a good accessibility while Liguria suffers from its peripheral location (see (Drewello and Weiß 2014)). Therefore, the objectives of performing the qualitative analysis in these two regions is, to identify specific bottlenecks that are responsible for a limited accessibility. On the other hand, it is important to identify local strengths, to be able to focus on these in the future.

Thus, firstly an overview about the regions will be given by describing their most important traffic infrastructure. Afterwards, the methodology of the performances will be described and subsequently the results will be presented.

The borders of the TRK cross the German states Baden-Württemberg and Rheinland Pfalz. The TRK is situated around the city of Karlsruhe and includes next to the metropolitan area and administrative district of Karlsruhe the city

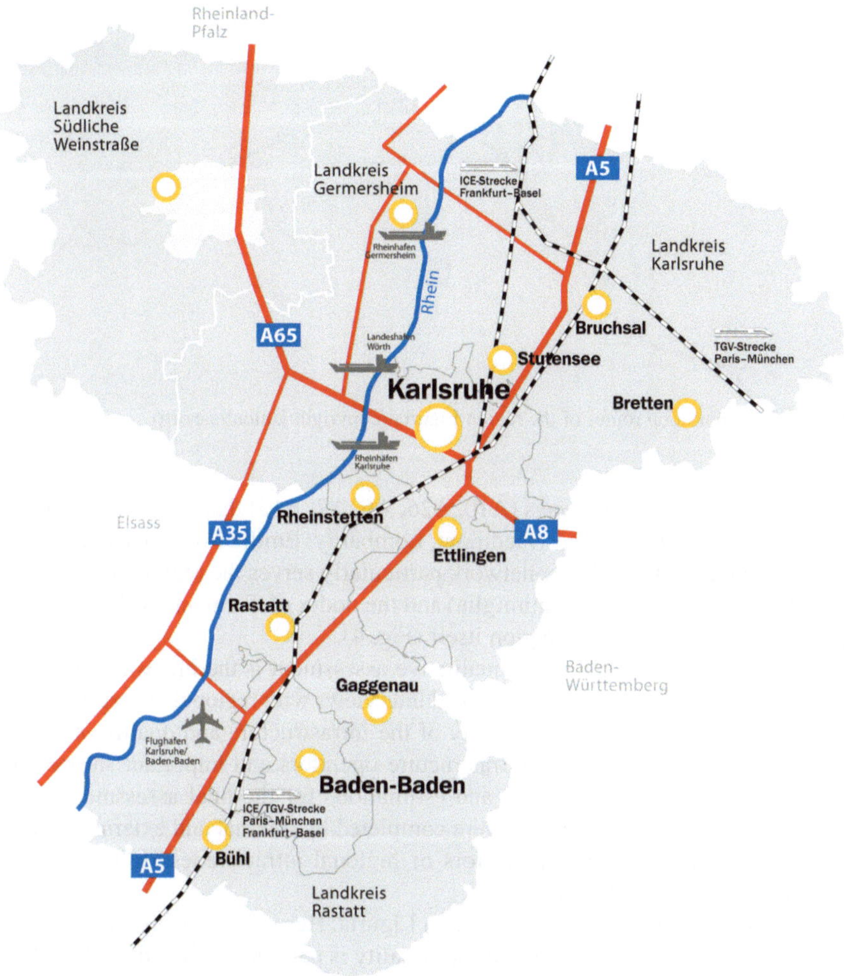

Fig. 3 Transportation routes of the Technology Region Karlsruhe (copyright TechnologieRegion Karlsruhe)

districts of Baden-Baden, Rastatt, Germersheim and Südliche Weinstraße. The available infrastructure of the TRK is characterized by its direct connection to the Rhine, the important highways (A5, North-South link and A8, link to Stuttgart) as well as its location on the intersection of the rail traffic axes 17 and 24 (Fig. 3).

The region Liguria is situated at the Mediterranean coast at the southern end of the corridor Genoa-Rotterdam, in the North-West of Italy. Its regional capital is Genoa and further includes the provinces of Imperia, La Spezia and Savona. Liguria's most important nodes traffic connections are via three international ports (Genoa, Savona and La Spezia) and the airport in Genoa. From the land

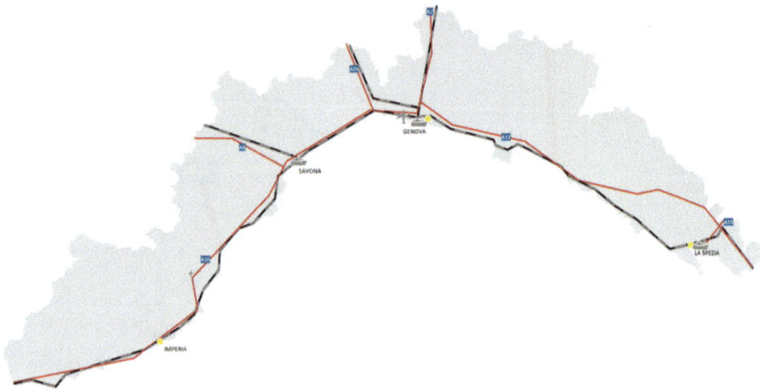

Fig. 4 Transportation routes of the region Liguria (copyright Uniontrasporti)

side, relevant are the highways (A10, A26, A6, A7, A15) that connect Liguria with neighbouring Italian regions (Piedmont, Lombardy, Emilia Romagna and Toscana) and with France. The railway network particularly serves the region in connections with France (line Genoa–Ventimiglia) and the nodes of Turin and Milan, as well as the Connections within the region itself (Fig. 4).

An important element of the qualitative assessment is the inclusion of regional experts. In the case study TRK, personal interviews with regional experts were held. The indicators concerning the quality of the infrastructure and quality of services were discussed with regional infrastructure operators and important shippers that could give valuable information and estimations for the final assessments of the indicators. These considerations were completed by internal and external available objective information. The indicators of regional attractiveness fully base upon available statistics and studies.

Slightly different was the approach in Liguria. Here, the analysis concerning the quality of infrastructure and the service quality is particularly based on the surveys to the entrepreneurs involved in the boards of directors of Ligurian Chambers of commerce and to infrastructure experts, otherwise on publications. At the same time, the indicators of regional attractiveness are implements starting from the statistical data available in literature.

In both performances, the indicators were covered per each transport mode (road, rail, sea/waterways, air) or infrastructure node (train station, port, airport) available in the region. That means for example that information for the *Indicator I. a) Quality of infrastructure* was separately covered and assessed per transport mode road, rail, sea/waterways and air. The final indicator result is based on the mean of these results. This structure provides more specific information for a later executed analysis of the indicators in order to identify bottlenecks.

Though the approach of retrieving data was slightly different in the two case studies, this didn't affect the methodological soundness and both results show a good level of comparability. This means for future case studies that both approaches

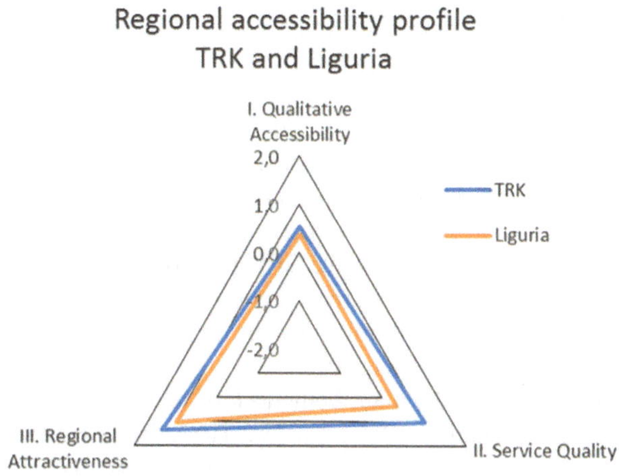

Fig. 5 Regional accessibility profiles. *Source*: Own illustration

Table 4 Qualitative accessibility analysis results

Results of the qualitative accessibility analysis			
	Indicator	TRK	Liguria
I. Qualitative accessibility	(a) Quality of existing infrastructure	0.6	−0.5
	(b) Connection to supra regional logistics nodes (rail, air, sea)	1	0.3
	(c) Extension of Infrastructure to remove infrastructure bottlenecks (capacity and functional) within the region	0.1	1
	(d) Up to dateness of infrastructure equipment	0.5	0.7
	Total	0.6	0.4
II. Service quality	(a) Delivery service and quality	1.1	0
	(b) Accessibility to service provider	1.0	1
	(c) Prohibitions and restrictions for goods transport	1.0	0
	Total	1.0	0.3
III. Regional attractiveness	(a) Growth of the region	1.5	1
	(b) Spending power	1.5	1
	(c) Recreational value	1.0	1
	Total	1.3	1.0

(either in personal interviews or with a questionnaire) can be advised and can be performed as the case may be.

The output of the analysis is shown in the following table where the indicator results are stated and of which the regional accessibility profiles developed (Fig. 5; Table 4).

8 Conclusions

Regarding the accessibility profiles, both regions perform over average in all indicators. However, especially the assessment of the indicator *Quality of infrastructure* detects weaknesses for both regions. For the region Liguria, this difficulty is higher in network infrastructure, compared to those of nodes. The indicator *Quality of logistics connection and infrastructure equipment* shows especially potential capability for Liguria while the TRK performs quite well. Regarding the *Regional attractiveness*, both regions reach good values and seem not to suffer from lack of logistics potential. In the following, some important indicator results of both case studies shall be points of discussion.

According to the results, the main bottleneck of the TRK's accessibility can be found by regarding the Quality of Infrastructure. Here, the TRK performs quite well considering available links to infrastructure nodes (especially due to its capable and multimodal equipped ports in Karlsruhe, Germersheim and Woerth) but shows stronger weaknesses in other fields. The main problems on infrastructures that the regional accessibility is suffering, concern congestions on road (especially motorways A5, A8) and rail capacity (Rheintalbahn) (see (Brenienek 2014)). To overcome those bottlenecks, investments in infrastructure are needed. Regarding the planned measures in the region (Ministerium für Verkehr und Infrastruktur, Baden-Württemberg 2013), investments on road and rail infrastructure are considered and planned but their finalizing is questionable. This is especially the case for the already long lasting construction sites on the Rheintalbahn.

Considering the indicators of assessing the logistics services of the TRK, these show an overall positive image. However, regarding the results in detail, they show that these good assessments base generally only on good marks for the transport modes on road and ship. The quality and availability of the transport mode rail was constantly only assessed averagely. This is—according to the discussions with local experts—owing to low frequencies and not enough direct train connections to logistic nodes and metropolitan areas. Additionally, also reliability problems and customer service were assessed as only average.

From Liguria side the existing infrastructure network shows some lacks in terms of quality. The road network is congested especially in the node of Genoa. While the rail network has tortuous lines and still single-track segments, in addition to interoperability problems with France.

In terms of accessibility, there are good connections with the Ligurian ports. Some difficulties are found with respect to the rail system. The airport instead has many weaknesses, especially regarding connections to major European cities. There are numerous initiatives and interventions planned to extend the infrastructure removing bottlenecks within the region, but few of these have a real financial chance in a medium-long-term.

Considering the evaluation indicators of logistics services in the region Liguria, these show fair results in the overall scenario. The positive results are more for road and sea transports. The equipment of rail transport is considered to be quite negative, especially in terms of quality and customer satisfaction: the two main issues are reliability and customer service.

All in all, both results allow a deeper insight into accessibility issues of the regions. On the one hand, regional strengths could be identified, which are for example in the TRK the availability of strong inland ports and in Liguria the regional sea ports equipment. On the other hand, the analysis especially helps to identify bottlenecks in the fields of regional infrastructure, service and attractiveness. Thus, the achieved results form an approach to identify measures that help to strengthen the regional economy by improving the locational factor accessibility.

References

Banister D, Berechman J (2000) Transport investment and economic development. University College London Press, London

Brenienek J (2014) Logistikcluster entlang des Transportkorridors Rotterdam-Genua. Universität Duisburg-Essen, Lehrstuhl für Wirtschaftsgeographie, insbes. Verkehr und Logistik, Essen

Bundesamt für Bauwesen und Raumordnung (2001) Kriterien für die räumliche Differenzierung des EU-Territoriums: Geographische Lage. Forschungen, Bonn, Heft 102.1

Dicken P, Lloyd PE (1990) Location in space, 3rd edn. Harper & Row, London

Drewello H, Weiß F (2014) Correlation analysis between regional accessibility and sector output for regions along the transport corridor Rotterdam-Genoa. 4th Code24 report of action 9. Kehl

Ministerium für Verkehr und Infrastruktur, Baden-Württemberg (2013) Priorisierungslisten für den BVWP 2015, Stuttgart

Porter ME (1993) Nationale Wettbewerbsvorteile: Erfolgreich konkurrieren auf dem Weltmarkt. Wirtschaftsverlag Carl Ueberreuter, Wien

Riley RC (1973) Industrial geography. Chatto and Windus, London

Rodrigue J-P, Comtois C, Slack B (2006) The geography of transport systems. Routledge, New York, NY

Spiekermann K, Wegener M (2006) Accessibility and spatial development in Europe. Scienze Regionali 5(2):15–46

Stewart JQ (1947) Empirical mathematical rules concerning the distribution and equilibrium of population. Geogr Rev 37:461–485

Weber A (1909) Über den Standort der Industrien. Erster Teil: Reine Theorie des Standorts. Mohr Siebeck, Tübingen

Wegener M, Eskelinen H, Fürst F, Schürmann C, Spiekermann K (2001) Kriterien der räumlichen Differenzierung des EU-Territoriums: Geografische Lage—Studienprogramm zur Europäischen Raumplanung des Bundesministeriums für Verkehr. Bau- und Wohnungswesen, Bonn

Alessandro Africani graduated in 2001 in Business Administration at the University Carlo Cattaneo—LIUC, he took a Master's Degree in "CFO: Chief Financial Officer". In 2004, he started as researcher in Uniontrasporti. In 2010, as project and finance manager of Uniontrasporti, he took part to the Interreg Project CODE24. Alessandro is a PhD candidate in Business Administration.

Roberta Delpiano graduated in 2006 in Mathematical Engineering at the Polytechnic University of Turin (Piedmont, Italy). In 2007 she gained the Master's Degree in "Transport and Logistics: systems, Networks and Infomobility" (COREP-Polytechnic University of Turin). In March 2008, she started as researcher in Uniontrasporti and she mainly cooperated to the evaluation of social and economical effects of new infrastructures on territorial and business systems. In 2010, as employee of Uniontrasporti, she took part to the Interreg Project CODE24.

Since 2009 **Hansjörg Drewello** has been a professor of economics, first at the University of Applied Sciences Göttingen and now at the University of Applied Sciences Kehl. He is the German director of the European Competence and Research Centre Cluster Management. His teaching and research deal with regional economics, in particular infrastructure development and the regional economic impact of cluster initiatives in a cross-border context. Until 2009 Hansjörg Drewello was the director of the Department of International Commerce, Transportation, Commerce and Tourism at the Chamber of Commerce and Industry Karlsruhe and a member of its management board. He was the lead partner of the study segment "Freight transport and logistics" of CODE24.

Antonello Fontanili graduated in 1998 in Civil Engineering (Transportation) at Polytechnic of Milan. After a brief experience for the Rinascente Group, he entered into Uniontrasporti in 1999 as a researcher, working—also as Project Manager—for project on seaport, on European Corridors, on trans-Alpine routes, logistics districts, technological innovation applied to transport. Between 2004 and 2009, he held the position of chief operating officer of Uniontrasporti, where is engaged in identifying, planning, organization and management of research and development in Italy and abroad. At the beginning of 2010 he became Director and took part to the Interreg Project CODE24.

Marcel Huschebeck received his Diploma in National Economics at the University of Freiburg, Germany. He has gained more than 3 years' experience in the field transport, logistics and transport policy as officer for the German association Bundesverband Güterkraftverkehr Entsorgung und Logistik (BGL). Since 1999 he is involved in research and consultancy projects in the field of freight transport and logistics IT systems at PTV Planung Transport Verkehr AG. Today he has the position of a Manager responsible for the Logistics Systems group within the Research & Innovation unit at PTV. As project co-ordinator he has managed several EU research projects: IRIS, IDIOMA, AIMS, BESTUFS, BESTUFS II, PROMIT, BESTFACT, MODULUSHCA.

Daniela Taake was student at the Fachhochschule Heilbronn, Germany. She finalised her Master Thesis "Entwicklung und Anwendung eines Indikatormodells zur Bewertung der logistischen Erreichbarkeit auf regionaler Ebene" as part of the CODE24 project.

Towards an Integrated Railway Network along the Genoa–Rotterdam Corridor

Maurizio Arnone, Tiziana Delmastro, Peter Endemann, Noriko Otsuka, Stefano Pensa, and Andrea Rosa

Abstract New scenarios arose with the development of high speed rail (HSR) services: their shorter travel times make them competitive against the car on short distances and the aeroplane on medium to long distances. HSR integration is best realised if a hierarchical system is conceived whereby cities not served by HSR lines are well connected by rail [Chen and Hall (J Transp Geogr 19:689–704, 2011)]. Such integration among the different railway services, e.g. HSR, long distance and regional trains, and freight, plays a crucial role in being able to take advantage of these new opportunities. Moreover, the integration with interregional and local services would help provide better regional accessibility to HSR, allowing people living in the hinterland along the corridor to travel easily between regions. Better use of existing tracks will also contribute to avoiding or alleviating the saturation of the lines, thus allowing railways to achieve a better level of service without large new infrastructural projects. This chapter focuses on the current provision of high speed and long distance services along the Rhine–Alpine Corridor and presents a new methodology developed to assess their integration.

M. Arnone (✉) • T. Delmastro • S. Pensa • A. Rosa
SiTI, Higher Institute on Territorial Systems for Innovation, Via Pier Carlo Boggio 61, Turin 10138, Italy
e-mail: maurizio.arnone@siti.polito.it; tiziana.delmastro@siti.polito.it; pensa.stefano@gmail.com; andrea.rosa@rocketmail.com

P. Endemann
Regionalverband FrankfurtRheinMain, Poststr. 16, 60329 Frankfurt am Main, Germany
e-mail: endemann@region-frankfurt.de

N. Otsuka
Research Institute for Regional and Urban Development gGmbH, Brüderweg 22-24, 44135 Dortmund, Germany
e-mail: norikootsuka.ils@gmail.com

H. Drewello, B. Scholl (eds.), *Integrated Spatial and Transport Infrastructure Development*, Contributions to Economics, DOI 10.1007/978-3-319-15708-5_9

1 Introduction

In recent years, an increasing number of high speed rail (HSR) operations have been witnessed at national and European levels, and the EU's commitment to developing high speed rail corridors has been highlighted in the TEN-T (Trans-European Transport Network) policy (EC 2010). Among other EU policies, the TEN-T policy has largely influenced major infrastructure developments in the field of transport (Marshall 2014). As a main planning methodology for future TEN-T development, the concept of a 'core network' was introduced to combine and integrate all transport modes and intelligent transport systems. Corridor 24 has become one of nine core network corridors, i.e. the Rhine–Alpine Corridor (EC 2013) and the integration of different railway services, e.g. high speed, long distance, regional and local trains, and freight, should play a crucial role in providing seamless travel along the Rhine–Alpine Corridor.

Although the initial focus of the CODE24 project was on improving the flow of rail freight and its implications for corridor development, there is a need for exploring better ways of maximising passenger rail services in parallel with rail freight, given the increasing importance of integrating HSR on existing tracks.

The Rhine–Alpine Corridor has been served by HSR over the last two decades and some high speed services are operated at a transnational level. Key challenges for HSR and long distance rail services along the corridor are to improve their service reliability and frequency as well as to integrate with other railway services. Since additional new infrastructure is difficult to envisage due to limited financial and spatial resources, a better use of existing rail tracks before building new infrastructure will help in easing congestion on existing lines.

This chapter focuses on discussing passenger rail development with a particular emphasis on HSR provision along the Rhine–Alpine Corridor, which is based on work performed for one of the action plans of the CODE24 initiative. The aim of the action is to develop guidelines for pursuing an integrated railway network along the corridor through the introduction of an integrated international timetable that encompasses HSR as part of its long distance services, regional trains and freight transport. The chapter starts with presenting a review of HSR experiences and the current flow of passengers along the corridor. After a brief summary about high speed (HS) lines and service structure on the corridor, the major outcomes of the analysis on the level of integration of high speed services within the corridor railway services are presented. Two types of integration have been considered: (1) corridor accessibility: integration between HS and long distance (LD) services to connect the main stations along the corridor; and (2) regional accessibility: integration between high speed and long distance (HS/LD) and interregional and local (IR/L) services to connect the main stations along the corridor with their hinterland. The chapter concludes by highlighting important factors in the integration among different railway services and suggesting open issues for future investigation.

In the following text, the original language is used for cities' names of railway stations as it is common in the railway sector.

2 High Speed Rail: A Review of Experiences

In many cases, HSR is focused on passenger long distance rail services with an operating maximum speed of 250 km per hour (kph) if new infrastructure is considered, but also of 200 kph and upwards if services are run on upgraded conventional rail lines (Council of the European Union 1996). However, "speed" as referring to the notion of maximum speed is not everything in terms of HSR and needs further specification (Chen and Hall 2011; Givoni 2006). Givoni and Banister (2012) argue: "While maximum speed of 350 kph is considered the new standard for HST (high speed trains), most HST services are provided at a much lower average speed, and the world's most successful HSR line in terms of passengers carried, between Tokyo and Osaka in Japan, operates at an average speed of less than 240 kph (for the fastest service)." Givoni (2006) provides a broader HSR definition: "...high capacity and frequency railway services achieving an average speed of over 200 kph". Table 1 summarises the average and maximum speed of HSR between relevant cities. It confirms to a certain extent that, with a maximum speed of 270 kph, the Tokyo–Osaka line attains a higher average speed than the fastest German line Köln–Frankfurt with a maximum of 300 kph. On the Milano–Roma line, the effect of additional stops can be appreciated. Overall, the notion of average speed appears more appropriate for assessing the effectiveness of rail performance than maximum speed.

It becomes apparent that designing HSR for maximum speed may reduce the number of stations served and thus requires longer station headway. In literature, a station headway of 150–200 km is proposed (Vickerman 2013) or even lower as an additional stop within a metropolitan area can be suitable as suggested by Garmendia et al. (2012). Thus, a trade-off between speed/travel time and potential ridership is required (Givoni 2006). Vickerman (2013) discusses the potential of HSR generating demand among commuters, as is the case for the Javelin HST,

Table 1 Comparison of HSR lines: speed and distance

Origin–destination	Distance (km)	Travel time (h)	Average speed (kph)	Maximum speed (kph)
Köln–Frankfurt	179	1.05	170.48	300
Brussel/Bruxelles–Paris	310	1.42	218.31	300
Tokyo–Osaka	515	2.42	212.81	270
Torino–Milano	125	0.90	138.89	300
Milano–Roma (nonstop)	515	3.00	171.67	300
Milano–Roma (with stops)	515	3.50	147.14	300
Madrid–Puertollano	209	1.08	192.92	270–300
Madrid–Toledo	75	0.50	150.00	240–260

Source: Adapted from Guirao (2013) and Germanwatch (2013)

which allows daily commuting from Kent to London, and the French TGV in the Nord-Pas de Calais region. Rebmann (2011) considers commuting as a less important trip purpose for long distance travelling. Therefore, a frequency of every 4 h appears to be sufficient in order to bundle those groups who mostly undertake planned trips. Analyses from the Roma–Napoli corridor suggest that around 6 % of trips are made for commuting purposes, but a high proportion of trips are made for business reasons (ranging from 38.7 % on Sundays to 57.4 % on weekdays), while education-related trips (percentages ranging from 3.4 % on Sundays to 6.2 % on weekdays) and "other purpose" trips (percentages ranging from 52.5 % on Sundays to 30.2 % on weekdays) show lower, but still very significant rates (Cascetta et al. 2013).

In addition to trip purpose, the mode shift effect for HSR needs to be assessed. Shifting demand from air to HSR is one aspect. This is confirmed by the substantial shift in the Paris–Lyon line and the Madrid–Sevilla line achieved 3 years after their opening in 1981 and respectively in 1991 (Givoni 2006). In Madrid–Sevilla, the market share of air travel was reduced from 40 to 13 %, while train ridership rose from 16 to 51 %. For Paris–Lyon, the share of air travel was reduced from 31 to 7 % and train travel increased from 40 to 72 % (Table 2). The impact on the amount of car travel is lower if one considers the overall increase of train trips of 37 % in the Paris–Lyon case and 35 % with respect to the Madrid–Sevilla line (Givoni 2006). Givoni and Dobruszkes (2013) stress the mode shift effect from air to rail while referring to HSR services such as London–Paris respectively Brussels/Bruxelles or lines in China, Taiwan or South Korea. Dobruszkes (2011) raises awareness for the supply side, since he has observed an overall increase in air traffic in Europe, though HSR is successful on some connections. The substitution effect on car use appears less evident, first because figures are not often available and second, because a HSR network with fewer stations may require more car use to get to the stations and thus increases the amount of car kilometres travelled. Moreover, it is often neglected that there are considerable shifts from conventional rail to HSR (Givoni and Dobruszkes 2013). Nonetheless, the car is an important competitor for HSR, especially for shorter distances. For the Barcelona–Madrid HSR line, opened in 2007, a survey carried out in 2009 revealed that 44 % of the customers used the car before shifting to rail, 8 % used the bus, 16 % made their trip by plane and

Table 2 Mode shift effects of HSR introduction in %

	Paris–Lyon (TGV)			Madrid–Sevilla (AVE)		
Mode	Before 1981	After 1984	Relative change (%)	Before 1991	After 1994	Relative change (%)
Air	31	7	−77	40	13	−68
Rail	40	72	80	16	51	219
Car/bus	29	21	−28	44	36	−18
Total	100	100		100	100	

Source: Adapted from European Commission (1996), quoted by Givoni (2006)

Table 3 Evolution of modal split in the entire Italian travel market

Mode	2009 (million trips)	2009 share (%)	2013 (million trips)	2013 share (%)	Δ 2013–2009 (abs.)	Δ 2013–2009 (%)
Car	38.7	57.3	31.4	45.2	−7.3	−19.0
Air	7.1	10.5	5.0	7.3	−2.1	−29.0
HSR	17.0	25.2	30.8	44.3	13.8	81.0
Intercity	4.7	7.0	2.3	3.2	−2.5	−52.0
Total	67.5	100.0	69.6	100.0	2.0	3.0

Source: Adapted from Cascetta and Coppola (2015)

another 23 % "moved" from other conventional trains to HSR. The remaining 10 % can be considered induced traffic (Frontier Economics et al. 2011).

A before-after study in Italy comparing the modal split between 2009 and 2013 made by Cascetta and Coppola (2015) proves that HSR can gain market shares also from the car, which was reduced by 19 % in relative terms, but started from a higher level than the airplane. The number of air trips was reduced by 29 %, but that share is relatively low. HSR increased by 81 % from 2009 to 2013, but conventional train travel lost about 52 % of its users within 4 years (Table 3). This latter case and the Madrid–Barcelona case reveal another aspect: the loss of customers for conventional rail services. Givoni and Dobruszkes (2013) report that up to 94 % of users in the case of Madrid–Sevilla did not use any more conventional trains once HSR was introduced. On the Sanyo Shinkansen in Japan, 55 % of the traffic was diverted to the new line from conventional rail lines, while the rest comes from other travel modes (23 % from air, 16 % from car and bus, and 6 % new induced demand) (Sands 1993b, quoted by Givoni 2006).

In view of this information, it should be discussed at what cost rail infrastructure is designed to deliver air substitution if the land-use transport nexus may be threatened as mentioned above. The emergence of new stations at the edge of towns or in greenfields, which the authors qualify as "TGV-generation stations",[1] are promoted by the European Commission, makes integration between rail and land-use and between conventional rail lines and their supply more difficult (EC 2010). Moreover, railways as a backbone for development loses its overall quality if a loss in conventional rail service takes place.

Integration into the existing rail network also becomes more difficult if regional accessibility with the possibility of transferring at relevant nodes to the long distance network is neglected. In essence, HSR is suitable to serve a wide range of purposes and is able to shift demand from other modes. This should not prevent

[1] This term is based on the French TGV-planning strategy with the creation of additional, predominantly peripheral stations for cities like Mâcon, Avignon or Belfort with a "TGV" label in order to distinguish them from the traditional "main" station mostly located in the respective city centre.

from neglecting the effect of induced traffic and cannibalisation of conventional rail.

3 An Analysis of Passenger Flow along the Rhine–Alpine Corridor

The Rhine–Alpine Corridor covers some of the most important economic regions in Europe. Its catchment area includes 70 million inhabitants. However, specific data on passenger mobility along the corridor are partial and not up-to-date. Eurostat publishes some data related to different modes of transport (air, rail and road), but there is a lack of information concerning origin-destination flows (OD) between the corridor zones. Road OD matrices are not available. The only data available are total passenger-km per country (travelling by car, motorbike and public transport) up to 2012. Railway OD matrices between NUTS2 regions are available for 2005 and 2010, but several OD are not included, especially for 2010. Finally, air OD matrices containing the number of passengers travelling between the main airports along the corridor are available from 1993 to 2013. However, not all airports are considered and transit passengers,[2] which are included, cannot be separated from the mobility directly originated or destined for those ODs. As a result, it is impossible to build modal OD matrices, e.g. road, rail, air, with a high level of detail, i.e. between NUTS3 zones, only using data coming from official statistics.

However, the ETIS+ project[3] provides modelled data with this level of detail. All means of transport are considered, but data are only available for 2005 and 2010 and, being modelled data, there could be a deviation with respect to observed data. Nevertheless, those data are useful to provide an overview on the corridor mobility and identify the main OD relations. Therefore, the Rhine–Alpine Corridor area was divided into zones and the corresponding demand data provided by ETIS+ were analysed.

In order to have comparable data, zones were defined on the basis of their population. The zones correspond to provinces (NUTS3) in Italy and France whereas in Germany, Switzerland and the Netherlands the zones correspond to regions (NUTS2). Only zones along the corridor were considered, e.g. the Bayern Region in south-east Germany was not included, and only interzonal demand (trips between different zones) was analysed. This is because the present research was aimed at analysing medium- to long distance mobility demand along the Rhine–Alpine Corridor that can be served by HS/LD services properly integrated with interregional and local services. However, it is noteworthy to highlight that internal demand (intrazonal demand) usually represents most of the mobility of each zone.

[2] People going by plane to an airport to catch another plane to go on to a different destination.

[3] ETIS+ (http://www.etisplus.eu) was a project funded by the EC through the Seventh Framework Programme for research with the aim of building a European database on the flow of goods and people.

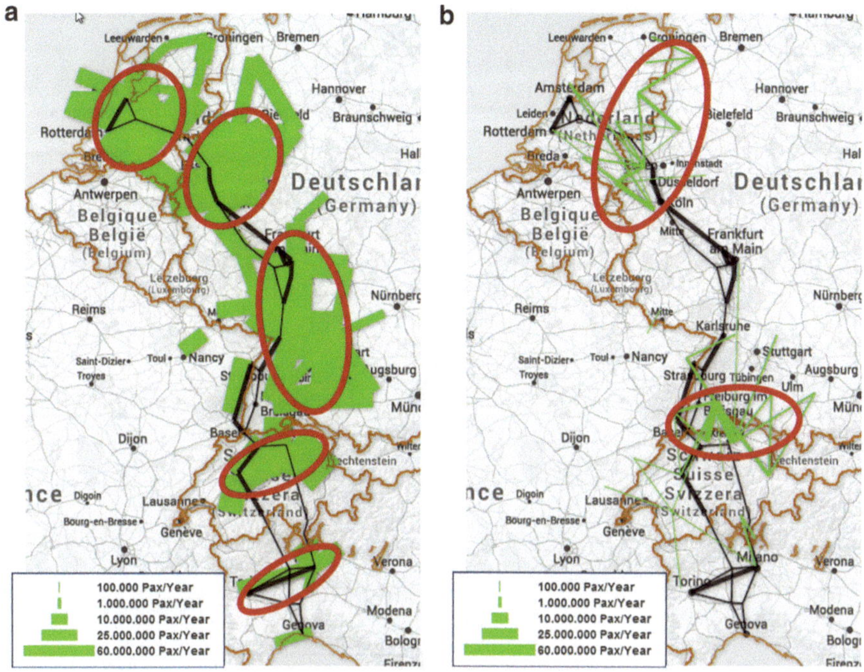

Fig. 1 (**a**) Most relevant interzonal OD relationship along the Rhine–Alpine Corridor for all modes of transport. (**b**) Most relevant transnational interzonal OD relationship along the Rhine–Alpine Corridor for all modes of transport. *Source*: Elaborations of ETIS+ data; base map: Google Maps, Google Inc.

The ODs with the highest passenger mobility (relationships with more than five million passengers per year) are represented in Fig. 1a. The most relevant OD relationships are at the national level: five large passenger demand clusters (indicated with circles in Fig. 1a) can be identified:

1. The Netherlands
2. North-West Germany
3. Central/Southern West Germany
4. Switzerland
5. The Piedmont-Lombardy axis in Italy

It is remarkable that most passengers travel between zones that are less than 100 km apart.

If only significant transnational mobility demand is considered (more than 0.2 million passengers per year), it becomes evident that there are fewer trips between different countries (Fig. 1b). These ODs are always characterised by less than 0.5 million passengers per year. Two main cross-border relationships can be identified: one between Northern Germany and the Netherlands and another between Southern Germany and Switzerland. The relationships between France-Germany, France-Switzerland and Italy-Switzerland appear to be less significant.

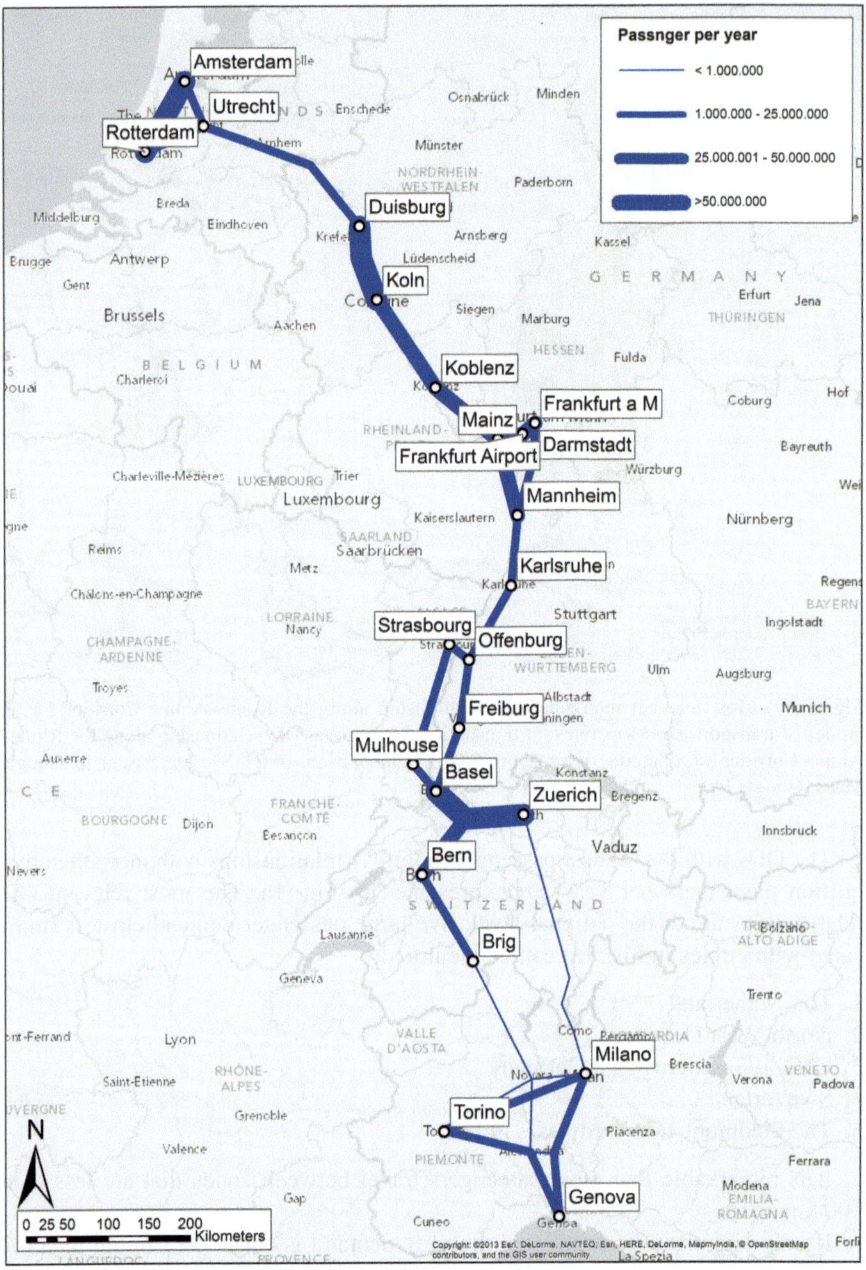

Fig. 2 Yearly interzonal mobility demand on the Rhine–Alpine Corridor for all modes of transport. *Source*: Elaborations of ETIS+ data; base map: Canvas/World_Light_Gray_Reference, Copyright: ©2013 Esri, DeLorme, NAVTEQ

The previous comments are confirmed by Fig. 2, which represents the annual interzonal mobility demand on the corridor for all the modes of transport. The figure reveals high passenger demand between Duisburg and Mannheim, and within the Netherlands and Switzerland. Lower, but still significant, demand exists for travel between the Netherlands and north-western Germany, between south-western Germany and Switzerland and within north-western Italy. Finally, demand is low between Italy and Switzerland.

4 High Speed and Long Distance Rail Services: The System along the Rhine–Alpine Corridor

Before moving to the analysis of service integration along the corridor, this section gives an overview of the rail axis between Rotterdam and Genoa. This comprises lines with different characteristics that are used by a mix of services, although mainly passenger services (Rhine–Alpine Corridor 2013). Figure 3 shows the railway lines forming the corridor and highlights those that allow trains travelling at a speed higher than 200 kph, consistent with the interest of the CODE24 project (and with some studies, see e.g., Givoni 2006). Only some sections of the corridor are equipped for trains travelling at 250 kph or more and may be considered high speed lines according to UIC (2013).

In the Netherlands, there exists a high speed line linking Amsterdam with Rotterdam and then providing an international link to Antwerp in Belgium. The rest of the network, also along the corridor, reflects the density of settlements in the country, which warranted fast services based on high frequency rather than very high speed, due to the frequent stops resulting from the settlement pattern.

Germany has both polycentric metropolitan regions and long intercity distances. There, high speed and long distance services together form the higher tier of rail connections. On the German section of the corridor, there are lines allowing speeds over 250 kph such as those between Köln and Frankfurt and between Mannheim and Stuttgart. In Switzerland, the dense railway network provides frequent integrated services not only between major cities, but also with their hinterlands. The pattern of services and lines was developed according to the Rail2000 strategy, which focused on connections and frequency.

In Italy, the area of the CODE24 project comprises the Torino–Milano high speed railway connection, whereas the corridor links between the Alpine crossings, Milano, and Genoa do not include HS lines.

The high speed lines along the corridor use conventional stations that facilitate the interchange with local services. Indeed, Italy and the Netherlands are the only corridor countries where HS trains travel on dedicated lines though, at present, they use conventional lines when approaching nodes.

The description above highlights the lack of high speed links across the borders of the corridor. This is consistent with the national focus of the original HSR

Fig. 3 Main rail lines along the corridor: *thick links* indicate lines allowing a speed over 200 kph, whereas thin links indicate *lines* with allowed speed not exceeding 200 kph. *Source*: CoDe24 Portal; base map: Google Maps, Google Inc.

network development plans, noted by Vickerman (1997), and only later expanded by a Europe-wide vision. Moreover, as mentioned above, services have a different organisation among the countries. In Germany, long distance and high speed services have similar functions and, in some cases, even services provided by local authorities cater to long distance travel (Beckers et al. 2009). The Netherlands and Switzerland have comparatively short intercity distances and national long distance services that do not rely on very high speed. In Italy, high speed services have replaced many long distance ones, which are left with a minor role, although this is only partly relevant along the corridor since, with the exception of the Milano–Torino link, it is formed by conventional lines where only intercity and regional services are provided.

It is remarkable that many HS lines have been developed to obviate capacity shortages, rather than with the single aim to provide HS links (Givoni and Banister 2012). Such a point is also noteworthy as one of the reasons that motivated the CODE24 project was to examine issues about the mix of different types of traffic and the relevance of the Rhine–Alpine Corridor for freight traffic. In fact, the corridor carries the largest logistics volume in Europe in tonnes-km (Rhine–Alpine Corridor 2013). Work to overcome capacity shortages and improve operations and interoperability between different rail services has recently been completed or is ongoing along the corridor. Projects of the former group include the Katzenberg Tunnel in Germany and the Lötschberg Tunnel in Switzerland (concerning one of the two tracks). Ongoing projects include new tracks between Karlsruhe and Basel, Emmerich and Oberhausen in Germany; the Gotthard Tunnel, and the Ceneri Tunnel in Switzerland; and enhancements on the lines between the Alps and Milano and the third Giovi line in Italy. Operations on existing and new infrastructure should also be eased and made increasingly interoperable by the deployment of the ERTMS.

5 High Speed and Long Distance Rail Services: Integration along the Rhine–Alpine Corridor

This section reports on the investigation on corridor accessibility aimed at understanding whether the entire Rhine–Alpine Corridor and, in particular, cross-border mobility, are suitably served by HS and LD trains. The analysis considered the main stations belonging to the OD pairs with highest demand (more than one million passengers). Data related to the supply of direct rail services were collected, with the support of the CODE24 partners, on a typical day (a weekday in October 2013) among those stations.

The number of services was used as a proxy of the seats provided because the number of seats of each train is not available in the train operators' publicised data and it is not easy to estimate since train capacity can vary greatly between different countries and lines.

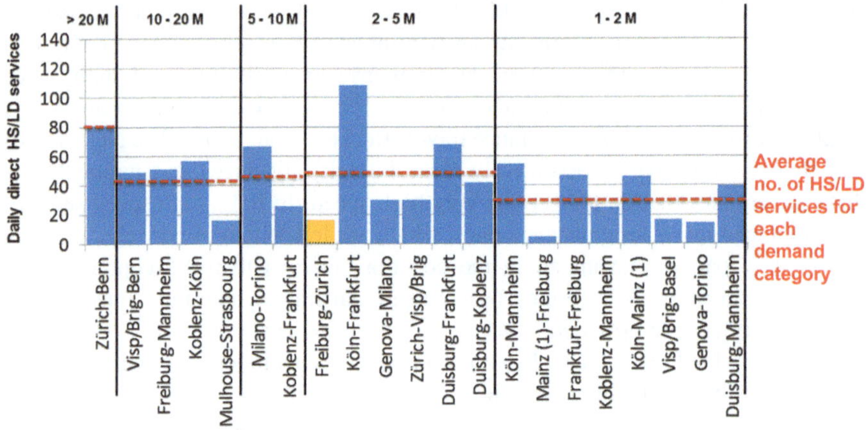

Fig. 4 Daily direct HS and LD services between ODs with a high mobility demand. *Source*: Elaborations of DB European timetables available on web. (1) The Region of Mainz, Rheinhessen-Pfalz, is also served by trains to/from Mannheim

Figure 4 shows the number of daily direct services (sum of both directions) between the most important OD pairs, grouped in five categories on the basis of the mobility demand estimated as discussed in paragraph 3 (>20 million, 10–20 million, 5–10 million, 2–5 million and 1–2 million passengers per year). For each demand category, the average number of HS/LD services is depicted by a dashed horizontal line. The only transnational OD (Freiburg-Zürich) is shown in a different colour.

The number of direct services within each demand category can vary significantly. In particular, for some ODs, a seemingly low number of direct services is provided, i.e. Mulhouse–Strasbourg, Koblenz–Frankfurt and, the only cross-border relation, Freiburg–Zürich: these could be "critical connections".

When interregional and local (IR and L) services are also included, the number of critical connections decreases. For short distance ODs (less than 200 km), in particular, IR/L trains are well integrated with HS/LD trains since they aim at serving not only the regional mobility demand, but also the medium to long distance demand that cannot be served by HS/LD services. In fact, the latter are characterised by a low number of stops in order to reduce the total travel time between the more distant stops.

However, some ODs, e.g. the only transnational OD, Freiburg–Zürich, appear critical even considering IR/L services. For those relationships, data collection was extended to include connecting services requiring transfers.

Figure 5 reports all the daily connections between Freiburg and Zürich. The graph shows direct connections on the top (eight direct services per direction with an average travel time of 1 h 54 min) whereas below are the other connections that require up to three transfers (22 connections/day: 17 with one transfer, 3 with two transfers and 2 with three transfers).

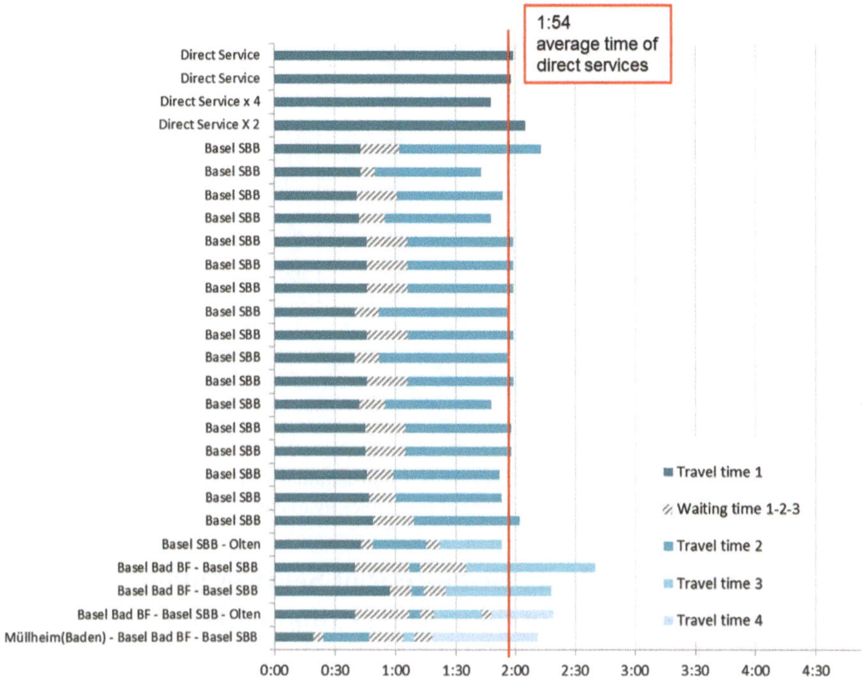

Fig. 5 Freiburg–Zürich daily connections and respective total travel time split into travel time and waiting time at the interchange stations. *Source*: Elaborations on DB European timetables available on the Internet

It is evident that connections with one transfer suitably integrate the direct services: total trip times are similar to trip times with direct services (in some cases even shorter) and the waiting times at the interchange stations are usually shorter than 20 min.

This situation recurs along the corridor. Some high-mobility ODs have few direct services compared to other relationships with a similar demand, although they are served by very good indirect connections with total travel times similar to those of direct services. A further analysis could be carried out to investigate what other factors, in addition to mobility demand, are considered in the choice between direct services and options including transfers.

The previous comments remain valid for other transnational ODs with lower demand as well (Fig. 6). The only exception is the relationship between Torino (Porta Susa station) and Brig. In that case, there is no direct service and changing trains usually requires waiting more than 30 min in Milano Centrale Station or travelling by underground between Milano Porta Garibaldi and Milano Centrale Station.

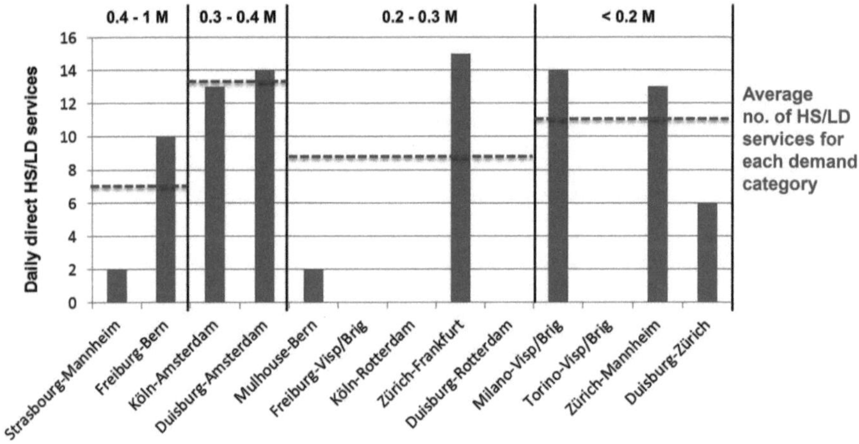

Fig. 6 Daily direct HS and LD services between transnational ODs with the highest mobility demand. *Source*: Elaborations of DB European timetables available on the Internet

6 High Speed and Long Distance Rail Services along the Rhine–Alpine Corridor: Integration with Local Connections

Regional accessibility to HS/LD services in the main nodes along the Corridor was also investigated: the analysis focused on the connections of HS stations to their hinterland and the integration of HS/LD trains with interregional (IR) and local (L) trains.

The level of integration was analysed in terms of timetables and efficient transfer times between two different services in a node. In particular, services were defined as:

- Integrated with short transfer time, 5–15 min
- Integrated with medium transfer time, 15–30 min
- Potentially integrated when transfer time is between −5 and 5 min (negative values mean that a service arrives a few minutes after another service that it could be integrated with; in such cases small timetable shifts could increase the number of possible transfers)
- Not integrated in all other cases

On the basis of the level of mobility demand attracted and generated by the zones along the corridor (number of passengers per year using all transport modes, see Sect. 3), eight main nodes were identified for the analysis. Since some of these nodes are served by more than one HS station, the timetables of services in 14 stations (those highlighted in Fig. 7) were collected and analysed in detail for a typical time slot (8:00–10:00 am) and a typical day (a Tuesday in October 2013).

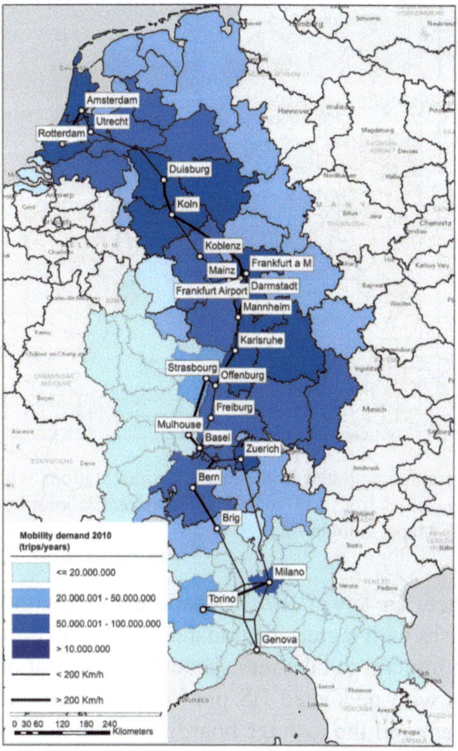

Main rail stations	
The Netherlands	Amsterdam
	Rotterdam
	Utrecht
Germany	Duisburg
	Darmstadt
	Koblenz
	Freiburg
	Offenburg
	Mainz
	Köln Hb
	Koln Messe Deutz
	Frankfurt Hbf
	Frankfurt Airport
	Mannheim
France	Strasbourg
	Mulhouse
Switzerland	Basel SBB
	Basel Bad Bhf
	Zuerich
	Visp/Brig
	Bern
Italy	Milano Centrale
	Milano P. Garibaldi
	Milano Rogoredo
	Torino Porta Nuova
	Torino Porta Susa
	Genova

Fig. 7 Main rail stations identified along the Genoa–Rotterdam Corridor on the basis of mobility demand in 2010. *Source*: Elaborations of ETIS+ data; base map: Canvas/World_Light_Gray_Reference, Copyright: ©2013 Esri, DeLorme, NAVTEQ

Starting from the arrival and departure times of the trains at the stations, transfer times for all possible service combinations were evaluated in order to assess integrations between HS/LD trains and IR/L trains serving the corridor hinterland. HS and LD services were considered as equivalent for the analysis.

This method allowed us to identify and represent, for each analysed station, all possible final destinations that can be reached arriving at that station from 8:00 to 9:00 am with a HS/LD service and transferring on a local train with an appropriate transfer time or, vice versa, all the possible locations from where one can leave with a local train in order to transfer on a HS/LD service.

Comparing some of the most important nodes along the corridor (Fig. 8), it is noticeable that in both the German and Swiss main stations, e.g. Frankfurt am Main, Köln, Zürich, HS/LD services are very well integrated with IR and L trains: the hinterland is efficiently connected and a high number of regional services (calling at many other stations along their path) are already available within 5 and 15 min transfers. The same considerations hold also for Dutch stations that were not considered in the analysis due to the very high number of LD services provided (all local trains are integrated with at least one such service). Milano Centrale,

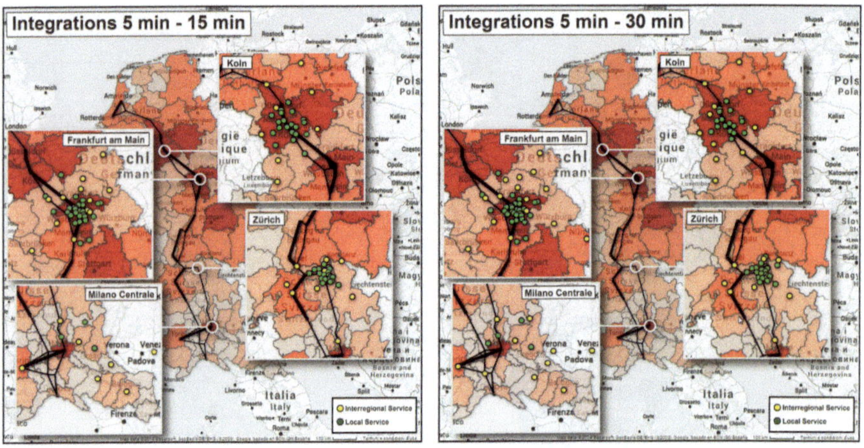

Fig. 8 Origins and destinations of IR and L trains integrated in the main corridor stations with HS/LD services having short or medium transfer times from 8:00 to 9:00 am. *Source*: Elaborations of DB European timetables; base map: Google Maps, Google Inc.

in Italy, presents a different picture: there are fewer IR/L trains integrated with a short transfer time. Their number increases with increasing transfer time (5–30 min) since 15 min is considered too short an interval to take into account the HS station size (usually large stations, sometimes with dedicated HS platforms that are not always near those used by local services) and the average boarding/disembarking time (people travelling on HS trains usually have luggage).

In terms of IR/L services provided, Milano Centrale is similar to smaller stations (Fig. 9), for example, Mannheim and Basel SBB, which provide fewer services, but are both 1-h away from a large station along the Corridor, Frankfurt and Zürich respectively. However, almost all the integrated trains in Milano are IR services, connecting the node with large cities further away and serving fewer stations in the proximity of Milano, while in Mannheim and Basel SBB more integrated local services are available.

This point is even more evident if we compare different nodes that are served by more than one HS station (Fig. 10). Two different service models for IR and L connections in such nodes can be observed. In Germany and Switzerland, the central station provides most of the IR/L services, as shown by the examples of Köln central station and Basel SBB compared with the Köln node (also served by Köln Messe Deutz) and the Basel node (also served by Basel Bad Station). In contrast, in Italy, different stations serving the same node have different functions: in the case of Milano, the Centrale Station connects the city with other important cities (providing more IR services), while Rogoredo and Garibaldi serve the hinterland (providing L trains). In such cases, it is sometimes necessary to move between two different stations in order to transfer from a local train to a HS/LD service and vice versa.

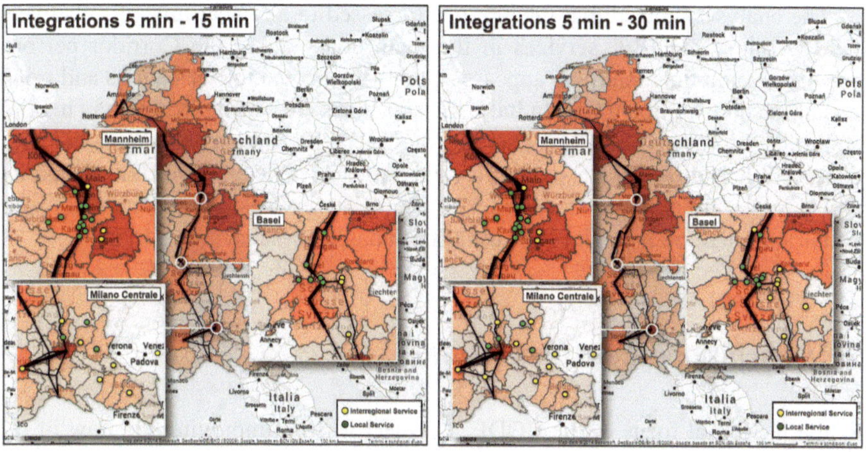

Fig. 9 Origins and destinations of IR and L trains integrated in other corridor stations with HS/LD services with short or medium transfer times from 8:00 to 9:00 am. *Source*: Elaborations of DB European timetables; base map: Google Maps, Google Inc.

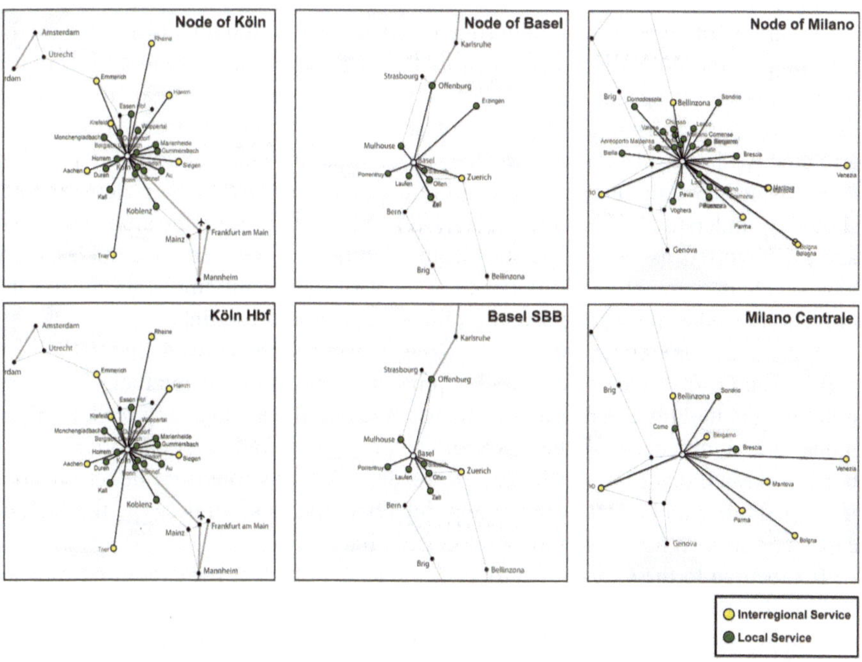

Fig. 10 Incoming integrated IR and L services in nodes served by more than one HS station and in the corresponding central station. *Source*: Elaborations of DB European timetables; base map: Google Maps, Google Inc.

The analysis showed that integrations (assessed in a typical time slot) between HS/LD trains and IR/L services in the main nodes along the Corridor perform adequately and the hinterland appears suitably connected to HS stations and transfers have short waiting times. In Italy, transfer times are usually longer than in other countries. Moreover, a different service model for IR and L connections has been observed since different HS stations serving the same node have a different function and provide either more L services or more IR services, compared with other countries where both services are usually available at the central station.

7 Conclusions and Open Issues

While the initial focus of the CODE24 project was on improving the flow of rail freight and its implications for corridor development, it was soon realised that there is a need to investigate how passenger rail services could be optimised in parallel with rail freight, and how to make the most of the opportunity provided by the increasing importance of HSR, which, on the corridor, typically shares the tracks with the other services.

A survey of experiences underlined which factors should always be in the forefront of the HSR discourse, such as the importance of average speed, rather than of maximum speed, the trade-off between speed and additional stops, which affects, in part, ridership, frequency of services and types of users expected or attracted, with the different experiences around the relevance of commuters rather than business people or people travelling for other purposes. Other elements that should be understood when considering HS and LD services are the attraction of air and car travellers as well as the shift from other non-HS rail services. Such factors have a clear commercial importance, but also significant implications for the provision of other rail services and on the transport land-use links.

A brief description of the system along the corridors showed that settlement patterns and rail service offers can be quite different from one another. The latter point recalls the initial, and current, focus on national passenger services. Furthermore, the significance of the corridor in terms of freight carried, often across borders is such that many technical efforts are underway to ensure better corridor interoperability and capacity for trains. Interestingly, in several cases, the lack of capacity was also the reason to develop HSR lines.

It appeared to the CODE24 project that one key element for exploiting railway technical capabilities, e.g. capacity and speed, for passenger transport is the integration among services. Integration is also relevant from the viewpoint of the land use/transport interaction to avoid losses in levels of service for locations not served by HS services and to ensure that the entire corridor may benefit from HS links thanks to feeder services.

The paper explored two aspects of integration: integration of services among the Rhine–Alpine Corridor nations and integration of LD and HS services with local services. The focus of the investigation was based on an assessment of the travel demand for all modes that exist along the corridor. The data presented show that the key OD pairs are national, whereas international travel demand plays a much smaller role. What is more, much passenger travel happens between zones that are less than 100 km apart. A comparison between demand and supply of HS and long distance services between the most significant OD pairs revealed that the number of direct services is not necessarily similar across ODs with a similar potential demand, although dissimilarities in supply decrease when IR/L services are also considered. Further, the investigation highlighted the important role of indirect—but connecting—services in ensuring a high level of service between important OD pairs.

The analysis of the integration between HS/LD and IR/L services focused on the availability of local connections within a given time at a selected set of stations. The results show a good integration in German and Swiss main stations in particular, with a longer transfer time in Italy. Milano experiences a different service model where three stations are served by HS services, with one right in the city centre, but not integrated with the other services and the other two benefitting from more local services. Moreover, HS trains calling at the first station do not serve the others.

The results summarised in this paper were presented during an experts' workshop held in Frankfurt in June 2014. The experts confirmed the importance of long distance services with the ensuing savings in travel time serving the backbone of the Corridor. They also agreed that, in the interest of regional accessibility, the focus should not be limited to HS and LD services. It was proposed to foster the existing multi-scale accessibility approach, which takes into consideration the numerous nodes along the corridor. This multi-scale accessibility brings a second advantage: the integration of regional feeder services in relevant regional and national transfer nodes, which results in better accessibility than focusing on high speed solutions. The fewer stops provided by the latter take the risk of losing customers "along the lines", notably when creating parallel lines, which make network accessibility for stations further away more difficult. Moreover, demand may be shifted from air towards rail, but to a lesser extent than from cars. This is justified by the density of the corridor, the polycentric character of the regions and the interregional relationships along the corridor, which again justify the multi-scale strategy.

Regular services along the corridor at intervals of 2 h may be suggested. Standardised transfer times of 15–30 min would ensure a seamless travel chain without unattractive waiting times. Additional faster services calling at fewer stations may be a top option for coping with considerable air demand levels, as found between Frankfurt and Zürich or Amsterdam, Zürich and Milano or Cologne and Zürich.

Considering this proposal, further work may include an assessment of the introduction of such services and the possible loss of customers on other trains of the proposed 2-h service structure. Another topic to investigate is whether passengers would prefer air travel within Europe if airports were more accessible and integrated. Furthermore, it would be interesting to evaluate how HSR should be part of an overall rail development strategy with the centres and their central stations as the backbone for more sustainable settlement patterns aimed at reducing the overall carbon footprint as a sum of predominantly air and car trips. This, especially, is an issue that Givoni and Dobruszkes (2013) raise in their paper on the effects of HSR.

Moreover, it would also be useful to extend the integration analysis to other factors that are as important as saving transfer time in order to improve service integration, such as:

- Service frequency, increasing the number of possible transfer choices
- Service reliability, reducing the risk of delays
- Integration of fares among the different operators, also considering interregional and local services so that it is possible to transfer to the next useful train in case of a delayed arrival in transfer nodes
- Information that allows users to share knowledge about available services
- Regulations to coordinate cooperation among both the public authorities and the operators

Further useful work on the matters discussed here would include investigations on the rationale and the effects of the choices between direct services and options including transfers, typically made by operators.

In closing, the lack of official data on passenger mobility along such an important corridor and the need to resort to modelled data should be underlined. Further work on reconstructing actual demand patterns in Europe is desirable.

References

Beckers T, Haunerland F, von Hirschhausen C, Walter M (2009) Long-distance passenger. Rail services in Europe: market access models and implications for Germany. Discussion Paper No. 2009-22, OECD International Transport Forum

Cascetta E, Coppola P (2015) New high-speed rail (HSR) lines and market competition: short term effects on services and demand in Italy. Transp Res Rec (forthcoming)

Cascetta E, Coppola P, Velardi V (2013) High-speed rail demand: before-and-after evidence from the Italian market, disP. Plann Rev 49(2):51–59

Chen C, Hall P (2011) The impacts of high-speed trains on British economic geography: a study of the UK's intercity 125/225 and its effects. J Transp Geogr 19:689–704

Corridor Rhine-Alpine (2013) Corridor implementation plan (CID Public Version 1.0—05.12.2013). www.corridor1.eu. Accessed 23 July 2014

Council of the European Union (1996) COUNCIL DIRECTIVE 96/48/EC of 23 July 1996 on the interoperability of the trans-European high-speed rail system. European commission, Brussels

Dobruszkes F (2011) High-speed rail and air transport competition in Western Europe: a supply-oriented perspective. Transp Policy 18:870–879

European Commission (EC) (2010) High-speed Europe—a sustainable link between citizens. Publications Office of the European Union, Luxembourg

European Commission (EC) (2013) Trans-European transport network—TEN-T core network corridors map. REGULATION (EU) No 1316/2013 O.J. L348—20/12/2013

Frontier Economics, Atkins, ITS Leeds (2011). Ex post evaluation of cohesion policy interventions 2000–2006 financed by the cohesion fund—work package B: cost-benefit analysis of selected transport projects. Appendix 1: High Speed Railway Madrid—Barcelona in Spain (March 2011)

Garmendia M, Romero V, de Ureña JM, Coronado JM, Vickerman R (2012) High-speed rail opportunities around metropolitan regions: Madrid and London. J Infrastruct Syst 18:305–313
Germanwatch (2013) Emissionsminderung durch Hochgeschwindigkeitszüge. Bonn/Berlin. www.germanwatch.org/de/7155. Accessed 27 Feb 2014
Givoni M (2006) Development and impact of the modern high-speed train: a review. Transp Rev 26(5):593–611
Givoni M, Banister D (2012) Speed: the less important element of the high-speed train. J Transp Geogr 22:205–206
Givoni M, Dobruszkes F (2013) A review of ex-post evidence for mode substitution and induced demand following the introduction of high-speed rail. Transp Rev Trans Transdisciplinary J 33(6):720–742
Guirao B (2013) Spain: highs and lows of 20 years of HSR operations. J Transp Geogr 31:201–206
Marshall T (2014) The European union and major infrastructure policies: the reforms of the trans-European networks programmes and the implications for spatial planning. Eur Plann Stud 22(7):1484–1506
Rebmann M (2011) Hochgeschwindigkeit und Klimaschutz: 3 ½ Stunden per Bahn zwischen deutschen Metropolregionen ermöglichen Verkehrsverlagerungen vom Flugzeug. Bahn-Report No. 2, 74-79
UIC (2013) High speed lines in the world. www.uic.org. Accessed 19 Feb 2014
Vickerman R (1997) High-speed rail in Europe: experience and issues for future development. Ann Reg Sci 31:21–38
Vickerman R (2013) High-speed rail and regional development: the case for intermediate stations. Paper for the Jean Monnet programme on the high-speed train and its intermodality in medium-sized cities in Spain and Europe, Barcelona, 14 March 2013

Maurizio Arnone holds a degree in Civil Engineering and a master in "Transport: systems, networks and infomobility" specialising in planning, management, control and monitoring of transport systems.

He has been working in SiTI Logistics and Transport Research Unit since 2007, initially as researcher and, since 2010, as unit manager. In the last years his research works have focused on the analysis and the reorganization of urban mobility systems and interurban public transport services.

Tiziana Delmastro holds a master degree in Mathematical Modelling in Engineering. Since 2007 she works as a researcher for the Logistics and Transport Research Unit in SiTI—Higher Institute on Territorial Systems for Innovation. Her studies deal with transportation planning, accessibility analysis, estimation and assessment of mobility demand, development and application of passenger and freight transport models and of discrete choice models.

Peter Endemann has a background in Geography, Spatial Planning and Urban Design. Since 2003 he is employed at Regionalverband FrankfurtRheinMain (Regional Authority Frankfurt-RheinMain). His activities comprise travel behaviour analysis and interaction with land-use, transport infrastructure development, rail freight and long distance passenger rail. Within CODE24 he was co-responsible for the development of the Online Rail Freight Exchange (ORFE) and the action on high speed rail integration.

Dr. **Noriko Otsuka** is a senior researcher at Research Institute for Regional and Urban Development (ILS) in Dortmund. She has conducted a number of research projects in the field of urban and transport planning, brownfield regeneration, real estate and construction management. Previously she worked in the Institute for Spatial and Landscape Planning at ETH Zürich where she was involved in the CODE24 project. She also worked as a Post-doctoral researcher at Oxford Brookes University and as a senior consultant at TRL (Transport Research Laboratory) Limited in the UK.

Stefano Pensa has completed his PhD in Architecture and Building Design in 2013, with a thesis concerning the use of geovisualisation tool as a support for spatial planning and decision-making processes. He currently works both as a researcher and as consultant. His scientific interests range from the visualisation of spatial data to interactive systems, from web mapping to VGI. He is the author of several national and international publications.

Andrea Rosa is a transport planning researcher and consultant with a background in academia and consultancy. He worked in transport planning as project manager and in academia on transport and traffic modelling and, later, on rail and air transport regulation. Andrea holds a degree in civil engineering (transport section) and a PhD in transport modelling.

Part III
Logistic Processes: Organizing Efficient Freight Transport in the Corridor Rotterdam-Genoa

Evaluation of Impacts of Logistics Clusters in the Corridor Rotterdam–Genoa

Rudolf Juchelka and Julian Brenienek

Abstract The emergence of logistics clusters is an economic development which could be observed in particular locations during the last decade. These locations are primarily linked to the global trade or have the function of a hinterland hub for these global links. They are characterized by a very high amount of handled goods, an excellent infrastructure of intermodal transfer facilities and a large pool of specialized labor. These factors lead to high economic growth, which can be read on specific indicators.

While the concept of industry or production clusters is well recognized in academic research, analyses of logistics clusters are marked by a definitional ambiguity of spatial and economic delineation of the study area. Authors identify global, national or regional clusters. There is, however, no complete list of identified clusters in academic literature concerning application-oriented cluster concepts. This is partly grounded in the definitional dispute over fundamental cluster structures, which determines the discussion on cluster concepts of recent years. While it still lacks an empirical concept to identify these clusters: Is a location already a cluster if an above-average amount of companies out of a particular industry branch is in a particular location? Or is it a cluster if actors denote "from inside out" the location as a cluster?

The paper gives an overview of the approach used in the project CODE24 with regional examples from case study areas.

1 Introduction

Since 1990, the EU pursues a concerted transport-policy with respective development and planning strategies in order to support the linking of European regions and national networks by an efficient transport infrastructure. This is a prerequisite for a functioning internal market and is related to the aim of the EU's economic, social and territorial cohesion (Europäische Kommission 2001).

R. Juchelka (✉) • J. Brenienek
Department Economic Geography, Transport and Logistics, University of Duisburg-Essen, Schützenbahn 70, 45127 Essen, Germany
e-mail: Rudolf.Juchelka@uni-due.de; Julian.Brenienek@uni-due.de

H. Drewello, B. Scholl (eds.), *Integrated Spatial and Transport Infrastructure Development*, Contributions to Economics, DOI 10.1007/978-3-319-15708-5_10

The core zone of economic growth in Europe, in spatial models referred to as "blue banana", extends as a corridor starting in South East England, via the BeNeLux countries, via Germany, along the Rhine, via Switzerland and northern Italy to the Mediterranean coast. The aim is to maintain this growth region of Europe due to the development and expansion of transport infrastructure. As part of the Trans-European Networks (TENs) the main traffic line Rotterdam-Genoa, also described as Corridor No. 24, is of central importance.

The forecasts for the sea port of Rotterdam are enormous. By the year 2030, according to an analysis of the port, the overall throughput numbers will double in size, solely by the completion of the newly built part of the port "Maasvlakte 2". The forecasts are based on several possible scenarios that include the current European trend, the global economy and the rising oil price and also assume a lower growth. By 2030, the port is to be upgraded from the leading European hub to *the* global hub, which increasingly serves as a hub for global trade towards the U.S., Japan and Australia (Port of Rotterdam 2012). But global ambitions can not only be observed at the northern end of the corridor 24. The port of Genoa is also a gateway for the connected corridor. Of particular importance is the impact that both ports exercise on their hinterland and thus also on the connected corridor 24 (Rodrigue et al. 2009).

The corridor no. 24, seen as traffic-, transportation- and trade-corridor, merges international traffic flows and aligns them to the respective global link in form of the seaports. The emergence of logistics network structures in large distances of these global gateways, their alignment along these traffic flows and their increasing importance within the corridor no. 24, is an observation of recent years. This paper will analyze these network structures using selected regional case studies which will then be compared at an international level. The analysis is based on the intermodal traffic, the transport- and economic structure of each area with special emphasis on the individual situation of each region.

2 Logistics Clusters

The emergence of logistics clusters is an economic development which could be observed in particular locations during the last decade. These locations are primarily linked to the global trade or have the function of hinterland hubs for these global links. They are characterized by a very high amount of handled goods, an excellent infrastructure of intermodal transfer facilities and a large pool of specialized labor. These factors lead to high economic growth, which can be read on specific indicators.

The first theoretical foundations for today's cluster concepts go down to the first concepts of Alfred Marshall in 1890 (Marshall 1890). Current cluster concepts and today's cluster research is marked by the concept of Michael Porter and his diamond model (Porter 1998). Many industry or production clusters have been analyzed and evaluated in multiple theoretical and empirical studies (Porter 2000). The methods vary and range from a primary survey, with interviews of selected stakeholders, to analyses of selected secondary data. In many cases, there are two issues to be addressed after this analysis: Is the selected location a cluster? What must be done for a positive economic development of the area?

While the concept of industry or production clusters is well recognized in academic research, analyses of logistics clusters are marked by a definitional ambiguity of spatial and economic delineation of the study area (Elsner et al. 2005). Authors identify global, national or regional clusters (Rivera and Sheffi 2012). There is, however, no complete list of identified clusters in academic literature concerning application-oriented cluster concepts. This is partly grounded in the definitional dispute over fundamental cluster structures, which determines the discussion on cluster concepts of recent years. While it still lacks an empirical concept to identify these clusters: Is a location already a cluster if an above-average amount of companies out of a particular industry branch is in a particular location? Or is it a cluster if actors denote "from inside out" the location as a cluster?

In the following considerations, the chosen methodology in the context of the selected case studies in the project CODE24 is described. The results of the case study areas will then be compared on an international level.

3 Methodology of the Case Studies

By the selection of eight areas within the corridor 24 and the associated international character of the case study areas, a sufficient cross-section comparison is ensured.[1] Nevertheless, this international character poses a challenge for the statistical approach to the respective regions. On the national state level, basic economic data such as gross domestic product or statements of import and export are available. Inquiries on micro level, such as statistical analyses for container traffic at NUTS 3 level, are not available on an international level and are sometimes not provided by the statistical offices of the states. Due to the different national collection of records the datasets had to be edited for a later comparison. Gaps are filled by interviews with experts in each case study region. Furthermore, this ensures that each case study region is adequately presented.

4 Definition of the Study Area

The relevant case study areas were identified at NUTS-3 level and present the respective municipal districts as core regions. As the case study areas form around a logistical nucleus, in most cases an inland port, the study area is extended to

[1] The case studies were carried out in Cologne, Frankfurt, Mannheim, Karlsruhe, Nijmegen, Basel, Milan and Turin.

adjacent counties. This is to ensure that the respective core regions are comparable to the extended area (on a statistical level) and that the investigated region covers a large enough area to represent the entire extent of the logistics cluster.

As in the empirical view of a logistics cluster no comparatives for the spatial extent were explored, a set of criteria for selection and specification of the case study areas was used. In a study of the logistics cluster Basel, for example, the area is analyzed as a "tri-national Euro district" (Stölzle et al. 2009) and thus is considered as a union of sub-regions of the three involved nation-states. In this context, the authors make their selection by considering various administrative areas, such as cantons and counties, individual municipalities and a variety of regions, who are members of a regional planning association (Stölzle et al. 2009, p. 7). Due to the inclusion of the adjacent administrative units and their diversified shape, the case study areas within the corridor no. 24 have a different spatial extent. It is nevertheless essential to analyze these regions on this level, because nation states collect the respective statistical data according to NUTS-3 regions. At micro level (below NUTS-3), these datasets are not available. Additionally, an assumed radius of 50 km is used to limit the study area (Rivera and Sheffi 2012, p. 14). This limitation is based on a comparative analysis of equidistance for road freight transport and proximity to major (intermodal) logistic locations for each area. Furthermore, it is important that logistics activities, due to lack of suitable land, are increasingly outsourced from urban areas.

The following points for the limitation of the study area can be summarized:

- A radius of 50 km as the boundary of an analysis of a logistics cluster seems appropriate, since the maximum extent of the cluster can be detected. Thus, all information relevant to the core region can be included in the analysis.
- For a detailed analysis compatible datasets according to NUTS-3 units have to be used.

5 Expert Interviews

In most cases, the required statistical data on the respective country is made available by land offices or by the federal statistical office. Nevertheless, it is necessary to unify the previously gathered data, due to the varying datasets on an international level. Special analysis criteria, such as handling data of individual container terminals, can give rise to gaps in the datasets, since the throughput numbers are not always completely published. To close these gaps within the records, personal contact to regional stakeholders was an important part of the analysis.

In addition, the perspective of the local stakeholders should be included in the analysis of the logistics cluster region. For this reason, expert interviews were conducted in all case study areas. These were conducted with representatives of the respective chamber of industry and commerce and the economic promotion of the region. Additionally, representatives of the busiest intermodal facilities were selected. Within the German and Swiss case studies, for example, the respective port authority was chosen, since the terminals are located in the port area. Due to project workshops along the corridor, the perspective of the logistics service providers of a region can be considered. These regional roundtables and workshops were conducted by the participating regional associations or chambers of industry and commerce and were explicitly directed to the local companies. The results of these events are reflected in the analysis of the respective case study area.

The expert interviews were conducted based on an individual guide, but an actual recording of the interview was not possible. Many contacted experts had the fear that their statements can quickly lose any validity in the context of the rapid development of the logistics sector. In addition, the requested datasets, such as for supply chains, trade relations, terminal capacity and available potential, etc. are subject to privacy policies. The requested data are closely linked to economic interests of companies and regions. For the purposes of the experts, we worked without a record of the interview and thus an explicit analysis and evaluation of the conversation.

6 Indicators of Regional Economic Structures and Analysis of Location Factors

Taking into account the individual indicators for regional economic structure, the economic situation of a region can be described and illustrated. The indicators provide information about the socio-economic situation of the population and the economic environment in which companies operate (Stölzle et al. 2009, p. 9). In the examined regions, the number of companies and employees, the gross domestic product and gross value added within the logistics sector were analyzed and evaluated at NUTS 3 level.

The following points for the analysis of the regional economic structure can be summarized:

- GDP and GVA can be used for the first approximation to the investigated area, but must be analyzed in the context of the particular research area (here: transport and logistics)
- A separate consideration of the branch code H (trade) was only possible in rare cases. Information will often be provided by summed up datasets for branch codes G-J.
- Analysis and identification data for the logistics sector should be carried out due to the presented logistics sector (see Fig. 1) and should not include branch codes for passenger traffic.

Warehousing, storage, cargo handling, forwarder, postal and courier activities		
WZ 2008 NACE Rev. 2	**description**	
52.10	**Warehousing and storage** if available: further subcategories	Warehousing, storage and cargo handling
52.24	**Cargo handling** if available: further subcategories	
52.29.1	**other transportation support activities, here: forwarder**	Forwarder
53.10	**Postal activities under universal service obligation** if available: further subcategories	postal and courier activities
53.20	**Other postal and courier activities** if available: further subcategories	

freight transport on different modes of transport and support activities for transportation		
WZ 2008 NACE Rev. 2	**description**	
49.20	Freight rail transport if available: further subcategories	freight transport on different modes of transport
49.41	Freight transport by road if available: further subcategories	
50.20	Sea and coastal freight water transport if available: further subcategories	
50.40	Inland freight water transport if available: further subcategories	
51.21	Freight air transport if available: further subcategories	
52.21	Service activities incidental to land transportation	Support activities for transportation
52.22	Service activities incidental to water transportation	
52.23	Service activities incidental to air transportation	
52.29	other transportation support activities	other transportation support activities

Fig. 1 Selected economic activities defined in section H (transport and storage) to define the hierarchical order of the selected logistics sector

- For a detailed analysis of the relevant economic sector, particularly companies and employees of selected sectors should be considered.

By analyzing the location factors for the logistics, statements for the potential of a region to form a logistics cluster can be determined. For this purpose it is necessary to review the various modes of transport on their usage and potential to eventually highlight the importance within the transport network. In the context of the case studies for the CODE24 project road, rail, inland waterway and air transport were analyzed as indicators. To ensure the high level of similarity of records, published data and information (relating to the transport network) of the statistical offices are used. The available commercial space in a region is archived partly in digital databases. However, these are bound to the administrative boundaries of the respective operators, such as planning associations or chambers of industry and commerce. In this context, it was necessary for the extensive coverage within the case study areas, to merge and unify them within a data form. The published information at the institutional level of research and teaching and logistical cooperation and networks could be collected at the level of the respective districts and were collected in own research. Flows of goods and calculations for modal split are exclusively collected at the level of nation-states (NUTS-1) or for selected modes of transport at the level of the German government districts. However, since these data can only be roughly selected, it was necessary work with appropriate publications of the respective chamber of industry and commerce or the regional economic promotion.

The following points for the analysis of the road, rail, inland watery transport can be summarized:

- A classification of the utilization of road transport is possible for the German case studies through publications of the Federal Statistical Office and the Federal Office for Roads. There is no equivalent representation for rail freight transport.
- For the German case studies destination and source data for rail freight traffic are available at NUTS-2 level (administrative districts). At NUTS-3 level the statistics lose their explanatory power. A spatially differentiated provision of data would be desirable.
- A specification for container traffic of the German inland ports is not included in the official statistics of the federal statistical office. Consistent information about the container throughput can currently be obtained via the magazine *Schifffahrt, Hafen, Bahn und Technik.*

7 Analysis of the Spatial Distribution of Logistics Service Providers

In order to investigate the spatial extent of a logistics cluster, relevant logistics service providers were located and analyzed. To ensure a high level of detail on a statistical level for the above-mentioned analysis of the logistics potential of a region, a logistics sector has been defined based on selected economic sectors (see Fig. 1). In a following step records for the selected companies were created in cooperation with 28 German chambers of industry and commerce within the corridor no. 24. These were located using the geographical information system ArcGIS. Due to privacy policies of the chambers of industry and commerce, it was only possible to process the actual location of the company without including additional data, such as name or size of the company. Due to the observed activity of the companies within the logistics sector, the logistics sector has been divided and hierarchized.

The list of these companies includes the sectors warehousing and storage (5210), cargo handling (5224), forwarder (52291), postal activities under universal service obligation (5310) and other postal and courier services (5320). In the detailed research on the determination of individual companies to specific sectors (which is carried out in each case at the company's entry into the chamber of industry and commerce) a strong similarity between the sector "forwarder" and "freight transport by road" could be found. When classifying in these sectors, the respective chamber makes a distinction: when a company offers a higher quality service, like storage or consolidation of cargo, it is assigned to the forwarders (52291). If it offers only the transport of goods from a source to a destination, this company is classified in the sector "freight transport by road" (4941). Accordingly, the above-mentioned economic sectors are in the higher-quality services and thus are of special importance within the logistics sector.

The other selected companies in the logistics sector, which do not fit into the range of higher-quality services, have been consolidated in the area of transport. These companies are allocated to the economic sectors 4920, 4941, 5020, 5040, 5121, 5221, 5222, 5223, and 5229.

The following points for the analysis of the location of logistics service providers can be summarized:

- At NUTS-3 level a value for the number of companies within the logistics sector can be determined by the federal statistical office.
- Economic developments on a national level, however, are not based in these administrative boundaries, but along major transport routes. In order to give a detailed analysis of the shape of a logistics cluster, a localization of the logistics service providers is essential.

8 Comparison of the Case Study Results

In the following, the summarized results of each case study area are shown in contrast to each other. The list is based on the topics presented above, followed by the generalized results for each topic in order to derive recommendations for appropriate action.

9 Definition of the Study Area

Cologne: A 50 km radius is ideal to analyze the region especially in comparison to Duisburg and the Ruhr area. In addition, the cities of Düsseldorf and Bonn are included in the analysis.

Frankfurt: A 50 km radius includes the state capital of Wiesbaden in the analysis. At the same time the state of Rhineland-Palatinate is taken into account with the city of Mainz and the state of Bavaria with the city of Aschaffenburg.

Mannheim: A 50 km radius includes the cities of Mannheim (Baden-Württemberg) and Ludwigshafen am Rhein (Rhineland-Palatinate). The radius does also include the cities of Karlsruhe, Darmstadt and Kaiserslautern.

Karlsruhe: A 50 km radius includes the surrounding administrative districts of the city of Karlsruhe and goes beyond the definition of the "Technology Region of Karlsruhe" and therefore also includes a part of the French NUTS-3 territorial unit Bas-Rhin. Due to this scale, the Karlsruhe region can be analyzed in comparison to the cities of Mannheim/Ludwigshafen am Rhein and Stuttgart.

Nijmegen: In the study area, the urban centers of Arnhem, Nijmegen, Ede, Kleve and Emmerich are represented with their respective catchment area. The shape of the study area allows an analysis of cross-national perspective.

Basel: A 50 km radius includes a cross-border (tri-national) area and enables the integration of locations near to the border. Due to this, the analysis includes transnational aspects.

Milan: A 50 km radius around the city of Milan includes large parts of the "logistics region Milan" (Regione Logistica Milanese), which includes important nodes of international trade.

Turin: The study area (the administrative province of Turin) should be extended to the provinces of Alessandria and Novara, which are key hubs for the logistic development of the Piedmont region.

9.1 Summary and Practical Application

- Logistics Clusters at the national level are not subject to administrative units or boundaries.
- At bilateral level, an orientation can be recognized.
- Transborder locations are of high functional importance for the cohesion of the entire Rotterdam-Genoa corridor and have to be analyzed according to their cluster potential.
- The superposition of the NUTS-3 territorial units through areas of responsibility of local authorities, chambers and associations leads to a wide range of available data, although these are not necessarily compatible. For an aggregation of data resources at a higher level, an orientation on a 50 km radius for publications on the economic characteristics of region would be desirable.

10 Regional Economic Structure

Cologne: The number of employees in the logistics sector is based mainly on the left bank of the Rhine from Duisburg via Düsseldorf, via the NUTS-3 district Rhein-Kreiss Neuss to Cologne and the Rhein-Erft-Kreis. The cities of Dortmund, Hagen and Unna are hotspots.

Frankfurt am Main: Compared to other case study regions an average amount of logistics companies is available. However, Frankfurt is a hotspot for employees in the logistics sector. Large values for the number of employees set to continue south on the Rhine axis. Aschaffenburg is a hotspot as well. The north of Frankfurt is characterized by low numbers of employees in the logistics sector.

Mannheim: The largest number of employees can be identified on the axis Gross-Gerau-Worms-Mannheim-Rhein-Neckar-Karlsruhe. High proportions of the total number of employees can be determined on the left side of the Rhine. High absolute employment records for the logistics sector are located on the right side of the Rhine.

Karlsruhe: The continuation of the axis described for Mannheim continues via the NUTS-3 units Rhein-Pfalz-Kreis and Germersheim to the city of Karlsruhe. The Rhein-Neckar-Kreis and the district of Karlsruhe are characterized by above-average values. The district of Ludwigsburg is a hotspot. In the counties around Stuttgart, a high distribution of large absolute numbers of employees in the logistics sector can be observed.

Nijmegen: Compared to the neighboring German NUTS-3 units Borken and Kleve the amount of logistics service providers is much higher. The region Arnhem-Nijmegen acts as a distribution hub for the hinterland traffic of the ports of the North Sea range. The region can be described as logistical hotspot.

Basel: Above-average gross domestic product. Basel has identified the logistics branch as a core industry with high added value.

Milan and Turin: The provinces of Milan and Turin take a significant role in terms of economic development of the country. The regions are characterized by high employment levels and an above-average GDP.

10.1 Summary and Practical Application

- The results of the evaluation of the number of employees are shown in Fig. 2 (excerpt from the case study Mannheim).
- An analysis must not only consider the context of national and international metropolitan areas, but also be based on important relations within the (logistics) network.
- Logistics Potential: An analysis of economic structure that is characterized by conditions of national and international metropolitan areas must be supplemented by important relations within the logistics network.

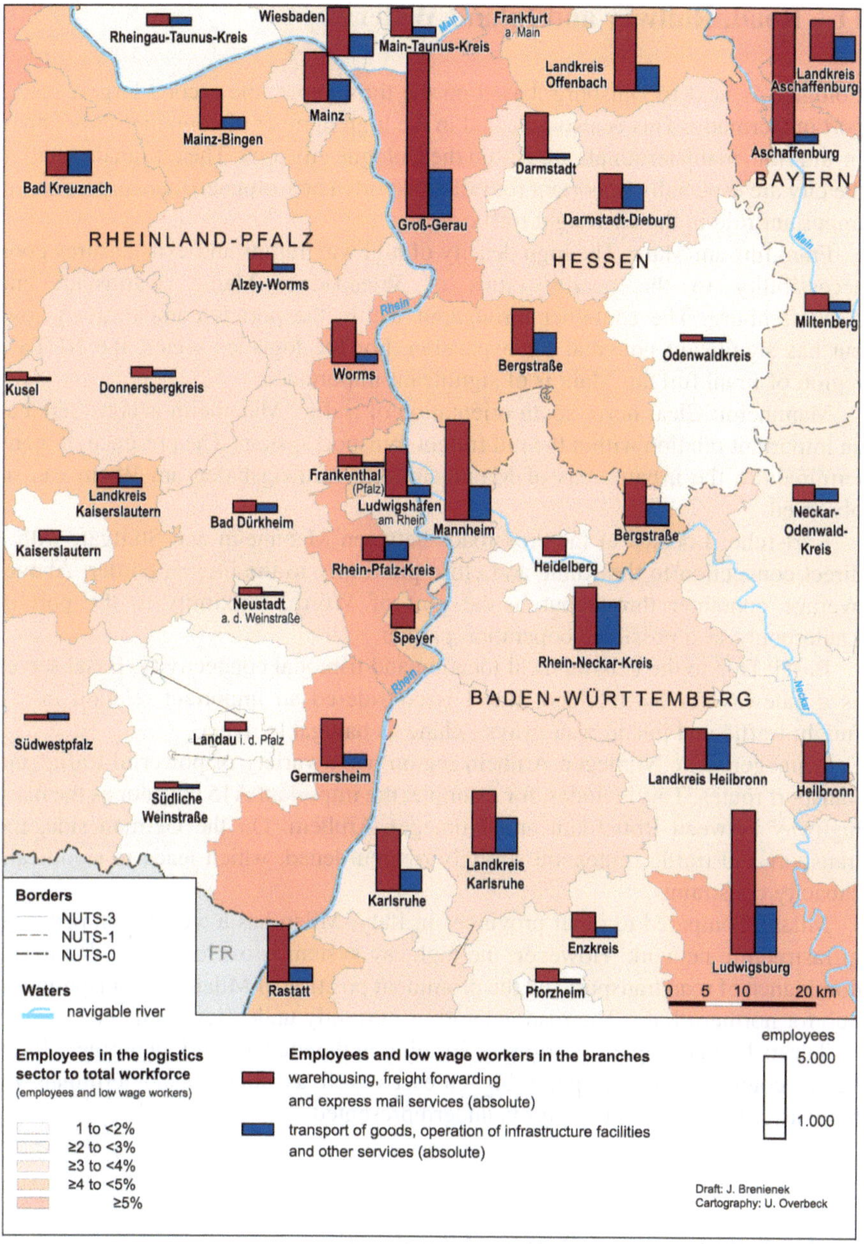

Fig. 2 Employees and low wage workers in the logistics sector in 2012 in the case study area Mannheim, *red*: branches 5210, 5224, 52291, 5310, 5320, *blue*: branches 4920, 4941, 5020, 5040, 5121, 52215, 52222, 52231, 52292, 52299 (data based on German Federal Employment Agency 2012)

11 Road, Railway and Inland Waterways

Cologne: Is next to Duisburg, based on the throughput, the second largest inland port in Germany. This is also reflected in the high number of container throughput. In addition, many terminals are within the Cologne city area. The terminals south of the city are especially important for rail transport. The Cologne/Bonn airport has an important role in the air freight traffic.

Frankfurt am Main: The high density of highway ramps and exits ensures good accessibility to the nearby cities of Wiesbaden, Mainz, Darmstadt and Aschaffenburg. The container throughput within the port is comparatively low, but has significant potential for expansion. For the logistics sector, the NUTS-3 region of Frankfurt am Main is of significant importance.

Mannheim: Clear north-south orientation of traffic. Mannheim is considered as an important relation within the rail freight transport system. The presence of many terminals in the urban areas of Mannheim and Ludwigshafen am Rhein can be observed.

Karlsruhe: Located at the crossroads between Mannheim and Stuttgart with a direct connection to the Rhine and close proximity to the French border. Above-average container throughput in the port of Wörth. Proximity to the port of Lauterbourg as a possible cooperation partner.

Basel: Due to the geographical location and trimodal connectivity, Basel serves as a gateway for Switzerland. Basel is considered an important relation in rail freight traffic and inland waterways (Rhine as navigable river).

Nijmegen: The Nijmegen-Arnhem region has a variety of powerful traffic and transport routes. This includes, for example, the important A15 corridor as the main highway between Rotterdam and Nijmegen-Arnhem. On the German side, the transport and traffic routes are strongly overburdened, which leads to significant capacity constraints.

Milan: Compared to other provinces in Italy, Milan has a well-developed road infrastructure network. However, the highway system is overburdened due to the dominance of road transport and the prominent position of Milan as a transportation hub for northern Italy. The road network is currently under development.

Turin: Turin occupies a peripheral location in the road network in northern Italy, but is connected to the major big cities as well as other inland terminals by highways. Rail freight transport is underrepresented.

11.1 Summary and Practical Application

- The logistics infrastructure and the determined container throughput are shown in Fig. 3 (excerpt from the case study Cologne).
- Logistics service providers are located around available transport infrastructure. This could also be detected on an international level due to the high density of data.

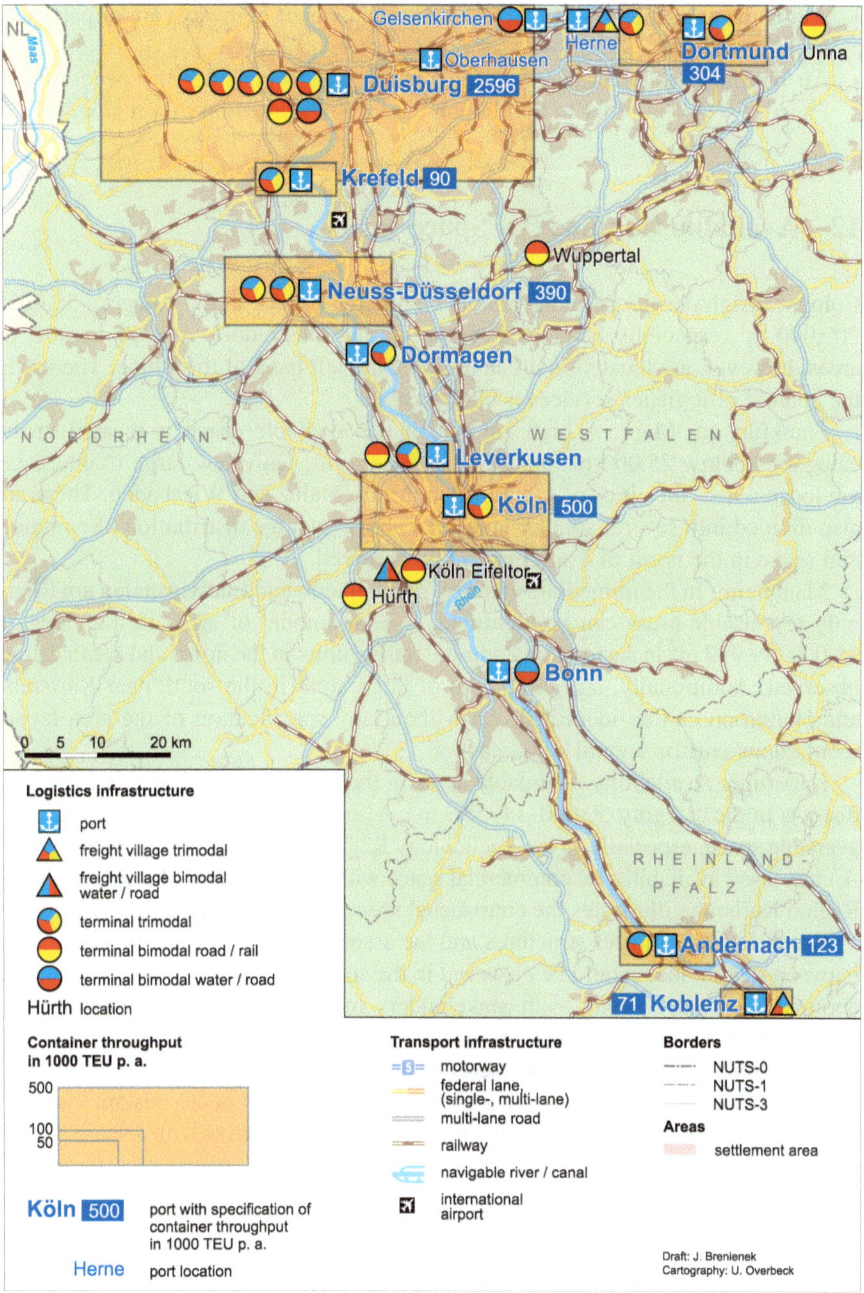

Fig. 3 Logistical infrastructure in the case study area Cologne (data based on SGKV 2010, DUSS 2013; Container throughput in 2012 by "containerticker" of the magazine "Schifffahrt, Hafen, Bahn und Technik" 2013; Information for Krefeld and Neuss-Düsseldorf: January to June 2012; Information for Koblenz based on an evaluation of Planco 2013)

- Along the available road network ramps and exits of highways are of particular importance.
- An optimization of transport infrastructure has to consider the needs of the logistics sector in order to build upon the logistics potential of a region.

12 Available Commercial Space

Cologne: High density of the available commercial space in the category 50,000–200,000 m^2, especially between Leverkusen and Düsseldorf. A large number of areas, however, are located in urban areas on the left bank of the Rhine. The use in the context of logistics services is questionable.

Frankfurt am Main: A high amount of the available space is located in the categories below 25,000 m^2 and 25,000–49,999 m^2. Only few large usable areas are located near the cities of Frankfurt am Main, Mainz and Wiesbaden. These are also divided into several smaller areas. An above average distribution of commercial space in the north of Frankfurt can be observed.

Mannheim: In the immediate vicinity of Mannheim and Ludwigshafen am Rhein only few usable areas can be identified. A high amount of areas in the category 25,000–49,999 m^2 in surrounding administrative units in the north and south can be observed. Additionally, a high amount of large areas in the south near Karlsruhe and Heilbronn can be identified, yet without any specification of the area layout (contiguous area or several sub-areas).

Karlsruhe: Availability of useable areas in the in the surrounding administrative districts in the category 50,000–149,999 m^2. A strong north-south orientation of the available space, especially in the Technology Region of Karlsruhe, can be observed. An increased availability of commercial space within the territory of the Technology Region Karlsruhe illustrates the constructed spaces in terms of data availability.

Basel: Existing cluster structures and the associated high degree of crosslinking between stakeholders could be observed in the solution of the shortage of available commercial space. Along with stakeholders in the region a regional plan for logistics is processed, which identifies usable commercial space.

Nijmegen: A shortage of commercial space for the logistics sector in business and logistics parks can be observed. Nevertheless, after years of constant increase, vacancy rates are recorded in the first half of 2010. This applies with a share of over 40 % to areas over 10,000 m^2.

12.1 Summary and Practical Application

- The available commercial spaces are shown in Fig. 4 (excerpt from the case study Frankfurt).
- The availability of suitable commercial space supports the growth of a logistics cluster.

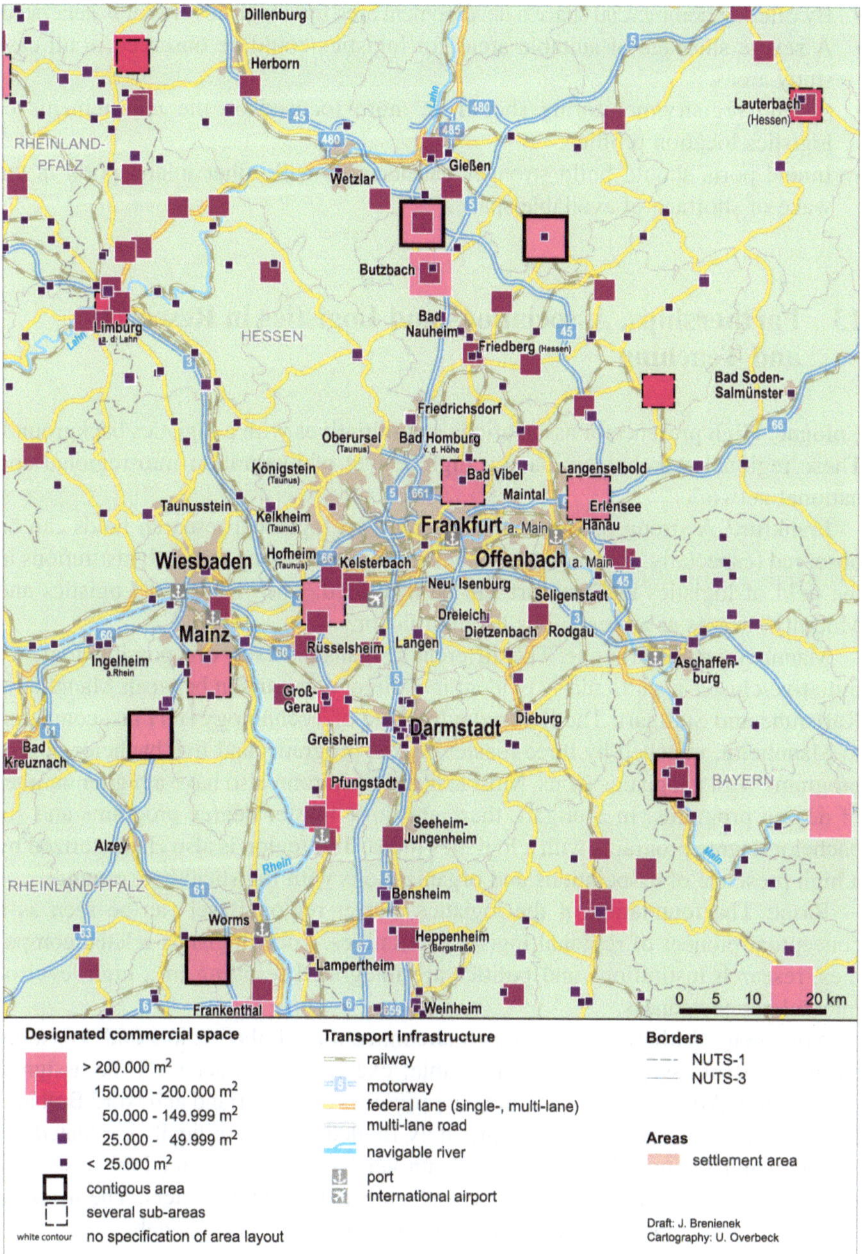

Fig. 4 Available commercial space in the case study area Frankfurt as of June 2013 (data based on Stadt Frankfurt am Main 2013, Standortportal Rhein Neckar 2013, ISB Rheinland Pfalz 2013, SIS Baden-Württemberg, SIS Hessen 2013, SIS Bayern 2013)

- By offering commercial space a development area for a logistics cluster is generated.
- A severe shortage of suitable areas for logistics could be observed in all case study areas.
- A land registry/monitoring should be more focused on the requirements of logistics location profiles.
- Inland ports should build stronger partnerships with other inland ports in the wake of shortage of available space.

13 Partnerships, Associations and Logistics in Research and Teaching

Cologne: High presence of associations and initiatives with a logistics background. These regional stakeholders have a high degree of integration into regional and national networks.

Frankfurt: A strong orientation of research on logistical research fields can be observed (especially TU Darmstadt). At the same time many research institutions in the field of logistics in Frankfurt can be identified. The "House of Logistics and Mobility" serves as a prestigious location factor.

Mannheim and Karlsruhe: The different "expansion state" of a possible logistics initiative/cluster is particularly evident in a direct comparison between Mannheim, Karlsruhe and Stuttgart. The Karlsruhe Institute of Technology (KIT) is, compared to Mannheim, matched by three master degree programs and five bachelor degree programs with a logistics focus. Mosbach and Heilbronn also have a higher number of degree programs. In Stuttgart, there are three master degree programs and six bachelor degree programs with a logistics focus. The region is also characterized by a high presence of associations and organizations with logistical background.

Basel: The foundation of the logistics cluster region Basel can be seen as a conceptual pioneer of regional logistics initiatives. Local logistics-related companies, research institutions and politics are integrated resulting in a high level of internal connectedness.

Nijmegen: With regards to the interconnection of the region and logistical partnerships in associations and academia, exemplary character can be assigned to the region Arnhem-Nijmegen and the provinces of Gelderland and Nord Brabant. Networks are politically initiated, privately funded and scientifically evaluated.

Milan and Turin: In both provinces, a university education in the logistics sector is possible. Collaborations take place with selected research centers. The network of logistics structures in local associations and academia is expandable.

13.1 Summary and Practical Application

- The offers for logistics in research and teaching are shown in Fig. 5 (excerpt from the case study Karlsruhe).

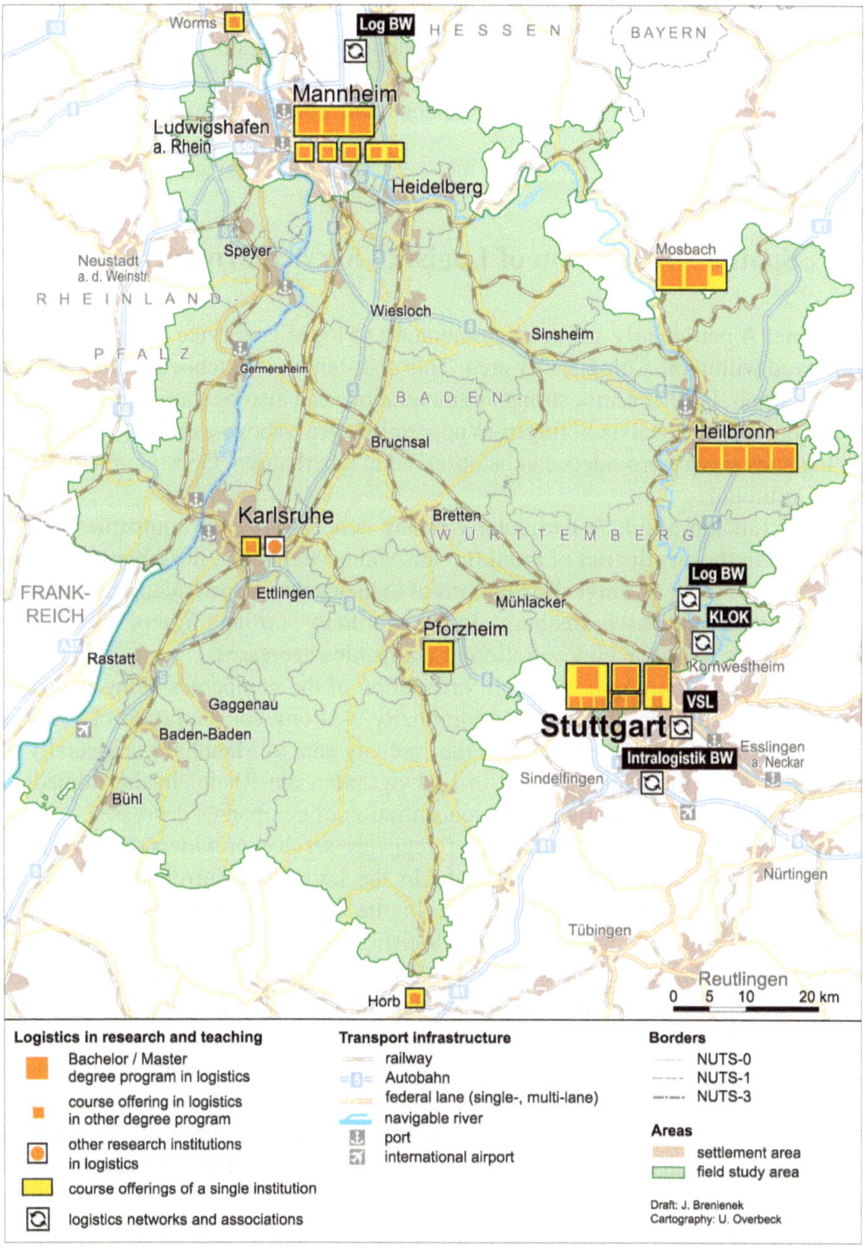

Fig. 5 Logistics in research and teaching in the areas of Karlsruhe, Mannheim and Stuttgart

- The availability of bachelor and master study programs with logistical orientation is an important indicator for the generation of a labor force for logistics companies.

- Collaborations of logistics companies, universities and research institutions are important factors for the continued existence of a logistics cluster in terms of knowledge transfer and innovation.
- Collaborations between industry, science and politics should be initiated and promoted.

14 Spatial Distribution of Logistics Service Providers

Cologne: A particularly strong agglomeration of logistics service providers can be observed within the Cologne city area. This distribution is extended to the southern city of Bonn. Further north, strong agglomerations can also be found in Düsseldorf, Duisburg and on the axis Solingen-Wuppertal-Hagen. Above-average distributions of logistics service providers can be observed in Oberhausen, Gelsenkirchen, Essen and Dortmund.

Frankfurt am Main: A particularly strong concentration of companies can be observed in the city district of Frankfurt am Main, extended to the direction Mainz/Wiesbaden. Above-average distributions of logistics service providers can be found in Darmstadt and Aschaffenburg. In the northern surroundings of the city of Frankfurt am Main the concentration of companies decreases.

Mannheim: A strong north-south orientation of the companies along the transport infrastructure can be found. Additionally, a strong concentration of logistics service providers can be observed in the core city area of Mannheim, especially on the left side of the Rhine located in Ludwigshafen am Rhein. In the immediate western and eastern hinterland, the concentration of companies decreases.

Karlsruhe: Strong concentration of logistics service providers in Karlsruhe, above-average concentration in Rastatt. In the immediate surrounding area, the concentration of companies decreases. In the immediate metropolitan area of Stuttgart a below average amount of logistics service providers can be found. Large concentrations can be observed in the northern towns of Ludwigsburg and Kornwestheim. The proximity to intermodal facilities is a possible explanation. At a distance of about 10 km from Stuttgart the concentration of companies decreases.

Nijmegen: The urban agglomerations of Nijmegen and Arnhem, and the towns of Tiel and Oss have the highest concentrations of logistics service providers. This also applies to the corridor along the highway A15 and along the Betuwelijn. Furthermore, it can be observed that locations close to the border are major logistics locations.

Milan: The distribution of logistics service providers is oriented along traffic routes and major logistic infrastructure such as airports and inland terminals. The structures grew endogenously.

Turin: The logistics service providers are mainly concentrated along the roads and in close proximity to the bimodal terminal and the airport Torino Caselle. Furthermore, a slight distribution in the direction of Lyon along the highway A 32 can be observed.

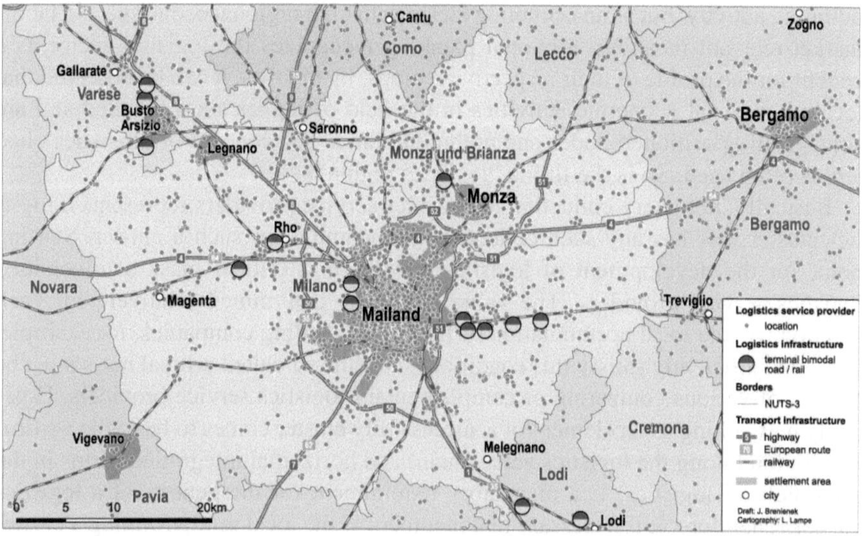

Fig. 6 Logistics service providers in the area of Milano, selected by branches 5210, 5224, 52291, 5310, 5320 and 4920, 4941, 5020, 5040, 5121, 52215, 52222, 52231, 52292, 52299 (data based on Camera di Commercio Milano 2013)

14.1 Summary and Practical Application

- The distribution of logistics service providers are shown in Fig. 6 (excerpt from the case study of Milan).
- The investigated logistics service providers are in close proximity to traffic routes.
- A high density of logistics service providers can be found in the respective core region of the case study area (city district), especially within a 5 km radius.
- Beyond the strong concentration within the core area, companies in the logistics sector show a disperse distribution.
- The selected locations of logistics service providers and other companies operating within the logistics sector should be involved in long-term monitoring in order to respond to changes in a prospectively-planning manner.

15 Practical Application and Implementation: Economic Promotion, Regional Planning, Cluster Management

In regional economic promotion and land use planning, bundled regional activities in the logistics sector gained importance in recent years. In the economic development, the logistics industry is no longer unilaterally recognized as traffic and polluting activity, but as an economic factor with high regional economic and labor-market-relevant meaning. Regional planning recognizes the logistics sector as a benefit for the profile of their respective regions. In addition, it can be observed that local or regional economic activities in the field of logistics are bundled and are strategically positioned under one umbrella brand. The concept of "logistics cluster" is often used. So is the titling "logistics region".

Basically, not every collection of logistics service providers represents a logistics cluster, nor has any such collection the potential for such a cluster. Starting point for the development of logistics clusters is a critical mass of diversified logistics service providers. The value has to be determined relatively for each location. A sole local accumulation of freight forwarding companies, for example, does not sufficiently satisfy this condition, since the so-called critical mass must be formed of various competing or complementary logistics service providers. Especially networking as an element of a successfully cluster comes to bear: the vertical integration (along the logistics value chain) and horizontal integration (firms in the same field of logistics) is a mandatory requirement for the genesis of a logistics cluster. Therefore, it is necessary that members of the local group forming a cluster are not only in the field of business enterprises, but rather relevant institutions, associations, local governments, public-private partnership organizations and research institutions.

The following facilities are necessarily involved in the development of a logistics cluster:

- Local and regional economic development agencies
- Chambers of industry and commerce
- Business associations
- Municipalities (staff positions, planning department, economic promotion, any existing specialist agencies)
- Universities and colleges (with appropriate logistics orientation in research and teaching)
- Professional (specialized) schools with training programs in logistics
- Outstanding local or regional actors or personalities with special relationship and/or experience in the field of logistics

Another key element for the development of a logistics cluster is an existing local or regional level of equipment in (transport) infrastructure. The logistics sector still needs—despite a major focus on telecommunicative networks—a top-class network of facilities with so-called hard location factors in traffic and infrastructure with existing and objectively verifiable advantages of geographical (not topographic) location.

The central control element of each cluster—and therefore also of the logistics cluster—is the aspect of networking: This is fundamentally accompanied by the regional restriction (small scale), so that a regional delimitation (but with demand-driven opening possibilities) for the logistics cluster space is necessary. Within this spatial limitation, the networking of actors, based on contact relationships and trust, plays a central role. Companies in competition with other commercial enterprises must first be aligned on one target, sometimes in a rather prolonged process, in order to gain synergistic benefits. Of course each company should be able to maintain critical internal information while they participate in the process. It is important that the respective actors of each company originate from the highest possible hierarchy level of enterprises or institutions and actively join in the process. The delegation of tasks to middle-levels of the hierarchy (these are mostly operationally executing and without their own decision-making) are often indicators of a lack of will to form a cluster structure.

A promising step in this process is an identification of a suitable speaker or promoter with a high (personality) awareness level and a broad political, economic and social acceptance. This step is hardly paid attention. Instead, a so-called cluster manager (at an early stage of cluster development) is used as operational and executive body. This, too, can often be assessed as an element of early failure in the cluster genesis, since strong and relevant decision-makers want to see administrative tasks bundled in such a person. The role of the cluster manager will become more significant in the maturity phase of the cluster formation process, when trust, networking and strategic direction lead into concrete implementations together with an increased amount of organizational work. During that phase, it appears particularly useful when an external cluster manager is involved, with experience from other cluster genesis processes, to accompany the regional developments and to provide critical comments based on his knowledge of other cluster development processes.

The overall conclusion is that logistics clusters cannot serve as a panacea for a regional profile. The substantive filling of the term and an appropriate logistics cluster development with focus on long-term goals can only be successful due to the analysis of the empiric as well as the "soft" starting conditions.

References

Elsner W, Hübscher JA, Zachial M (2005) Regionale Logistik-Cluster. Statistische Erfassung, Stärken und Schwächen, Handlungspotentiale. Peter Lang, Frankfurt am Main

Europäische Kommission (2001) Weissbuch—Die europäische Verkehrspolitik bis 2010. Amt für amtliche Veröffentlichungen der Europäischen Gemeinschaften, Luxemburg

Marshall A (1890) Principles of Economics, vol I. Macmillan, London

Port of Rotterdam (2012) Hafenkompass. Hafenperspektive 2030. Deutsche Übersetzung der Zusammenfassung des Fortschrittsberichts 2011

Porter ME (1998) The competitive advantage of nations. Free Press, New York, NY, 1st edition 1990

Porter ME (2000) Location, competition, and economic development: local clusters in a global economy. Econ Dev Q 12:15–42

Rivera L, Sheffi Y (2012) Logistics clusters in the US: an empirical study. Draft

Rodrigue J-P, Comtois C, Slack B (2009) The geography of transport systems, 2nd edn. Routledge, London, p 57f
Stölzle W, Hofmann E, Wessely P (2009) Logistikcluster region Basel: Ergebnisse der SWOT-Analyse, Bewertung der Stärken und Schwächen sowie der Chancen und Risiken des Logistikstandortes Basel—final report, Phase I, St. Gallen

Rudolf Juchelka is since 2007 a full professor for Economic Geography, esp. Transport and Logistics at Duisburg-Essen University. In research and teaching he is dedicated to the field of economic space and locations of commerce, spanning from a local to regional and to a global scale while exploring spatial structures, functions and interdependent complexities. He owns the single one Geography Chair in Germany incorporating a specific focus on transport and logistics into the economic geographical research.

Julian Brenienek did his state examination in English (TU Dortmund) and Geography (Uni DUE). From 2010 to 2014 he was research associate in the Interreg-Project Corridor Development Rotterdam-Genoa at the Department of Geography at University of Duisburg-Essen. As project officer of Action 7 he was in charge of developing a research approach for the topic of logistics clusters and its application in case studies along the corridor no. 24. He is member of the German Association of Applied Geography (DVAG) and the Essen Society of Geography and Geology (EGGG).

Linking Terminal Ports to the Hinterland

Alessandro Africani, Roberta Delpiano, Antonello Fontanili, Andreas Deutsch, and Johannes Kohlschütter

Abstract The purpose of this article is to show a simplified but effective reading of Genoa's and Rotterdam's port performances and their potential hinterland economic power, depicting port catchment area size at the NUTS 2 and NUTS 3 levels. Developing a multi-criteria approach, a proper methodology is designed to produce a Port Performance Indicator (PPI) as a synthetic tool to assess port performance at each end of the CODE24 axis with reference to a panel of Mediterranean ports (La Spezia, Marseille and Barcelona) and a panel of North European ports (Antwerp, Bremen and Hamburg), where the two ports of Genoa and Rotterdam are placed, respectively. The analysis demonstrates how the combination of the rail and road transport mode may extend the hinterland's potential and its competitiveness over both the short and the long haul. The strength of the hinterlands of Genoa and Rotterdam is the extent of their intensive industrial and commercial activity and their dense infrastructural network.

Today, the greater flow along the corridor is southbound, while considerable northbound capacities are available. Further analyses were put in place to check for suitable conditions to render the setting up of a shuttle linking the Ligurian ports to the European intermodal network feasible. Due to their geographical location, transport chains via Ligurian ports can have time, flexibility and environmental advantages compared to other transport chains.

1 Port Hinterland Analysis

For a long time, the analysis of ports has been based on gathering quantitative data, referred to as 'throughput traffic', on the availability of infrastructure facilities, e.g. yard areas, cranes, berths, etc., or other minor performance indicators (de Langen et al. 2007). However, these are not at all useful for carrying out a

A. Africani (✉) • R. Delpiano • A. Fontanili
Uniontrasporti scarl, Via E.Oldofredi, 23, 20124 Milan, Italy
e-mail: africani@uniontrasporti.it; delpiano@uniontrasporti.it; fontanili@uniontrasporti.it

A. Deutsch • J. Kohlschütter
TransCare GmbH, Danziger Straße 14, 65191 Wiesbaden, Germany
e-mail: a.deutsch@transcare.de; j.kohlschuetter@transcare.de

H. Drewello, B. Scholl (eds.), *Integrated Spatial and Transport Infrastructure Development*, Contributions to Economics, DOI 10.1007/978-3-319-15708-5_11

comprehensive evaluation, since they do not fully depict the role a port plays in the region where it operates.

In order to provide the true economic impact of a port, the analysis should go further and ought to include other dimensions. This is not an easy task, since the data required are different and more comprehensive, given the economic dimension involved.

1.1 Port Hinterland Definition: The Methodology

The methodology identified as suitable for the port hinterland is related to the concept of generalised cost of transport (GC henceforth) and the use of the freight transport model TRANS-TOOLS.[1]

In detail, the evaluation considers road and rail modes.[2] Each mode has a type-specific catchment area that is concentric with the others; their borders depend on travel time and GC functions.

Output derived from the model is useful to determine mode-specific catchment areas, the share of traffic attracted by the ports from their neighbouring areas and to/from districts along the Genoa–Rotterdam transport corridor.

1.2 Genoa and Rotterdam Hinterlands

The TRANS-TOOLS model has returned the following variables regarding both road and rail modes and container traffic:

- Distance in kilometres
- Travel time for the main mode (FreeTime for road and OnBoard for rail)
- Travel time to ancillary modes (FerrySailing[3] for road and Access/Egress for rail[4])
- Travel time for crossing borders (BorderCrossing for both modes)
- Fees paid in €/ton

[1] TRANS-TOOLS is a European transport network model, developed in a research project co-funded by the European Commission within the sixth Framework programme activities. TRANS-TOOLS is similar to a traditional four-stage model that simulates both passenger and freight transports; it has a modular structure that includes the Computable General Equilibrium economic model, which produces as output an estimate of impact on the GDP of a given region of the policies implemented in the scenario analysis, and a module able to calculate energy consumption, emissions, external costs, accidents according to the output of the assignment model. Output of version v2 was used.

[2] A third mode for Rotterdam, with respect to the inland waterway, is not considered in this study.

[3] When a section by sea is required.

[4] *Feeder* section.

Table 1 Relationship between GC thresholds and distance from ports by mode (average distance ranges)

Road	Distance (km)		Rail	Distance (km)	
Thresholds (€/ton)	Genoa	Rotterdam	Thresholds (€/ton)	Genoa	Rotterdam
0–60	0–260	0–310	0–30	0–320	0–290
60–100	260–440	310–490	30–50	320–520	290–510
100–150	440–670	490–785	50–100	520–1060	510–1080
>150	>670	>785	>100	>1060	>1080

Source: Author's elaboration on TRANS-TOOLS data

- Road transport cost (driver and other perceived VOCs[5]) in €/ton
- Rail transport cost in €/ton

Based on these data, the port hinterland is defined with respect to GC thresholds,[6] which show classes of distances to the ports. A last step before the calculation of GCs entails a set of suitable time-values. Once the GCs for both modes were obtained, they were sorted according to their increasing magnitude.

The ranges of distance from the port are functional and reasonably representative of GC thresholds. In detail, the analysis identified four ranges and for each of these determined the average, minimum and maximum values (Table 1).

Based on this methodology, hinterland dimensions were defined for the two ports of Genoa and Rotterdam.[7] To get a homogeneous measure of the port hinterland, this relationship was defined with a 60 €/ton threshold of GC for road transport and a 30 €/ton threshold of GC for rail combined transport, which corresponds to the 'inner' hinterland.

In the NUTS 2 partition, the road hinterland of the port of Genoa corresponds to the northwestern Italian regions (Liguria, Lombardy, Piedmont and Valle d'Aosta), while the rail mode includes a larger portion (Liguria, Lombardy, Piedmont, Valle d'Aosta, Emilia Romagna and Tuscany), reaching as far as the Canton of Ticino in Switzerland. As to the port of Rotterdam, both its road and rail hinterlands encompass the Netherlands, Belgium, the bordering regions of Germany and the district of Calais in France.

A more in-depth analysis was finally dedicated to the hinterlands of Genoa and Rotterdam ports by applying another network model that allows elaboration at the NUTS 3 level: the TRACC (Transport Accessibility) model.

This model was developed as part of the ESPON[8] TRACC project, which had the objectives of (a) gathering and updating the results to be found in existing

[5] Vehicle Operating Costs.

[6] Suitable ranges have been selected arbitrarily.

[7] This methodology was applied to evaluate Cluster 4, related to the port hinterland performances. The catchment area dimension was defined for all eight ports that make up the selected sample of this analysis (four ports in the Mediterranean Sea Range and four ports in the North Sea Range, including Genoa and Rotterdam).

[8] European Observation Network for Territorial Development and Cohesion.

studies on accessibility at a European scale, (b) extending the range of accessibility indicators by adding further indicators that respond to new policy issues, (c) extending the spatial resolution of accessibility indicators and (d) exploring the likely impact of policies at the European and national scales to improve global, European and regional accessibility in the light of new challenges, such as globalisation, energy scarcity and climate change. The model, implemented in the MEPLAN context, aims to compute travel time and costs for freight transport. Times and costs are therefore used to estimate an accessibility index throughout the European area.

Inter-modality is explicitly dealt with in the network model. Rail for unit freight (combined transport) is simulated based on the probable use of a truck feeder to reach the closest intermodal terminal. The road and rail networks are derived from the TRANS-TOOLS model. The location of the intermodal terminals is added, based on the information included in the RRG database, which includes various European sources.

The model calculates the time and cost as a function of parameters associated with the network links, e.g. tolls, loading times, and/or the different modes of transport, e.g. variable operating costs.

For each mode of transport (road, rail, inland navigation, maritime, air), the model computes the shortest path between each pair of zones. The shortest haul is defined on the basis of generalised time, defined as the sum of travel time, plus cost, divided by the value of travel time. The algorithm used is stochastic (Stochastic User Equilibrium) and allows the assignment of part of the application to sub-optimal paths. The Origin/Destination time and cost are the average of the various hauls used. The model is without capacity constraints, but link-speed is exogenously set to provide realistic speeds rather than free-flow speed. The cost of the non-road modes derives from a cost function that encompasses a fixed cost and a variable cost (function of the distance). These functions were obtained from the parameters included in Trans-Tools and adapted in order to reflect the competition between the modes of transport according to different distance ranges. The fixed cost share is higher for modes used mainly on longer distances and lower for the others.

A high density of intermodal terminals may be noted, especially around Milan (see Fig. 1) with respect to the Genoa hinterland, and in the Netherlands' main logistic areas as well as Belgium and Germany with respect to the Rotterdam map (see Fig. 2). It is clearly evident why these regions represent a high potential for each of the two ports considered.

Other considerations were drawn from the analysis that related to the overall GDP of the region, within the lower threshold set and in terms of GDP per inhabitant. The elaboration showed clearly that, among the reasons for the success of these two ports, both leaders in their respective ranges, the economic dimension of the hinterland plays a major role.

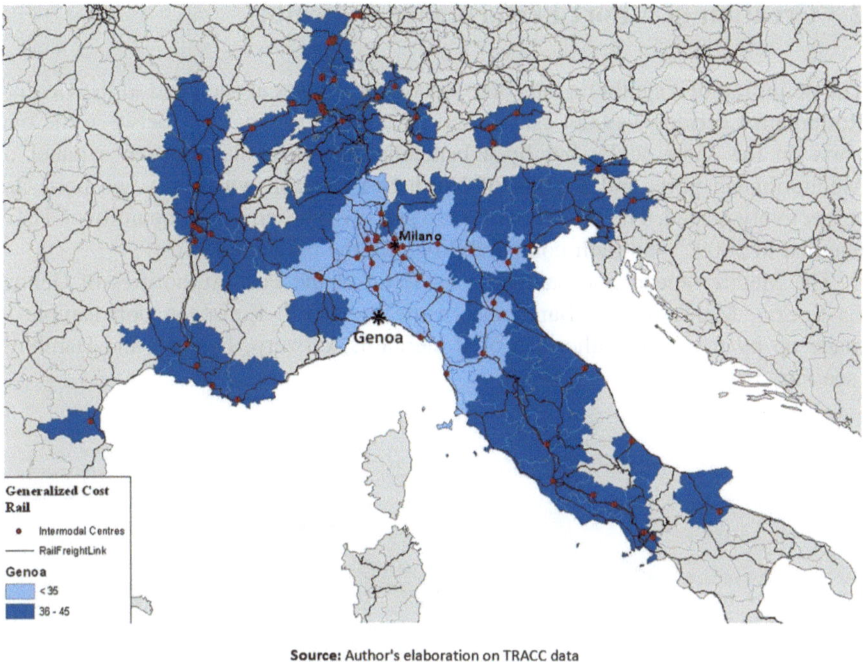

Fig. 1 Example of TRACC output: Genoa rail combined hinterland (35 and 45 €/ton GC threshold). *Source*: Author's elaboration on TRACC data

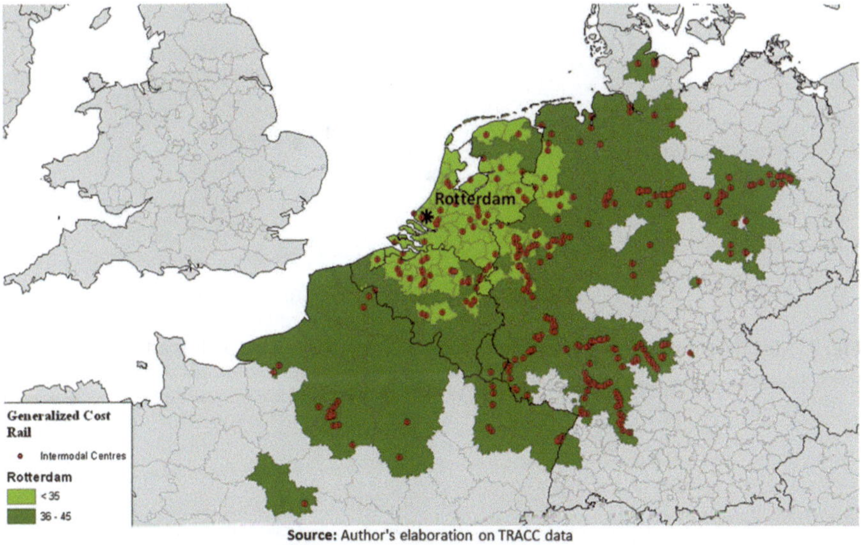

Fig. 2 Example of TRACC output: Rotterdam rail combined hinterland (35 and 45 €/ton GC threshold). *Source*: Author's elaboration on TRACC data

2 Creation of a Port Performance Indicator

The definition of a synthetic indicator to assess the ports at each end of the Genoa–Rotterdam Corridor was not established by a direct comparison between the two ports on the same scale, but by a benchmarking analysis between a panel of Mediterranean ports and a panel of North Sea ports in which the two ports are respectively placed. The selection of these ports was based on the quantity of TEU[9] handled in 2009[10] and their comparability with the ports of interest.

The Mediterranean port sample is made up of the Port Authorities of Genoa, La Spezia, Marseille and Barcelona; while the North European port sample is made up by the Port Authorities of Rotterdam, Antwerp, Bremen and Hamburg (see Fig. 3).

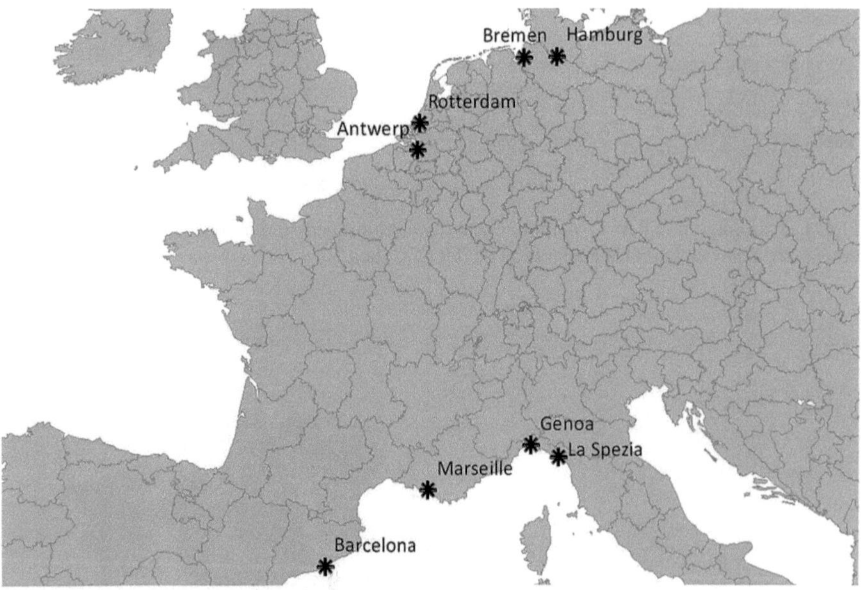

Fig. 3 Mediterranean and North Sea port samples. *Source*: Uniontrasporti scarl

[9] Twenty-foot equivalent units.

[10] The study was concluded in September 2011.

2.1 The Methodology at a Glance

For the purpose of the analysis, a multi-criteria approach has been chosen and five CLUSTERS have been considered in order to take a qualitative and quantitative measure of these ports' performances.

Cluster 1 Port size concerns the dimension of ports, typically in terms of *throughput*, direct employment, and vessel size allowed to enter the port and the maximum draught.

It is important to notice that the dimension is not itself a standalone measure of the port's efficiency; smaller ports may treat less volume, but may create higher added value and greater local employment. All ports, both large and small, are of great relevance and this relative importance is not necessarily determined by turnover. The main problem for the small- and medium-sized ports, as well as for the larger ports, is its port–hinterland connections. Various degrees of congestion exist, depending on the capacity of the hinterland infrastructure, which is why this aspect is considered specifically in Cluster 4.

Cluster 2 The efficiency of container terminals represents the core of port activity. Commonly, the performance of ports has been associated with attempts to calculate and improve, and, where possible, to optimise the operational productivity of cargo handling at the berth, with the particular focus, in recent years, on the container yard. Even within a container terminal itself, there are a very large number of variables that may be taken into consideration. The choice of considering only the container category is also linked to the role of ports in relation to the Genoa–Rotterdam Corridor and also to the type of cargo that is suitable for forwarding along the same Corridor. To analyse container terminal efficiency in the eight selected ports, the most important terminal has been considered for each of them in terms of capacity, number of containers handled and its equipment.[11]

Cluster 3 The use of modern technology has had a great impact on the productivity and efficiency of port operations, reducing times and costs. Significant progress has been made in recent years, thanks to technological innovation processes found in the maritime and port industries and improvements in the organisation and administration of ports. In order to cut down on transport costs, container traffic requires ultra-large containerships and thus terminals with facilities and technology suited to handle them and a maritime transport system, leading to a reduction in transport costs. This cluster has been evaluated, taking into account the basic computerisation of services, ICT integration (within the port, for operators and by the customs authorities), identification technology (access to the port,

[11] Voltri Terminal Europa (VTE), Genoa; La Spezia Container Terminal (LSCT); Terminal de Contenidors de Barcelona SL (TCB); Terminal à conteneurs de Fos, Marseille; ECT Delta Terminal, Rotterdam; Terminal MSC Home, Antwerp; HHLA Container Terminal Burchardkai, Hamburg; Terminal Eurogate Bremen.

tracking and security), the more recent sophisticated facilities, such as the availability of shore-to-ship power supply.

Cluster 4 Port hinterland connections are a basic element in the smooth running and development of a terminal since they guarantee the inland transport of goods to and from the port. Road, rail, and river connections can also connect a port with specialised cargo hubs located in suitable inland depots. These stations may also serve to smooth over the peaks in demand and the supply of goods to any port having limited storage areas.

It is a very hard task to produce a one-size-fits-all definition of what constitutes a port hinterland. A lot will naturally depend on geographical location, type of commodity (bulk freight versus containers), seasonal impact, economic cycles, technological changes, transport policy changes or the availability of feeder transport modes. In addition, an evaluation of this cluster implies the need to draw, in a systematic way and with barely sufficient detail, the borders of the geographical catchment area of all eight ports involved in this study. To do so as objectively possible, and not arbitrarily, a proper transport model tool has to be considered, as stated in the previous section.

Cluster 5 Port development strategies is a cluster that gives a framework for efficiently handling the increasing freight volumes expected at the ports. It analyses the aspects related to the investments in port infrastructures, modern technology, and training. Furthermore, it includes a greater respect of environmental policies and sustainability and the level of managerial autonomy of Port Authorities. Governance factors play an important role in the performance of ports, but these are certainly not the only ones and, perhaps, are not even among the most important elements. The financial capability of a port authority is one of the key factors that determine the extent to which it can achieve its objectives and perform its functions.

Given the arguments presented above, an analysis of the current situation of the two ports and their hinterlands requires the use of a large number of variables. A proper selection is therefore mandatory and it has to be noted that data are not always comparable or standardised across countries and time, since they are gathered by various institutions using different methods. After a first scan of the existing literature, a survey was carried out with Port Authorities in both ranges as the only effective way to understand whether existing data are reliable and retrievable, and, in addition, to identify the cause and extent of existing problems and how they might be monitored.

Altogether, 30 elements and indicators were selected and divided among the five clusters, and every item had a weight assigned in order to establish the level of relative importance among the clusters and between the single elements of each cluster. Each element was assigned a score, which was based on a scale of values. The scores are from 1 to 5, depending on the units of measure and the thresholds linked to each element. For qualitative elements (Modern technology and Development strategies clusters), the scale provided a qualitative judgement.

The thresholds were defined according to the characteristics of each element, considering the relevance of the threshold values on the basis of literature review and on effective observations.

The scoring, together with the relative weighting for each element and cluster, allowed the final calculation of the total scores for each port, which correspond to the value of the performance index for each port, both in respect to each cluster and in respect to all the clusters (synthetic Port Performance Indicator).

2.2 *Application of PPI to Genoa and Rotterdam*

The analysis is different for the two ranges, Mediterranean and North European ports, as expected, considering the different characteristics and performances between the two ranges.

The overall objective is a synthetic evaluation of the ports' performance and therefore an identification of the extent to which ports may improve their efficiency and the leverages that might be exploited.

The sensitivity analysis, finalised to verify and corroborate the results and validate the multi-criteria analysis, was finally applied to the synthetic Port Performance Indicator. Sensitivity was also used to stress the overall results in two directions. On one hand, one cluster was switched off (not considered), on the other, it was weighted by applying a double weight with respect to the reference situation (doubling). In both cases, the weights of the other clusters were re-sized accordingly.[12]

The main considerations from all the results are summarised as follows:

- The strongest difference between the two port areas, besides the Port size cluster, is represented by the performances in the Port hinterland, one of the excellence aspects for all the North European ports.
- In northern Europe, the port of Rotterdam gets the best results in almost every cluster, while, in the Mediterranean area, the best cluster ranking is distributed among all the ports: Genoa comes first for its use of modern technology and its port hinterland, La Spezia for container terminal efficiency, Barcelona for development strategy and Marseille for port size.
- Some port's performances are more balanced among the clusters, while others have higher variation between clusters, i.e. Marseille and Hamburg.

[12] For more in-depth insights, please refer to the final publication *Port assessment tools: hinterland definition and performances analysis* by Uniontrasporti, September 2011.

2.2.1 Genoa and the Mediterranean Ports

The port of Genoa is the leading port in the Mediterranean range. Setting an ideal satisfaction threshold at a score of 3 out of 5, the port of Genoa shows a higher score for all the clusters, thus indicating that the overall port performance is the result of a homogeneous development trend. The clusters that contribute most to this result are those related to Modern technology and Port size.

Other remarkable results for the Mediterranean ports were:

- The high terminal efficiency recorded by La Spezia, which is the result of a great effort to exploit the full capacity of the container terminal, even when faced with a scarcity of land.
- Both Barcelona and Marseille had a relatively low hinterland score. For Barcelona, this could be explained as a consequence of its peripheral position and to the relative weakness of the Spanish economy over the last few years. For Marseille, it is due to the distance from the wealthier French regions of the Rhone Alps and Ile de France.

Calculating the arithmetic mean of the cluster values for all four ports in the Mediterranean range, an average value of the synthetic Port Performance Indicator was obtained, equal to 2.86 (see Fig. 4). This value is mainly influenced by the poor performances of clusters related to Development strategies, Port size and Container Terminal efficiency. With respect to this synthetic value, the port of Genoa is 0.35 higher, the port of La Spezia is just 0.02 higher, while the Barcelona and Marseille ports are slightly lower.

The Development strategies cluster is penalized by the deficiencies in the autonomy of the port authorities, which plays an important but limited role in the

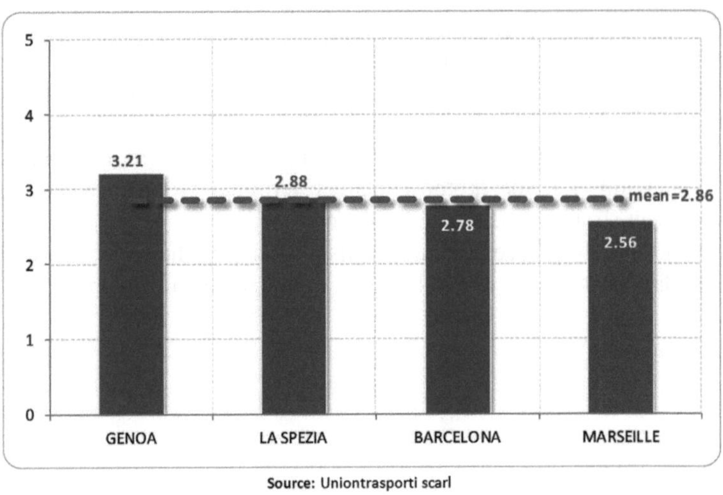

Fig. 4 The synthetic port performance indicator for Mediterranean ports sample. *Source*: Uniontrasporti scarl

Mediterranean range. The lack of financial autonomy for the Italian port authorities, whose governance fundamentals still depend on the 1994 law, is a well-known fact. In recent years, several initiatives to improve this law have been discussed, but without arriving at any significant changes; this may be included among the reasons for the gap in evaluation compared to the North Sea range ports.

Focusing on "*the best*" and "*the improvable*" cluster (see Table 2) is obviously a simplification, but used here to get a quick impression of the port performances, one by one. In actual fact, the remarks give a more exhaustive picture of the specific context by highlighting points of excellence and improvement opportunities for each port.

Focusing on the port of Genoa results, some specific remarks may be made regarding the individual elements of each cluster:

- Port size: The cluster reaches a very good level and the only element with a moderate score is the employment sector measured by the element "Direct employees" (however, in the Mediterranean area only Barcelona gets a higher score).
- Container terminal efficiency: The cluster needs improvement in order to get a higher final score (this cluster weight is the most important as it deals directly with efficiency). In particular, container terminal productivity (in terms of

Table 2 Mediterranean ports—best and improvable clusters

	Genoa	La Spezia	Barcelona	Marseille
Best cluster (score)	Modern technology (3.90)	Container terminal efficiency (3.49)	Development strategies (3.58)	Port size (3.40)
Improvable cluster (score)	CT efficiency (3.05)	Modern technology (2.00)	Port hinterland (2.13)	Port hinterland (1.65)
Main remarks	Genoa comes first with regard to the synthetic indicator, with excellent performances in Modern technology and good results in Development strategies, Port size and Port hinterland. CT efficiency is its worst performing cluster, even if it is the second best result after La Spezia.	La Spezia has a very good performance in CT efficiency and a good result in Port hinterland. Improvements could be made, in particular, in Modern technology, currently the worst performer in the Mediterranean area.	Barcelona gets a very good result regarding Development strategies and performs well in Port size and Modern technology. The main improvement effort could be made in extending its hinterland.	Good results for Marseille in Port size and Development strategies. Modern technology and Port hinterland are the clusters that need improving. Its ambitious development plan may bring the necessary benefits to Marseille in the future.

Source: Uniontrasporti scarl

TEU/m^2) may be improved; the score is lower than its competitors in the Mediterranean ports, because of their larger work-surface, while for special areas, such as cranes, equipment and quay crane productivity, Genoa is in line with the other Mediterranean ports.[13]

- Modern technology: The cluster gets the highest score among all the clusters and Genoa distinguishes itself for its very high level of technology, i.e. equipment, integration, and innovation. The only element with room for improvement is that the environmental measure expressing the availability of the shore-to-ship power supply is missing.[14]
- Port hinterland: With a very good final result, the best in the Mediterranean area, the improvable elements are related to the modal split: The rail share is only 14 %. The difficulties of managing rail transport from Genoa are well known. The recent agreement with the inland terminal in Rivalta Scrivia and the connections thus managed, together with the shunting operator, could mean an improvement in the near future.
- Development strategies: A very good result (after Modern technology), even if lower than Barcelona and Marseille. To close the gap, an effort is needed in managerial strategy to ease investments in infrastructure and environmental policies.

A further interesting performance target for Genoa could be reaching the minimum score of the North European ports, e.g. Bremen's score, which would mean an increase of 17.4 % of its total value. A combination of a more dynamic development strategy, together with further growth of CT efficiency, and improvement in its connections with the hinterland might, in the short to medium-term, favour this type of performance improvement. It is evident that concentrating the improvement efforts on the most weighted cluster will give the best result, but the same result could be reached acting on the other clusters, leading to small improvements in every cluster. That would be a more realistic strategy, combining the effects deriving from various individual improvements.

2.2.2 Rotterdam and the North European Ports

The overall scores attributed to the North European ports are higher than those attributed to the Mediterranean ones. All the ports' scores are higher than a theoretical satisfactory threshold of 3 out of 5. This may also be seen as a result of an improvement in performance achieved over the last few decades.

[13] The feedback provided by Genoa port operators about the infrastructure equipment inside the port, outlines a prevalence of "inadequate" evaluations concerning: proper availability of parking area for heavy goods vehicles, proper road and rail network in the port and availability of covered areas for goods.

[14] Considering the results of the survey on the Genoa site, general satisfaction was registered regarding the computerisation of services and ICT integration available in the port.

Research, innovation, and political support, planning and investments are all factors that brought the northern range ports to figure among the best port systems in the world. This is reflected in the calculation of the Port Performance Indicator where different aspects, including either the size of the port or their efficiency in dealing with the richer sectors of shipping and logistics, are weighted properly. It is not surprising that a port like Rotterdam ranks number one in almost all the clusters taken into consideration and it may even further improve its performance by, for example, increasing its rail connections with the hinterland. The clusters that contribute most significantly to this result are those related to Port size, Container Terminal efficiency and Modern technology.

Rotterdam is, undoubtedly, port number one in the range, while the other ports show overall scores that are quite similar, which means that competition among them may bring each port to improve its performance to a level that is superior to its current assessment.

Calculating the arithmetic mean of the cluster values of all four ports, an average value of the synthetic Port Performance Indicator can be obtained that is representative of the North European range, 4.02 (see Fig. 5). This value is mainly influenced by the clusters related to Port size, Container Terminal efficiency and Development strategies. With respect to this synthetic value, the port of Rotterdam is 0.46 higher than the other northern ports, which are still higher than the Mediterranean ports average value (2.86).

Table 3 summarises the "*best cluster*" and the "*improvable cluster*" for the North European ports, with the principal observations regarding each individual port.

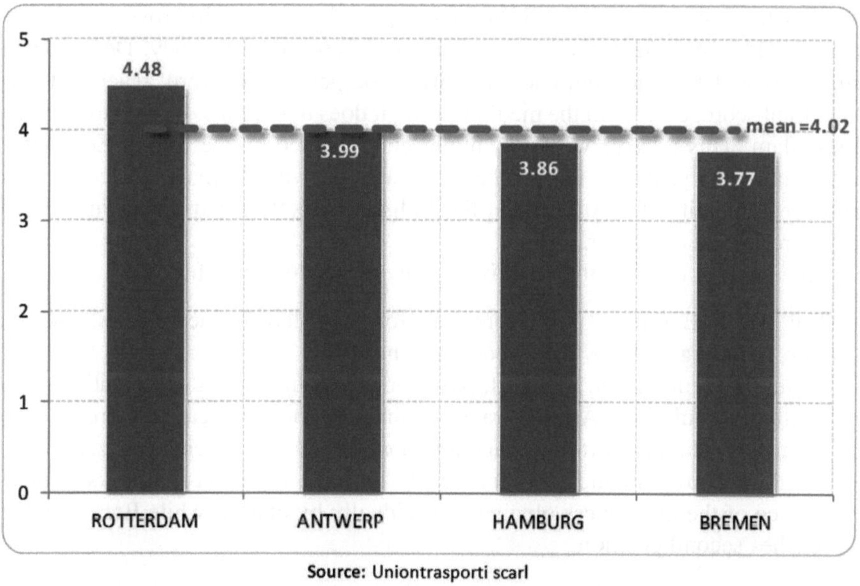

Fig. 5 The synthetic port performance indicator for North European ports sample. *Source*: Uniontrasporti scarl

Table 3 North European ports—best and improvable clusters

	Rotterdam	Antwerp	Hamburg	Bremen
Best cluster (score)	Port size (5.00)	Port size (4.39)	Port size (4.40)	Modern technology (4.00)
Improvable cluster (score)	Port hinterland (3.86)	Development strategies (3.86)	Modern technology (3.45)	Container terminal efficiency (3.53)
Main remarks	Rotterdam, the leading port according to the synthetic indicator, performs excellently in all the clusters. The main area needing improvement is the Port hinterland cluster (second in the ranking, after Antwerp).	Antwerp comes second among the North European ports; its performance is very good in Port size, CT efficiency and Port hinterland. Modern technology and Development strategies are the two clusters that need greater efforts to improve the results.	Hamburg gets very good results in Port size and development strategies (in this cluster, Hamburg is the best port); the main improvements are expected in the Modern technology and Port hinterland clusters.	Bremen is the port with the most homogeneous results, without any particular point of excellence compared to the other ports, but also without any particular weak points.

Source: Uniontrasporti scarl

Focusing again on Rotterdam's results, some specific remarks may be made in relation to the single elements of each cluster. Concerning Port size, this cluster reaches the top score (5) in all the elements and needs no improvement in this cluster. Other excellent clusters are Container Terminal efficiency, Development strategies and New technologies. Its direct competitor, Antwerp, reaches a very high total score, even if, in the medium-term, it does not seem to be able to overtake Rotterdam. This relative weakness might be explained by the role of Zeebrugge, which in recent years has recorded considerable growth and may have come to be another competitor for Antwerp in the Belgian market, in particular, in the segments ro-ro and container.

The main remarks on the sensitivity analysis results are as follows:

- The North European port scores, as for the Mediterranean ports, and the corresponding ranking, show moderate changes.
- Rotterdam remains the best port, based on any combination of weighting
- In all the simulations, Antwerp ranks second, except in two cases where it comes in third (excluding Port hinterland and doubling Development strategies).
- Hamburg shows the most variable results (second, third and fourth) as a consequence of the differences obtained individually by cluster, while Bremen never reaches second position.

For the port of Rotterdam, with an already outstanding result in absolute terms, it might be an interesting exercise to understand how it could attain the maximum score (5), which would mean an improvement of around 12 %. To reach this target, as shown above in the case of Genoa, some strategic initiatives could be suggested, improving every single cluster result, even if, most clusters already perform better than the competing ports. One hypothesis is that improvement in all the clusters (except that of the Port size, which already has the maximum score) with a combination of interventions for each cluster could bring the total target of 5 as the synthetic indicator. This particularly implies significant improvements in Port hinterland; this cluster should increase its performance by about 30 %.

3 Improving Rail Connection Capability from the Ligurian Ports to the Hinterland

Over recent decades, the North Sea ports have emerged as the standard entry gate for containers sent from Asia to Germany and adjacent countries. The Mediterranean ports have only played a minor role in these import volumes, despite the fact that they could offer significantly shorter transport distances for several destination regions. Figure 6 and Table 4 compare two alternative routes for transport from Shanghai to the Upper Rhine region (Basel). In this example, the route via Genoa is approximately 20 % shorter than the standard route via Rotterdam, and it would entail a reduction of lead times by about 4 days.

Excellent international hinterland connectivity is one reason for the strongly competitive position of the North Sea ports despite suboptimal transport distances. For instance, the port of Hamburg has achieved a high share of rail use in its

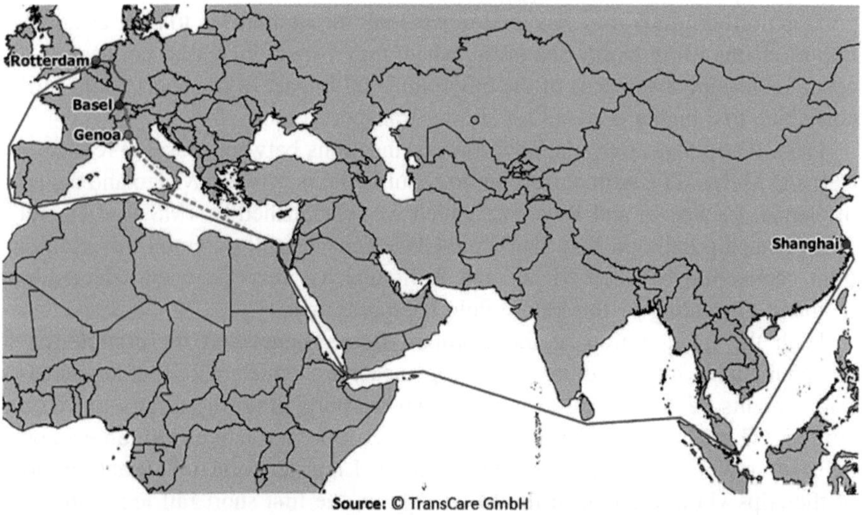

Fig. 6 Comparison of sea-rail connections between Shanghai and Basel. *Straight line*: via Rotterdam; *Dotted line*: alternative Mediterranean route via Genoa. Map created using QGIS/ Natural Earth data. *Source*: © TransCare GmbH

Table 4 Comparison of distances and approximate lead times for a sea-rail transport chain from Shanghai to Basel via Rotterdam and via Genoa

			Distance (km)			Lead time estimate (days)			
Origin	Port	Destination	Sea	Rail	Total	Sea	Port	Rail	Total
Shanghai	Rotterdam	Basel (Weil)	19,364	849	20,213	27	1	1	29
Shanghai	Genoa	Basel (Weil)	15,986	539	16,525	23	1	1	25

Source: Sea Rates LP, DB Schenker (TPS/DIUM), lead times estimated based on Maersk line transit time information, assuming improved transit times for Genoa. (Actual lead times may differ depending on vessel routing (port calls), etc.) © TransCare GmbH

hinterland, thanks to the numerous trains serving its terminals, and it ensures control of rail connectivity through HHLA's affiliated rail transport companies Polzug and Metrans. The port of Rotterdam extends to the hinterland via both rail and barge connections to Europe's largest inland port in Duisburg.

Several steps towards improving the competitive position of the Mediterranean ports concerning international hinterland connectivity have been taken in the past. Northern Italy is one example. In principle, connectivity between the Milan region and the rest of Europe has been at a satisfying level for many years: A 2011 analysis commissioned by the Milan Chamber of Commerce (Dallari et al. 2011) identified 551 weekly trains linking intermodal terminals in the vicinity of Milan to nine other European countries. One possible approach towards improving the international hinterland connectivity of the ports in the region is to consolidate flows to/from the ports in one terminal using shuttle trains and specific train connections from the terminal to the hinterland destination regions. Several operators have established such hinterland concepts based on shuttle trains in the past, but not all of them remained operational for a long time. However, demand for import routes via the ports in northern Italy may rise as shippers look for alternatives to their established import chains using North Sea ports, where they expect increased congestion and costs. Increasing awareness of the environmental impact of transport could further contribute to a higher demand for shorter transport routes.

From that perspective, the new train connections between Basel (Frenkendorf, CH) and Melzo (IT), with corresponding shuttle trains between Melzo and the ports of Genoa, La Spezia and Ravenna, which was established by IMS Rail/Hannibal SpA (Contship Italia) in May 2013 could have considerable potential. An additional train connection between Melzo and Karlsruhe has recently been selected as a recipient of funding by the Marco Polo II programme.

With the inauguration of the Gotthard Base Tunnel and the completion of complementary infrastructure upgrading measures, which will enable increased train weights, hinterland rail transport from the ports in northern Italy to Switzerland and Germany is expected to become ever more attractive from an economic point of view. In a rail transport chain from the Ligurian ports to destinations north of the Alps via a terminal in the Milan region, the first short rail leg is the most likely to face strong competition from road transport. However, a rough model calculation of the estimated costs for a shuttle train linking several Ligurian ports to the terminal in Busto Arsizio/Gallarate (see Table 5) suggests that one-way

Table 5 Rough estimate of costs for container shuttle trains linking the ports of Vado Ligure/Savona, Genoa and La Spezia to the intermodal terminal in Busto Arsizio/Gallarate[a]

Origin port (Routing via Genoa/Pavia)	Distance to Busto Arsizio (km)	Cost estimate before overhead/margin (€)							Costs incl. 20 % overhead + 5 % margin (€)	
		Personnel	Locomotive	Energy	Track access charge	Wagon rental	Shunting etc.	Total	Per train (capacity: 36 containers)	Per container at 80 % utilisation
Vado Ligure	218	230	1179	881	601	555	500	3946	4971	173
Genoa	185	207	1137	760	459	555	500	3618	4559	158
La Spezia	258	258	1231	1069	674	555	500	4287	5402	188

Source: © TransCare GmbH

[a]Train configuration: 18 container wagons (Sggrss 80 type, 6 axles), double electric traction (DC). Energy costs calculated based on train weight of 1.195 ton, maximum speed of 100 km/h, a maximum terrain gradient of 16 % and an energy price of 0.15 €/kWh. Track access charges calculated according to network operator information (RFI). Assumptions on personnel, locomotive, wagon rental and shunting costs: TransCare assumptions/expert information. Handling charges not included

transport prices below 200 € per container are feasible for transport legs of about 180–260 km.

Still, costs and lead times are not the only challenge. Transport providers need to ensure that their connections are reliable and meet shippers' requirements. Winning the support of a major vessel operator, e.g. by volume commitments, could be one important step towards the further development of hinterland transport to and from the Ligurian ports.

4 Conclusions

Defining a port hinterland is a challenging task. The choice of a model-based analysis has led to an approach that is consistent, replicable and adaptable to different scales. The strength of the hinterlands of both Genoa and Rotterdam is the extent of their intense industrial and commercial activity and of their dense infrastructural network.

The Port Performance Indicator (PPI), based on a sound theoretical background analysis, exploited the hinterland definition that was used to gather all the available information for an assessment of the performances of the two ports of Genoa and Rotterdam with reference to their geographical areas: the Mediterranean range and the North Sea range.

Based on a multi-criteria approach, the Port Performance Indicator calculation produced some very interesting results that encompassed several aspects of a port's performance. What emerges is how the strength, size and efficiency of a port are strictly related to its hinterland on one hand and how they, on the other, rely on its capacity to satisfy the demands and the evolution of the market strategically.

Both Genoa and Rotterdam play a leading role within their respective port ranges, thus reflecting a high capacity for innovation and dialogue with their hinterlands. The PPI application also identified possible areas of improvement. This result is broadly confirmed and validated by the sensitivity analysis, performed to highlight the changes in scores, varying the weight assigned to the single clusters. At the same time, the results of the sensitivity analysis showed the weakest point of each port more clearly.

Today, the major cargo flows along the corridor are southbound leaving considerable northbound capacities available. Due to their geographical location, transport chains via the Ligurian ports show, in fact, flexibility and environmental advantages compared to alternative transport chains via the North Sea ports. A rough cost estimate indicates that shuttle trains linking the Ligurian ports to the European intermodal network can be viable. Cooperation with a shipping company is recommended with regard to such shuttle trains in order to be in a position to offer attractive prices and to attract sufficient volumes. To support the development of rail hinterland connectivity, the Italian railway infrastructure should be upgraded for longer, heavier and higher trains.

References

Dallari F, Curi S, C-Log Università Liuc (2011) Il sistema logistico in Lombardia: trasformazioni in atto e scenari evolutivi. http://www.mi.camcom.it/web/guest/archivio-ricerche. Accessed 25 June 2014

de Langen P, Nijdam M, van der Horst M (2007) New indicators to measure port performance. J Marit Res 4(1):23–36

Alessandro Africani graduated in 2001 in Business Administration at the University Carlo Cattaneo—LIUC, he took a Master's Degree in "CFO: Chief Financial Officer". In 2004, he started as researcher in Uniontrasporti. In 2010, as project and finance manager of Uniontrasporti, he took part to the Interreg Project CODE24. Alessandro is a PhD candidate in Business Administration.

Roberta Delpiano graduated in 2006 in Mathematical Engineering at the Polytechnic University of Turin (Piedmont, Italy). In 2007 she gained the Master's Degree in "Transport and Logistics: Systems, Networks and Infomobility" (COREP-Polytechnic University of Turin). In March 2008, she started as researcher in Uniontrasporti and she mainly cooperated to the evaluation of social and economical effects of new infrastructures on territorial and business systems. In 2010, as employee of Uniontrasporti, she took part to the Interreg Project CODE24.

Antonello Fontanili graduated in 1998 in Civil Engineering (Transportation) at Polytechnic of Milan. After a brief experience for the Rinascente Group, he entered into Uniontrasporti in 1999 as a researcher, working—also as Project Manager—for project on seaport, on European Corridors, on trans-Alpine routes, logistics districts, technological innovation applied to transport. Between 2004 and 2009, he held the position of chief operating officer of Uniontrasporti, where is engaged in identifying, planning, organization and management of research and development in Italy and abroad. At the beginning of 2010 he became Director and took part to the Interreg Project CODE24.

Dr.rer.pol. **Andreas Deutsch** is a Senior Consultant at TransCare GmbH, Wiesbaden. With a background in economics, he joined the company in 2006. The focus of his projects is on rail and intermodal transport. In his PhD thesis, which was completed in 2013, he addresses capacity issues and shifting effects in container-based hinterland traffic in Continental Europe.

Johannes Kohlschütter is a Project Leader at TransCare GmbH, Wiesbaden. With a background in science, he joined TransCare in 2010 after a postgraduate program in European studies. The focus of his project work is on rail and intermodal transport.

References

Online Rail Freight Exchange (ORFE): Better Rail Competitiveness and Acceptance

Peter Endemann

Abstract In many European countries, rail freight demand has increased considerably during the last 10–15 years. A closer look at the German case reveals that—given a considerable amount of empty wagons and an untransparent market—there is still some potential to increase efficiency and competitiveness of rail freight and to overcome knowledge and information barriers in the rail freight market.

Embedded in the CODE24 strategy, an online rail freight exchange (ORFE) is a useful instrument to cope with these challenges. ORFE is a market platform where rail freight services are offered and requested on-line.

Besides some initial business and case studies, the market's relevance of such a tool and the requirements of potential users—predominantly shipping companies, transport logistics providers and rail operators—had to be explored. Therefore, a workshop was held with these groups and other experts in 2011. Based on the outcome, potential operators had to be identified.

In September 2012, CODE24 identified two companies who were interested in cooperating with the CODE24 network and developing such an online rail market ready for commercialisation. CODE24 decided to support both companies and emphasize their neutrality. The first platform is "railcargo-online" (http://www.railcargo-online.com), the second platform is "freit-one" (http://www.freit-one.de). The latter uses the software prototype developed in CODE24.

Both platforms are available since October 2013 and were presented at a joint workshop with the INTERREG-IVB programme in November 2013. A preliminary appraisal by February 2014 showed that both platforms are working well and count up to 80 registered companies. This paper uses metric tons throughout.

Preferred citation: Endemann, P. (2014). Online Rail Freight Exchange (ORFE)—Better Rail Competitiveness and Acceptance, Drewello, H. and B. Scholl (ed.), Integrated Spatial and Transport Infrastructure Development. The Case of the European North-South Corridor Rotterdam-Genoa, Springer-Verlag, Cham.

P. Endemann (✉)
Regionalverband FrankfurtRheinMain, Poststr. 16, 60329 Frankfurt am Main, Germany
e-mail: endemann@region-frankfurt.de

H. Drewello, B. Scholl (eds.), *Integrated Spatial and Transport Infrastructure Development*, Contributions to Economics, DOI 10.1007/978-3-319-15708-5_12

1 Goal of this Contribution

This paper has two goals. The first is to briefly consider the relevance of economic efficiency to the freight market by highlighting a number of real-world developments in freight practices and regulation as well as by identifying the barriers to further efficiency gains in the market. It then presents the concept of an *Online Rail Freight Exchange* (English language abbreviation "ORFE") as a means of addressing the deficits in the rail freight market outlined below and helping to realise further efficiencies. To demonstrate the market relevance of ORFE, experts that included potential clients have been inquired. The development and implementation of such an online market for rail freight is part of the EU-project CODE24.

2 Booming Rail Freight in Europe?

Using a Porter's five forces analysis, Wittenbrink (2007) was able to demonstrate that the liberalised rail freight market in Europe has a number of market characteristics in common with other modern transport markets in Europe. Despite relatively high entry costs, the growing demand for freight transport has incentivised many firms to enter the rail freight market. The new market entrants have mainly focussed on the intermodal and unit train segments, whereas the national incumbents still provide single wagon services, as observations from Germany and Sweden suggest (Endemann and Blees 2012; Vierth 2011; Vogt 2011).

At the European and national level, the political framework continues to prioritise rail freight whereas the road freight sector has been target of several restricting measures such as the digital tachograph, new regulation of driving times, rest and break periods (EEC No 561/2006) and the German national motorway-charging scheme for heavy good vehicles. These circumstances have led to a noticeable trend towards increasing rail freight volumes, both in absolute and relative terms, and partially explain the rail boom in Germany and other countries such as the United Kingdom, Sweden or the Netherlands (Fig. 1). This observation holds true even after taking into account the dampening effect experienced across all modes since the financial crisis of 2008 and 2009 (Fig. 1). According to a 2011 survey of German rail logistics companies, almost 80 % were able to achieve pre-crisis freight performance levels in 2010 (Wittenbrink 2011).

However, a closer look at the rail freight performance in some EU member countries reveals that only a few countries such as United Kingdom, Sweden, the Netherlands and Germany have achieved long-term success in the years since liberalisation. In other countries such as France or Italy figures continued to decrease (Fig. 1).

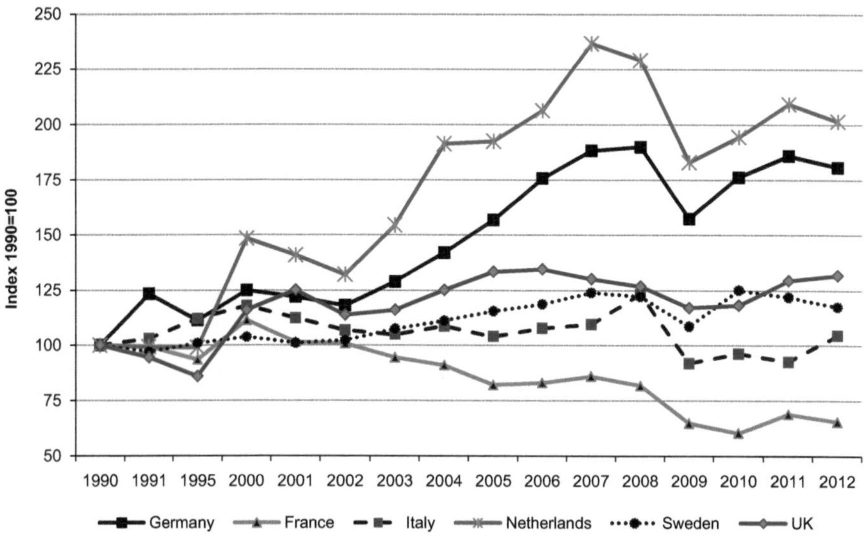

Fig. 1 Evolution of rail freight transport in Europe (in ton-km). *Source*: Eurostat (Retrieved from http://epp.eurostat.ec.europa.eu/portal/page/portal/eurostat/home/, accessed 16 July 2014)

There is widespread recognition that the open market has been responsible for the considerable increase in the number of rail freight operating companies (Jahncke 2008, 10 f). In this respect, Germany, along with Britain, Sweden and the Netherlands, are the countries that have applied the EU policy of liberalisation according to EEC Directive 91/440, most notably with tangible results. In Germany, one effect lies in the considerable market share that has been gained by non-incumbent rail freight companies. In comparison to other European markets such as France and Italy, again the Netherlands, Sweden, United Kingdom and Germany achieve higher shares of transported rail freight volumes the by non-incumbent rail operators (European Commission 2012; Table 1).

Experiences from overseas also stress the positive effect of liberalisation, which can also be observed in the United States of America, where the rail share of ton-km grew considerably from almost 30 % in 1980 to 38 % in the year 2000 following deregulation (Vassallo and Fagan 2007).

Another noteworthy aspect of liberalisation is the number of registered rail freight operators. In July 2014, the German Federal Railway Authority listed almost 372 German based rail freight operating companies having access to the rail network plus another 17 companies with headquarters abroad.[1] According to Vogt (2011), some 100 companies are currently competing nationwide. The other "successful" liberalised market candidates Sweden, the Netherlands and to a lesser extent United Kingdom all show similar results. Larger countries such as France

[1] Information retrieved from the website of the German federal Railway Agency; http://www.eba.bund.de (accessed 15 July 2014).

Table 1 Number of rail freight operators, transport performance and share of non-incumbent rail freight operators (2010–2011)[a]

	Germany	The Netherlands	Sweden	United Kingdom	France	Italy
Rail freight operators	362	17	17	26	4–6	11
Transported goods (bn ton-km)	113	6	23	21	34	18
Market share of non-incumbents (%)	26	31	28	51	20	22

Sources: Eisenbahnbundesamt[b], Office of Rail and Road[c], Deutsche Bahn (2012), Vierth (2011, 2012), European Railway Agency[d], Trafikanalys[e], European Commission 2012, based on enquiries made by the author

[a]According to Vierth (2011) the share may be higher as DB Schenker Rail may be considered as a national incumbent having overtaking the former Dutch incumbent. However, in the United Kingdom the share may be lower since there DB Schenker acts as a private competitor

[b]Now there are 372 licensed companies + companies with headquarters abroad; information retrieved from the website of the German federal Railway Agency; http://www.eba.bund.de (accessed 15 July 2014)

[c]Data retrieved from: http://dataportal.orr.gov.uk/, accessed 18 June 2015

[d]https://eradis.era.europa.eu/safety_docs/licences/default.aspx, accessed 15 August 2012

[e]http://www.trafa.se/sv/Statistik/Transportarbete/, accessed 29 October 2012

and Italy are still confined to a small number of rail operators when taking their smaller market size and a slightly lower rail market share into account (Table 1).

A fully functioning and competitive freight market has yet to be accomplished not only in Europe, but also in Germany. When comparing road and rail systems in Germany—and regardless of the implementation of the road pricing instruments mentioned earlier—there are still some significant imbalances in the market. While rail infrastructure needs to prove its economic performance permanently, road infrastructure is no longer questioned once it has been built (Clausen and Schwarz 2002).

Despite this, the following observations suggest that there is some potential for further increases in rail freight: reasons lie in an insufficient capacity utilisation (Weigand 2009), problem of loadings on return journeys and an overall lack of knowledge regarding the basic workings of the rail freight market. Other issues include:

– The use of existing loading capacities, which is still low in many European countries. In the average, rail freight trains can load 600–750 ton. However, according to Weigand (2009), the actual average loadings comprise 300–500 ton, e.g. French SNCF[2] 305 ton, Italian FS[3] 315 ton, German DB[4] 365 ton

[2] Société Nationale des Chemins de fer Français (French national railway company).

[3] Ferrovie dello Stato (Italian State railway company).

[4] Deutsche Bahn (German national railway company).

and Swiss SBB[5] 513 ton. Consequently, there is a potential of 20–50 % additional vehicle loading capacity.

- Several studies having suggested that there is a potential to shift additional freight volumes from road to rail, especially when better information is available and consolidation of volumes can be achieved (cf. Endemann and Blees 2012; Endemann 2009; Jahncke 2008; Wittenbrink 2008, Allianz pro Schiene 2007; TransCare 2006).
- Many smaller train companies either only operating regionally or not having knowledge of the destination market area in order to organise an additional transport on return (Clausen and Schwarz 2002). They furthermore often lack manpower to undertake marketing.
- Some businesses not being conscious of rail freight as an alternative having adjusted their supply chain management in order to meet road freight requirements (Endemann 2009).
- Many customers having been turned off by the rail system due to the monopoly market situation before starting liberalisation process in 1994.
- Some others had to switch from rail to road when the former national DB Cargo (predecessor of German DB Schenker Rail) introduced its market consolidation programme MORA C[6] in 2001 (Endemann 2009).
- Many shipping companies—amongst other reasons—have not envisaged the rail option as it appeared inflexible, unpunctual and the transit times was still perceived as being too long. Having said this, most of them still have—based on experiences in former years—the one national carrier in mind (Endemann 2009).
- Other shippers arguing against rail transport by evoking overloaded capacities of the rail network. According to Jahncke (2008), there are many good experiences to show that these justifications are—for the most part—no longer relevant.
- Single-wagon transport being highly cost-intensive, e.g. providing infrastructure and staff for marshalling services and maintenance of rolling stock (Vogt 2011).

Consequently, the rail freight market bears some further growth potential, even once the gains achieved under liberalisation are taken into account. This leads to the following central question to be addressed here: What are potential barriers to increase rail freight volumes in the market and how can an information tool help to overcome these barriers?

[5] Schweizerische Bundesbahnen (Swiss federal railway company).

[6] Marktorientiertes Angebot Cargo (market consolidation programme by the German freight rail incumbent company).

3 What is an Online Rail Freight Exchange?

An online rail freight exchange (ORFE) is an interactive online market platform where rail freight services are offered and requested. It thus enables market participants such as rail operators, shippers, forwarders and logistics service providers to better communicate by means of the internet in a user-friendly and interactive way. ORFE is suitable for the spot market as well as to establish new services and services for shippers with or without a private rail siding. ORFE can thus contribute to shift freight volumes from road to rail and to overcome information barriers in the rail freight market identified previously.

4 Objectives of ORFE within CODE24

ORFE is one CODE24 action that has been launched by the Institute for Production und Industrial Information Management (PIM) of the University of Duisburg-Essen, Germany. Other German partners involved in this action include the Regionalverband Ruhr, a regional planning association of the Ruhr Area, TransCare, a rail logistics consultant and the Regionalverband Frankfurt-RheinMain, an inter-municipal association responsible for regional development and land-use planning for the Rhine-Main region, which encompasses 2.2 million inhabitants living in 75 municipalities. Furthermore, a rail logistics service provider was embedded as a sub-partner to the Regionalverband and provides relevant insight on rail operating issues.[7] Some cooperation arose with CODE24-partner Port of Genoa in Italy.

The motivation to participate in the development of this online tool derived from the different activities and the experiences of Regionalverband Frankfurt-RheinMain and TransCare including a number of studies already carried out, as well as significant engagement with market participants in the Frankfurt Rhine-Main Area, as summarised here. One of the key conclusions arising from the studies lack of information on potential clients. In one market study, logistics service providers expressed the need for an online rail freight market that helped them to access to potential clients (Endemann 2009; TransCare 2006). ORFE is primarily designed with the intention to provide this missing information to the market. Bruns et al. (2012, p. 34) stress the need for a market: “The main focus of an online rail freight exchange is the provision of information between buyers and sellers in the rail freight sector via a ‘spot market’, where relatively small volumes of freight,

[7] Sub-partnership has started with Transpetrol, Hamburg and has recently continued with Klaus G. Becker, rail logistics consultant who formerly worked for Transpetrol.

moving predominantly in single wagons and heterogeneous wagon groups, can be bought and sold as freight options [for specific train paths[8]]".[9]

From the common perspective of the project CODE24 partners and the concerns put forward by the stakeholders, ORFE should fulfil the following goals:

- Provide greater transparency of pricing, leading to more competitive price structures and hence an increased competitiveness of rail freight transport
- Overcome the lack of information within the rail freight transport market
- Ensure better access to information for smaller and medium sized rail freight companies
- Ensure better utilisation of existing capacities in order to reduce unnecessary growth in freight rail traffic along the Trans-European Corridor 24, particularly for those municipalities that are subject to adverse effects (predominately noise) arising from freight trains.
- Contribute to the long-term longevity of local rail infrastructure (including rail sidings)
- Contribute to the long-term longevity of the broader rail freight transport network
- Contribute to the reduction of volumes on motorway and road networks

5 Method

A central component of the EU-funded project CODE24 is the examination of the rail freight situation along the aforementioned trans-European transport corridor 24 (TEN-T 24), Rotterdam-Genoa. Though the project's perspective is a transnational one, Germany has been chosen as an example for the analysis of existing deficiencies and the potential for an online rail freight exchange to improve rail freight efficiency. This can be justified on the grounds that Germany is one of the most liberalised rail markets in Europe and exhibits one of the highest levels of competition. Market barriers can thus be explored more clearly and also from different angles. Furthermore, the structure of the companies and the branches used for rail freight is broad. In essence, Germany is the biggest country on TEN-T 24 and a fundamental component of it.

In the market overview, however, experiences from other countries were included. In order to gain a greater understanding of real-world issues in the

[8] A train path is defined as the timing of an allocated movement of a train along a given route. Altogether this creates a timetable.

[9] Original citation: "Im Fokus der Online Frachtenbörse steht die Vermittlung zwischen Transportangeboten und Transportnachfragen des Schienengüterverkehrs auf dem so genannten 'Spot Market', auf dem relative kleine Gütermengen—vor allem in der Größenordnung von Einzelwagen—und Wagengruppenverkehren—gehandelt werden." Bruns et al. 2012, p. 34): translated by the author.

German freight market, an expert workshop was held in March 2011 with 30 delegates from peak bodies such as VDV[10] for the rail operators, IBS[11] for the rail freight forwarders and BME[12] for the shippers as well as various representatives from the research community and the relevant government authorities. The advantage of such an approach was that it allowed for a direct exchange on the pros and cons of the status quo, whilst simultaneously providing the basis for an evaluation of the different criteria required to establish an online tool.

The workshop was facilitated by an external moderator. The following aspects were analysed:

- Current situation of the rail freight market: where do deficiencies in the market exist?
- Market requirements for an online platform: what functionality must the platform offer?
- Elements of such an online freight market: what information must it contain?
- Framework for the business and operation model: how will users interact with the platform?

6 Market Contribution of an Online Rail Freight Platform

The description of the market situation in the first part of this paper suggests that there is some potential to increase the competitiveness of the rail system. The rail freight industry has been the benefactor of a number of significant microeconomic reforms since the early 1990s in Europe, most of which have been chiefly targeted at improving the productivity and profitability of rail operators (cf. Merkert et al. 2010). Furthermore, there are numerous examples of modernisation of rail freight infrastructure and operations, as well as new forms of co-operation designed to increase operational efficiency. Examples include the railport-system by DB Schenker Rail, the Xrail single wagon network established by some central European national incumbents, the establishment of numerous shuttle services to improve the hinterland access of sea ports—especially by new market entrants—as well as the enlargement of combined transport by Kombiverkehr or HUPAC), and more sector-specific rail based logistics services by firms such as Transpetrol or Crossrail. These developments have the potential to encourage further development and ensure rail services for the last mile of journeys.

Germany is well known for its concerted efforts in the areas of environmental, transport and energy policy since reunification in order to try and tackle issues of energy security and climate change. Such policies have also acted as a catalyst for a fundamental rethinking of attitudes towards rail freight (Endemann 2009).

[10] Verband Deutscher Verkehrsunternehmen (Association of German Transport Companies).

[11] Interessengemeinschaft der Bahnspeditionen (Interest Group of Railway Forwarders).

[12] Bundesverband Materialwirtschaft, Einkauf und Logistik e.V. (German Association of Materials Management, Purchasing and Logistics).

A closer look at the German case reveals that—given the tremendous increase of rail freight operators and increasingly restrictive and costly measures for the road freight sector (e. g. digital tachograph, EU–regulation on driving times, rest and break periods and motorway charging schemes for heavy goods vehicles)—there is still significant potential to expand the role of rail freight and meet latent demand. This can be achieved by increasing freight train loads to their capacity (see Sect. 1) and increasing the attractiveness of service offerings against road-based competitors. Herein lies the challenge for the rail freight and its contribution to the overall welfare: continue to boost the image of rail freight, to bundle volumes and increase the use of existing capacities before developing new services whilst decoupling them from a commensurate growth in rail traffic.

Increased productive efficiency, when translated to the freight market, can be thought of in its simplest form as 'moving more with less'. This kind of development in the market has the potential to increase profitability and to contribute to environmental policy outcomes if it is executed correctly and complemented by correct price and information signals to the market. This is best summarised in the OECD's (2006) general objective for the rail sector:

> The general objective of governments with respect to the rail sector is to for the end-user prices to be at an efficient level (taking into account the price of substitute services) with an optimal level of service quality and variety, a high level of productive efficiency (and therefore a minimum level of subsidy), and an on-going efficient level of investment and innovation in the rail sector (OECD 2006, p. 85).

ORFE can be thought of as an enhancement to the market that seeks to spread the benefits of rail freight by increasing the attractiveness of existing and potential new service offerings through the provision of timely and correct information to the market. The CODE24 partners envisage that this will lead to better rail freight operational efficiency and competitiveness along the entire corridor but is not restricted to it.

ORFE can further contribute to overcome other problems in the rail freight sector as stated by the experts who attended the workshop held in March 2011:

- At present there is no Europe-wide uniform and fair access to rail networks (more specifically no uniform operating language, difficult access to terminals and freight yards, differing national operating legislative requirements and rules).
- Cross-border shipments are subject to differing quality of service.
- Reduction of single wagon transport in some European countries, most recently in Italy. However, single wagon transport still constitutes of some 50 % of rail freight transport in Europe.
- In particular, forwarders claimed a high fluctuation in lorry prices, where price dumping is common. The response time for freight enquiries is considered too long among the rail operators. The price structure in the rail sector is not as transparent as in the road sector. Especially when goods are transported over borders, pricing becomes very complicated very quickly.
- Rail sidings are threatened but are still the most effective form for the door-to-door transport.

- Access to information on rail freight companies that are currently operating is difficult.

At the same time, there is a general recognition that taxation regimes, particularly in the area of indirect taxes (such as fuel and fringe benefits), still seem to favour road transport, rendering a ‘level playing field’ with all other modes a distant goal for policymakers. Nevertheless, for a few specific freight tasks encompassing dangerous goods and other heavy cargo, the industry as a whole has learnt to become accustomed to keep costs low in order to maintain and expand market share and win business from competing modes (Wittenbrink 2009).

7 ORFE Requirements in View of Potential Users

The workshop participants welcomed the idea of an online rail freight exchange as a tool to overcome the aforementioned market barriers and simultaneously increase the competitiveness of rail over the road sector. To achieve this, the following requirements needed to be fulfilled by ORFE:

- ORFE should be fundamentally oriented towards the whole of Europe in order to facilitate an increase in the international freight market; however, during the development of the prototype and testing phase there should be an emphasis on its ability to serve the entire Rotterdam-Genoa corridor.
- The platform operator should be neutral (an absence of preferential treatment) and thus not belong to any potential user group and related to a company using ORFE.
- The platform should also be able to serve not only the spot market in order to use up any existing short-term capacities identified, but also be a platform that can host longer-term jobs and thus facilitate tendering opportunities. If it is only confined to spontaneous freight tasks, then its potential for success is considered limited.
- The platform has the potential to address available capacities on existing rail back haulage quite well, which would have the effect of improving profitability of both existing operators and incentivising companies to transfer from road to rail.
- The opportunity to acquire new business using a platform is particularly welcomed by rail freight companies, as it would allow them to better augment spot capacities with existing scheduled services.
- An important benefit of the platform for shippers is the ability for them to have their details saved in a central repository, including their potential freight tonnage. This would allow shippers to build up new transport chains (conventional freight trains as well as combined transport) as well as identifying which freight pairs are suited for journeys.
- The aforementioned neutrality of the platform operator, with its freight capacity and volumes clearly identifiable, also allows for enhanced tendering opportunities for customers.

- The platform must encompass conventional wagon and intermodal transport. In case of the latter, the prototype needs to depict pre- and end-haulage by lorry as well as a list of intermodal terminals.
- The platform should also include information concerning existing freight patterns and the availability of rail sidings as well as a connection to the information system of DB Netz AG,[13] which contains data on other network operators.
- Another option would be to provide information concerning available freight wagons for hire.
- The ability to exchange information and to simplify communication is also seen as a prerequisite for more and stronger partnership.
- The 24/7 availability of a platform also allows enhanced communication opportunities, which is expected to foster closer partnerships and relations between the users as well as go a long way to overcoming the current information barriers in the rail freight market.

Based on the aforementioned aspects and although the rail freight market is smaller than the road freight market, the participants considered an online rail freight platform feasible if sufficient confidentiality of information lodged for statutory and commercial purposes can be guaranteed and if the platform is run by a neutral platform operator who is independent from any potential user group.

8 Quest for Operator and Market Entrance

Project partner University of Duisburg-Essen has developed a prototype based on own studies and including the requirements stated by the other project partners and which are based on the workshop outcomes presented previously and market observations made by each project partner. A first testing of the prototype has been made by some partners as well as by VDV, the German association of rail freight operators.

As it became evident in the workshop held with the potential user groups, an operator has to be identified which allows for a better and more suitable development of an online platform ready for the market.

A potential operator should be able to demonstrate an appropriate technical understanding of the project idea, fulfil the market requirements and respect the goals defined in a separate document and summarised hereafter:

- Cover the entire TEN-T 24 area or all of Europe
- Enable interaction and matching of demand and supply
- Ensure confidentiality of users' data
- Enable contact between ORFE users
- Enable monitoring of the platform in order to measure the success of reducing empty wagons

[13] Main German rail network operator.

- Ensure neutrality of potential operator (no shipper, no rail operator, no logistics service provider)
- Acknowledge the contribution of the CODE24 partners to the further ORFE development.

The prototype developed during the project phase was offered to any potential cooperating operator. In September 2012, the two private Germany based companies Bargelink from Xanten and MWP from Hamburg stated their interest to cooperate with CODE24 to prepare such an ORFE tool for the market entrance. In the following, both parties developed their market tool and launched their websites in October 2013. The first online market platform "railcargo-online" is owned by Axel Götze-Rohen from Bargelink. Bargelink benefits from its experience of 12 years' operating the inland shipping platform bargelink.com. Internet access for this platform is http://www.railcargo-online.com (Fig. 2).

Fig. 2 Illustration of the cover page of www.railcargo-online.com including locomotive and wagon search engines

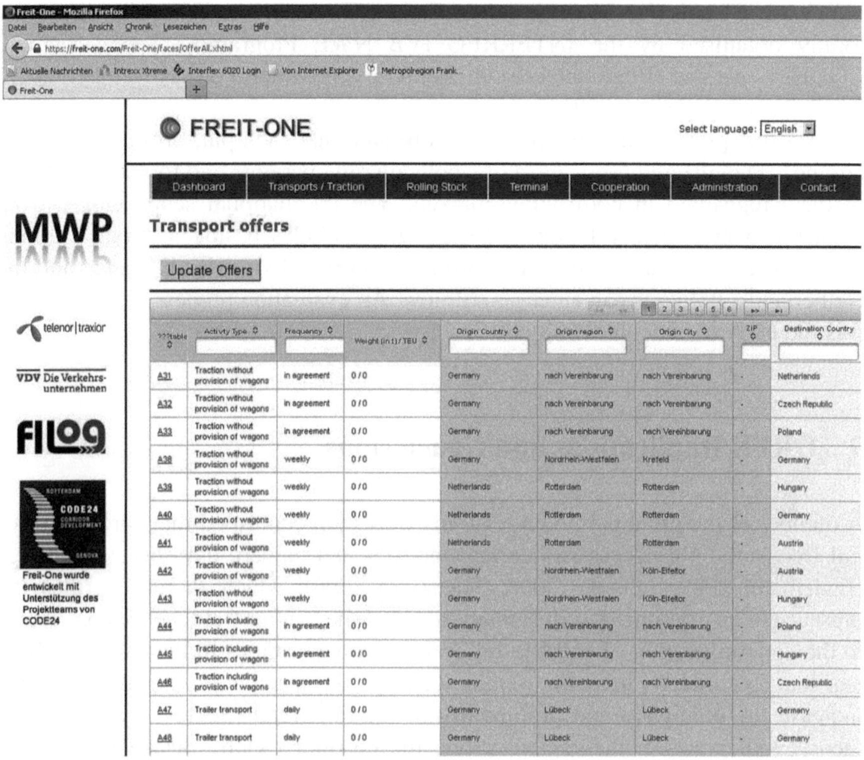

Fig. 3 Illustration of request/offer overview with the possibility to match in www.freit-one.com

Dr. Martin Makait from MWP runs the second platform, "Freit-One". MWP uses the software prototype developed in CODE24 and brings its long experience in rail consulting into a practice-oriented further development. Internet access is http://www.freit-one.com (Fig. 3).

Both platforms and their respective operator fulfil most of the criteria established by CODE24 and comprehend the neutrality of the operator, the confidentiality of user's data, the possibility to match demand and supply, the establishment of contact between users within the respective platform, the coverage of the whole corridor Rotterdam-Genoa area and the fact that cooperation with CODE24 is mentioned. While both platforms can be used all over Europe, another criterion was not met by both platforms: the idea was to include a monitoring option in order to appraise how ORFE can contribute to improve rail usage and to reduce freight movements on the road.

CODE24 emphasises its neutrality and offers its knowledge and network to both operators. CODE24 undertook some activities to push the idea of an online-market and to make it available to the market participants. Among them, there was a presentation on TransportLogistic 2013, one of the famous events in the logistic and freight sector. Both operators also presented their platforms on the workshop

"Innovative Tools for More Efficient Freight Transport in Europe" which was jointly organised by the INTERREG IVB NWE Programme Secretariat and CODE24 partner Regionalverband FrankfurtRheinMain in November 2013 (cf. Dörr and Endemann 2014).

As a first preliminary appraisal, it can be said that each platform benefits from around 100 registered users, most of whom are rail operators and forwarders. Some of them registered in both online-markets. The international scope ranges from Romania over Hungary to Great Britain and includes beside a strong German involvement many users from the corridor countries Switzerland, France and the Netherlands. Italian involvement is still low. An evaluation process has started but it is too early to conclude on it at this stage.

9 ORFE Implications and Conclusions

A number of the principal findings of this workshop confirmed that perceptions of rail freight did not necessarily correspond to actual modern day service offerings (Endemann 2009; Jahncke 2008; TransCare 2006), which in turn verifies the presence of an asymmetry of information within the market[14] and hence contributes to the underutilisation of existing rail freight train capacities in the market. ORFE has the potential to increase freight loadings on specific train paths and thus increase the efficiency and profitability of the rail freight sector. By combining theory with practice, the regions along the corridor can be placed in a strong position to realise substantial economic benefits whilst also improving land-use and amenity for residents and industry alike. Given the objective of increasing the use of freight train capacity, ORFE may contribute to improve the acceptance of rail freight among citizens, especially in those areas affected by rail noise. The fuller a train is, the clearer the awareness that freight is not only shifted in transit, the more rail freight may be publicly accepted as a modal alternative to road freight.

Market analysis reveals the economic imperative of implementing an online rail freight exchange: it is aimed at increasing transparency in the rail freight market and its competitiveness towards its counterparts in the road sector. Especially for smaller and medium-sized enterprises (SME) it has the potential to intensify marketing activities more efficiently and at lower cost for companies from both the demand and supply sides of the market.

The more the shipping companies use such an online tool, the higher the expected level of acceptance for rail freight. Nonetheless, the success of such an online market will be proven in the long-run. Consequently, CODE24 started the

[14] In economics, information asymmetry is a term that describes a situation in which buyers are not fully aware of what their purchase options and their respective costs are. Hence, they are inclined to make less-than-optimal purchase decisions.

search for an operator quite early in order to accompany the initial phase of market entrance with marketing and promotion measures.

In order to improve the usefulness and success of ORFE in the long run, the following issues need to be addressed on the European and the national level and by the respective network or terminal operator:

- Ensured access to the rail network, e.g. via funding schemes for rail infrastructure (terminals, railway sidings, shunting yards).
- Incentives for rail usage: funding schemes such as Connecting Europe Facility, regulatory measures.
- Preservation of (local) rail infrastructure, rail sidings, rail yards.
- Optimisation of transhipment and links to port infrastructure.
- Network operator(s): guaranteed open access & train paths.
- Provision of funding/support to implement noise reduction measures.
- Required use of ORFE as a prerequisite for any subsidy or for funding eligibility.

References

Allianz pro Schiene (2007) From truck to train. Allianz pro Schiene e.V., Berlin

Bruns AS, Föhring R, Zelewski S, Günes N (2012) Online-Frachtenbörse für Verlader und Güterbahnen. Güterbahnen 11(1):33–37

Clausen U, Schwarz M (2002) Gleisanschlüsse: Der einfachste Zugang zum Schienennetz. Der Nahverkehr 20(5):45–48

Deutsche Bahn (2012) Wettbewerbsbericht 2012. Deutsche Bahn AG, Berlin

Dörr H, Endemann P (2014) Moving forward freight mobility innovations. Internationales Verkehrswesen 66(2):60–63

Endemann P (2009) Booming rail freight market and collapsing road traffic—what can the regions do? Proceedings of the European transport conference, London

Endemann P, Blees V (2012) Mehr Kooperation und Koordination im Schienengüterverkehr gewünscht. Güterbahnen 11(2):26–31

European Commission (2012) Third report on monitoring development of the rail market. COM (2012) 459 final, Brussels

Jahncke R (2008) Branchenspezifische Leistungsangebote im Schienengüterverkehr, Schriftenreihe Wirtschaft & Logistik der BVL e. V. Deutscher Verkehrs-Verlag, Hamburg

Merkert R, Smith A, Nash C (2010) Benchmarking of train operating firms—a transaction cost efficiency analysis. Transp Plann Technol 33(1):35–53

OECD (2006) Structural reform in the rail industry. OECD J Compet Law Policy 8(2):67–175

TransCare (2006) Schienengüterverkehrskonzept für das Rhein-Main-Gebiet – Chancen für NE-Bahnen. Unpublished final report, Wiesbaden, summary. http://www.region-frankfurt.de/media/custom/1169_1363_1.PDF. Accessed 2 Feb 2012

Vassallo J, Fagan M (2007) Nature or nurture. Why railroads carry greater freight share in the United States than in Europe? Transportation 34(2):177–193

Vierth I (2011) Evaluation of the outcome of the opening market in Sweden—15 years deregulated rail freight market—lessons from Sweden. Proceedings of the European transport conference, London

Vierth I (2012) Uppföljning av avregleringen av godstrafiken på järnväg. VTI rapport 741, Linköping

Vogt A (2011) Zukunft des Einzelwagenverkehrs aus Sicht der NE-Bahnen. Güterbahnen 10 (4):26–30
Weigand W (2009) Mehr Kapazität für den Schienenverkehr. Eisenbahntechnische Rundschau 58 (12):722–727
Wittenbrink P (2007) Wo steht der Wettbewerb in der Güterbahnbranche? Güterbahnen 6 (4):13–15
Wittenbrink P (2008) Aufbruch und Handlungsbedarf bei den Güterbahnen. Güterbahnen 1 (5):34–39
Wittenbrink P (2009) Anpassungsstrategien von Güterbahnen in der Krise. Güterbahnen 8 (3):21–23
Wittenbrink P (2011) Ausblick Logistik 2015. Güterbahnen 10(1):20–23

Peter Endemann has a background in Geography, Spatial Planning and Urban Design. Since 2003 he is employed at Regionalverband FrankfurtRheinMain (Regional Authority Frankfurt-RheinMain). His activities comprise travel behaviour analysis and interaction with land-use, transport infrastructure development, rail freight and long-distance passenger rail. Within CODE24 he was co-responsible for the development of the Online Rail Freight Exchange (ORFE) and the action on high speed rail integration.

Inland Port Development

Manfred Rausch, Nina Marzioch, and Kerstin Ruppenthal

Abstract In this article, the results of two studies aimed at improving coordination of inland port activities are presented. The first study analyses port development in urban areas, focusing on the case of the port and city of Mannheim. The second study deals with optimising capacity use within a network of ports. The Upper Rhine ports are analysed in terms of market development, capacity needs and the complementarities of the platforms. The study has recently led to a common investment master plan.

Both studies draw the conclusion that inland ports are an important factor in the further development of multimodal transport, providing essential services for an intelligent transport system. Needed support and development opportunities can be found by a cooperation approach.

1 Introduction

Inland ports are key regional transport players, linking their hinterland to world-wide destinations. However, in their relationships with major players such as sea ports, national and local administrations and global economic partners, they are characterized by a relatively small size. The networking of inland ports and the pooling of common strategies make them more visible to those players and more efficient in developing a comprehensive logistics offer. Several reasons have recently been put forward to encourage some inland platforms to cooperate with each other on a strategic level in order to benefit from the ensuing reinforcement of their position. A parallel development is the upcoming cooperation between inland

M. Rausch (✉)
Port Autonome de Strasbourg, 25 rue de la Nuée Bleue – CS 80407, 67002 Strasbourg Cedex, France
e-mail: M.RAUSCH@strasbourg.port.fr

N. Marzioch
Staatliche Rhein-Neckar Hafengesellschaft Mannheim mbH, Rheinvorlandstr. 5, 68159 Mannheim, Germany

K. Ruppenthal
Stadt Mannheim Fachbereich Stadtplanung, Collinistr. 1, 68161 Mannheim, Germany

H. Drewello, B. Scholl (eds.), *Integrated Spatial and Transport Infrastructure Development*, Contributions to Economics, DOI 10.1007/978-3-319-15708-5_13

ports in urban areas and their cities, intended to resolve common difficulties by optimising space use for future development projects.

In this article, the results of two studies aimed at development of inland port activities are presented. The first study analyses port development in urban areas, focusing on the case of the port and city of Mannheim. The second study deals with optimising capacity use within a network of ports. The Upper Rhine ports are analysed in terms of market development, capacity needs and the complementarities of the platforms. The study has recently led to a common investment master plan.

Both forms of cooperation can currently be found in the Upper Rhine region where port issues are dealt within the EU-funded projects "Upper Rhine, a Connected Corridor" and "Corridor 24 Development Rotterdam—Genoa" (CODE24). On the one hand, collaboration has been initiated between an important inland port and its municipality, dealing with the issue of optimising space use for both port and urban development. On the other hand, since 2012, collaboration has been set up between nine inland ports aiming at elaborating a common strategy.

The common issue in both forms of cooperation is to obviate competition by working together and investigating win-win solutions for everyone involved. Both studies draw the conclusion that inland ports are an important factor in the further development of multimodal transport, providing essential services for an intelligent transport system. Needed support and development opportunities can be found by a cooperation approach.

The following article describes these two collaboration initiatives, providing information about context, methodology, outputs and implementation.

2 Master Plan Port.City.Mannheim 2035+[1]

Port cities often face the challenge of striking a balance between port and urban development. Port development is essential to foster regional development. However, port sites are often located near attractive inner city locations. Hence, urban planners aim to enhance the value of the city as recreational space and business or housing real estate by redeveloping industrial sites into ports. Urban development in ports usually goes hand in hand with increasing regional wealth. However, urban development on port land reduces the port area available for logistics progress and other port-related business and restricts economic development. These issues require balanced development, which considers port requirements alongside the interests of urban development (MWEBWV 2010).

[1] This article is an extract of the ongoing study on a port development plan for Mannheim. The study is carried out by a consortium of Drees & Sommer Infra Consult und Entwicklungsmanagement GmbH (Drees & Sommer) and Planco Consulting GmbH (PLANCO).

2.1 *Land Use and Site Potential*

The current situation of the port area is the starting point for port development. The characterisation of current land use and site potential is important in analysing options for future developments. This will help to identify available land for growing traffic flows and port-related business. The possibility of land use changes, both for port-related use and urban development, and is also influenced by current use and site characteristics. The analysis should consider a sufficiently large area around the port, so that all land potential for port activities and all conflicts between neighbouring uses can be identified and reflected in the Master Plan. Sites are classified in different categories of use, both for port-related and non-port-related use (PLANCO 2001, 2003, 2008a, 2012a).

2.1.1 Land Use Analysis

The approach outlined has been applied for the Master Plan Port.City.Mannheim 2035+ (Drees & Sommer/PLANCO). The Master Plan considers a large area including the port area and neighbouring areas as well as land along the banks of the Rhine and Neckar Rivers. Considering the requirement to identify additional land for port development and balance port and urban interests, the two sections B and E have been selected for further investigation (Fig. 1).

The analysis of site use applied in Mannheim distinguishes port-related use and non-port-related use. As a status quo projection it considers new developments and relocations, which have to be chosen. Port-related site use is classified into the following categories:

- Waterside container handling
- Waterside dry cargo handling
- Rail freight handling
- Logistics
- Port operation and port-related services

Non-port-related use is classified as commercial use with no relation to the port, non-commercial use including residential, recreational and green areas, and unused sites including vacant lots and brownfield (Fig. 2).

2.1.2 Site Potential Analysis

The site potential for port-related use and urban development requires separate analysis.

Site potential for port activities is important in determining additional land requirements. Sites within the analysed area are evaluated based on a set of indicators such as location, infrastructure connections, and noise and other

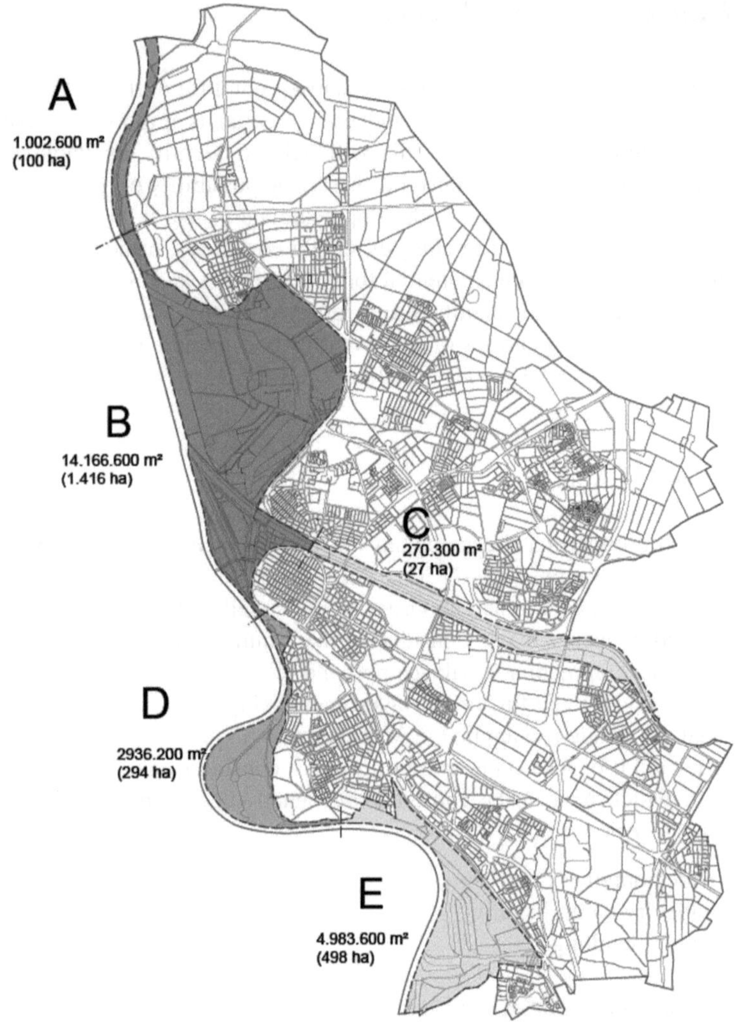

Fig. 1 Investigated area in Mannheim

restrictions, which underline the demand on a site for port activities. A site with water access is required for operations including waterside handling. Trimodality including good access by road and rail is an important factor as well. Generally speaking, activities should not be restricted by noise or other restrictions (PLANCO 2001, 2003, 2008a, 2012a).

The potential of sites located in the port area for urban development is determined by the city's objectives and related planning documents. They outline ideas and show areas or sites respectively in consideration for urban use. Different measures and sites are ranked according to the city's priorities. This determines the potential of sites for urban development (MWEBWV 2010).

Fig. 2 Port related land use south of Mannheim

2.1.3 Site Potential Port of Mannheim

The Mannheim analysis shows that potential sites with water access are limited. The use of this limited number of sites without waterside handling activity would, apart from a few vacant lots, require the relocation of current settlements to other sites. More room is available for settlements which do not require water access. In particular, green areas and brownfield, as well as sites without port-related use, could be used for logistics activities. However, the distance to the container terminals may limit the potential for container-related logistics activities.

A limited number of sites have been identified as having potential for urban development. The overlay of potential sites shows conflicts in the port area. A number of sites with potential for urban development are either used by the port or support the settlement of port-related activities (Drees & Sommer/PLANCO; MWEBWV 2010).

2.2 *Forecast of Port Throughput and Land Requirement*

The land requirement for port activities to accommodate cargo throughput in the long-term future is an important criterion for port development. The encouragement of port throughput growth is essential to maintain and strengthen the regional economic impact of the port. The Master Plan study shows substantial regional employment and added value related to port business. Insufficient land use and facilities jeopardise the contribution of the port to improved regional wealth.

A forecast on barge and rail volumes by load category up to 2035 has been made to determine anticipated throughput. The demand on land is determined by considering the average space needed for the handling of one unit of cargo for different load categories. It allows the comparison of the additional land needed with the projected land available for handling by loading category. The comparison shows land deficits by load category. These deficits should be addressed by port development scenarios (Drees & Sommer/PLANCO; PLANCO 2008b) (Fig. 3).

Cargo throughput at the Port of Mannheim will grow differently according to load category. This is the result of the forecast, which applies recently updated projections by the Federal German Forecast—to consider the impact of the global economic downturn 2009/2010—regarding conventional cargo to extrapolate the latest available figures until 2035. Container throughput projections are the result of a separate analysis considering the latest outlook for global maritime container

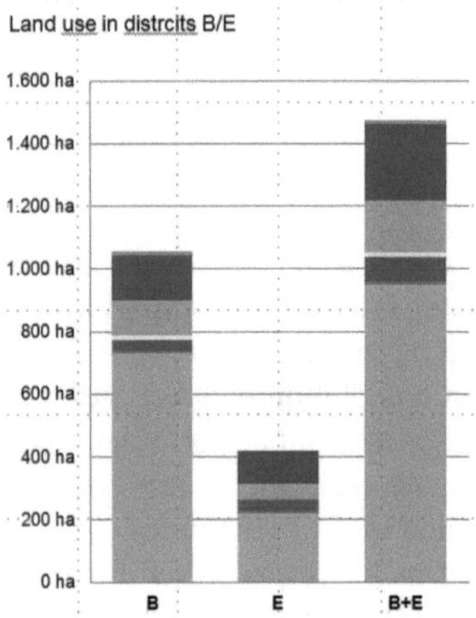

Use	District B	District E	Total
Waterside container handling	12 ha	--	**12 ha**
Waterside dry cargo handling	143 ha	104 ha	**247 ha**
Waterside liquid cargo handling	113 ha	50 ha	**163 ha**
Rail freight handling	14 ha	--	**14 ha**
Logistics	38 ha	41 ha	**79 ha**
Port operation and related services	5 ha	4 ha	**9 ha**
Other sites	731 ha	220 ha	**951 ha**

Fig. 3 Current land use by loading category

transport and development in hinterland transport of the seaports (PLANCO 2012b; OSC 2012; Port of Rotterdam 2011).

Conventional cargo volumes will grow moderately, except for the substantial growth in coal supply for the expanded power plant. However, the power plant uses private facilities, which are capable of handling the required volumes of coal. The land used for handling conventional cargo is sufficient and no additional land is required until 2035 (Drees & Sommer/PLANCO; PLANCO 2012b).

Container throughput will grow substantially until 2035 and lead to the requirement of additional land for container handling and related value added logistics. Overall container traffic growth will triple container throughput in Mannheim from 300,000 to 900,000 TEU in 2035. Barge containers will be in the majority, resulting in a greater requirement for trimodal terminal capacity with water access (Drees & Sommer/PLANCO; PLANCO 2012b; OSC 2012; Port of Rotterdam 2011).

2.3 *Scenarios*

The design of alternative scenarios, which illustrate an anticipated use of port sites, facilitates the decision for a preferred setting of land use in the port area. The alternative scenarios put the emphasis on port and urban development measures and show different aims. The evaluation of alternative scenarios will lead to a range of preferred land use structure in the port area. The evaluation criteria consider the interests of port and urban development. The involvement of stakeholders from port and city further strengthens the commitment to the chosen scenarios. (Drees & Sommer/PLANCO in preparation; MWEBWV 2010).

Demand on land and site potential are important aspects for the design of the scenarios. In the Port of Mannheim the selection of sites to satisfy land requirements for container handling is the main difference between scenarios (Drees & Sommer/PLANCO).

In addition to the port development options, the scenarios consider alternative settings of urban development in the port area. Options vary with respect to extent and forecast use of sites. The scenarios consider urban development priorities and illustrate alternatives ranging from a limited extension of urban development to a substantial extension into the port along with various levels of relocation of port activity and reduction in land potential for port-related use. Another dimension for the differentiation of urban development is the planned use of sites. This is critical for planning as the implementation of offices, services and, in particular, housing, lead to increased noise restrictions for neighbouring port sites. The coexistence of port and urban activity is as far as possible taken into account by the implementation of buffer zones. These zones ensure a distance between conflicting land use and contribute to a soft transition from port to city (Drees & Sommer/PLANCO).

2.4 Next Step: Evaluation of Scenarios

The scenarios will be evaluated based on criteria reflecting port and urban interests. The objective is to identify a preferred structure of land use in the port area. This scenario will be the focus of elaboration of the Master Plan Port.City.Mannheim 2035+. The criteria include:

- Compliance with port objectives
- Compliance with city objectives
- Traffic
- Restrictions for implementation
- Flexibility of planning
- Conflicting land interests/Risk of realisation
- Financial investment

The selection of the preferred scenario will be based on a qualitative evaluation of the criteria. An assessment of the feasibility of the preferred scenario considering regional economic impacts and investment in financial terms is scheduled in preparation for the Master Plan (Drees & Sommer/PLANCO in preparation; PLANCO 2001).

3 Cooperation between the Nine Upper Rhine Inland Ports

Within the framework of the project "Upper Rhine, a connected corridor", the nine ports of the Upper Rhine: Strasbourg, Colmar, Kehl, Karlsruhe, Ludwigshafen, Mannheim and the consortium of Rhine-Ports Basel-Mulhouse-Weil have conducted a set of studies to define the investments required and the relevant cooperation projects throughout the cross-border region, in order to optimise the performance of the network of port platforms of the Upper Rhine. Evaluating both the evolution of traffic by 2025–2035 and the capacity of transport infrastructure (cross-regional railway network, port platforms), these studies have culminated in late 2014 into the formalization of an investment masterplan involving a common program of investments and cooperation.

The project has been supported by the European Commission in the frame of the TEN-T programme, which aims to shift 50 % of road freight over distances of more than 300 km to other modes of transport and thus reduce CO_2 emissions by 60 % between now and 2050 (European Commission 2011).

This approach is innovative insofar as, even though initial coordination campaigns existed between certain ports in the zone and within Europe, there was no cross-border coordination between inland ports on this scale or any recognition of inland port networks within the TEN-T. This demonstrates the ambition of the undertaking, which is expected to find answers to the limits on the territory's capacities and connections but is also intended to expand service levels by setting up common actions.

3.1 The Inland Ports in a Growing Market

Current market surveys (40 comparative surveys of multimodal transport in the Upper Rhine region carried out between 2003 and 2013, including regional French and German port development surveys (CTS-Planco 2013; IVT 2010)) that have been analysed and compared by the ports, estimate that flows could increase by 25–40 % by 2025. In order to establish exactly what those flows are in the Upper Rhine region as part of an integrated, forward-looking vision, a market survey was conducted by the consortium of Upper Rhine Ports in order to make the most of an up-to-date diagnosis and detailed trends to design the development of the infrastructures and the range of services offered (CTS-ProgTrans 2014).

This survey takes as its starting point an analysis of traffic data and a series of interviews with the major economic players to assess market trends and offer a clearer understanding of the development of flows on the Upper Rhine.

The Etis+ 2010 database has been used for the data analysis as it provides comparable data on a European level. Etis+ 2010 was constructed for the European Union by a consortium of 18 international research organisations, using the same methodology as Etis+ 2005 and the TRANS-TOOLS model, which was created by the Joint Research Centre of the European Union and the Institute for Prospective Technology and Sciences in Seville. The data provided was cross-referenced with traffic statistics for the nine inland ports and the market development expectations for each sector provided by relevant inter-trade organisations.

The interview campaign conducted for the survey addressed development expectations of relevant economic stakeholders. It involved 89 economic players with port-related activities, such as shippers, river and rail carriers, and logistics firms representing all industrial sectors and port platforms concerned, as well as 14 players outside the corridor: seaports, inland port cooperation and a major inland port.

A major result of the survey reveals an increase in river traffic compared with 2013, estimated at 16 million tonnes in 2025 and 32 million tonnes in 2035, with two sectors (containers and metallurgy/siderurgy) predominantly increasing and one sector (coal) predominantly decreasing.

The development of flows of goods expected between now and 2015 and 2035 in the Upper Rhine Ports (Figs. 4 and 5):

The impact of this growth on the ports is closely related to the part of the growing sectors in each port's traffic. Some ports have a predominant activity sector (Ludwigshafen, Karlsruhe, Kehl, Basel), others are better outbalanced (Mannheim, Strasbourg, Mulhouse). However, even ports with outbalanced activity sectors will be impacted by the capacity limits of their equipment for the growing sectors.

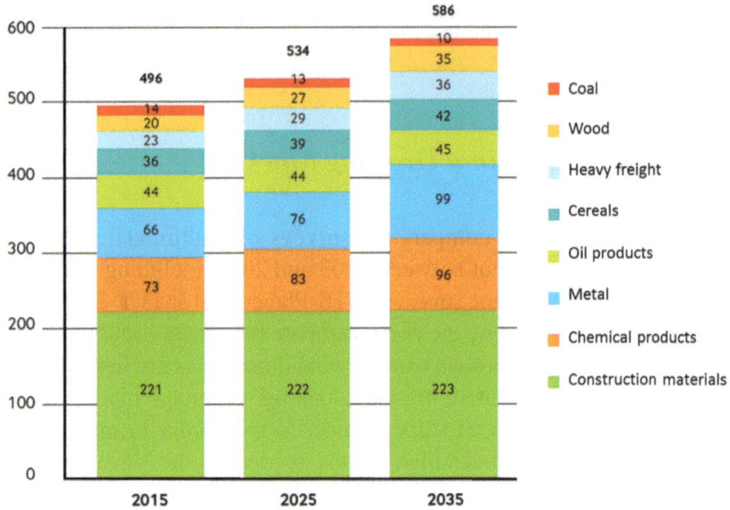

Fig. 4 River port traffic in the Upper Rhine Ports from 2015 to 2035 (Mensia/Steinbeis 2014, p. 31)

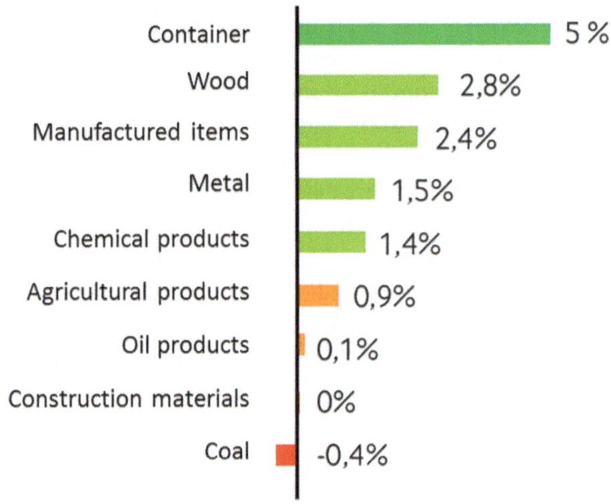

Fig. 5 Traffic Growth in the Upper Rhine Ports in 2025 (Mensia/Steinbeis 2014, p. 46)

3.2 *Potential of Modal Shift*

Based on the analysis of Etis+-data of intra- and extra-regional flows (characterized by origins-destinations and industrial sectors), the market survey identified the potential for modal shift between the Upper Rhine and other economic hubs in Europe. Major road flows to those hubs could, owing to their volume, their growth outlook and their organisation, potentially be shifted to railways, and thus transit via the ports network.

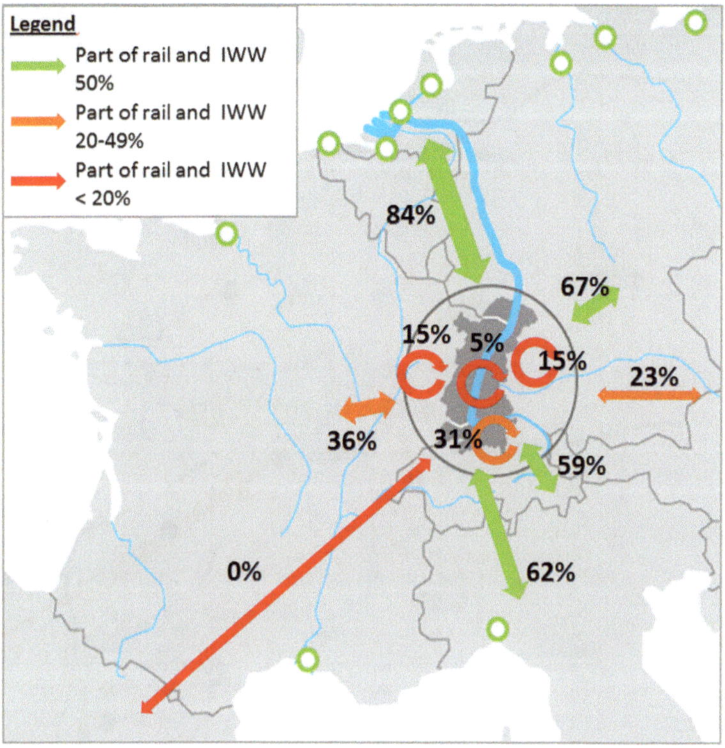

Fig. 6 Modal share of extra-regional traffics (Mensia/Steinbeis 2014, p. 37)

The survey showed that the modal share of railway and waterway transport is particularly high towards the ports of the northern range, Italy, Switzerland and Germany, but that flows to Eastern Europe, France and Spain consist essentially on road transport (Fig. 6).

3.3 *Infrastructure Capacities*

To have the capacity to respond to the forecasted increase in flows, the port hubs must adapt their facilities. Over and above each hub's own optimisation and development, the common line of thought sponsored by the ports within the framework of the TEN-T project should make it possible to identify projects that will enhance the inter-connection of the hubs and meet the challenges in the Upper Rhine region.

A capacity survey was initiated by the consortium of Upper Rhine Ports to establish the nature of the main limitations in terms of infrastructures, networks and

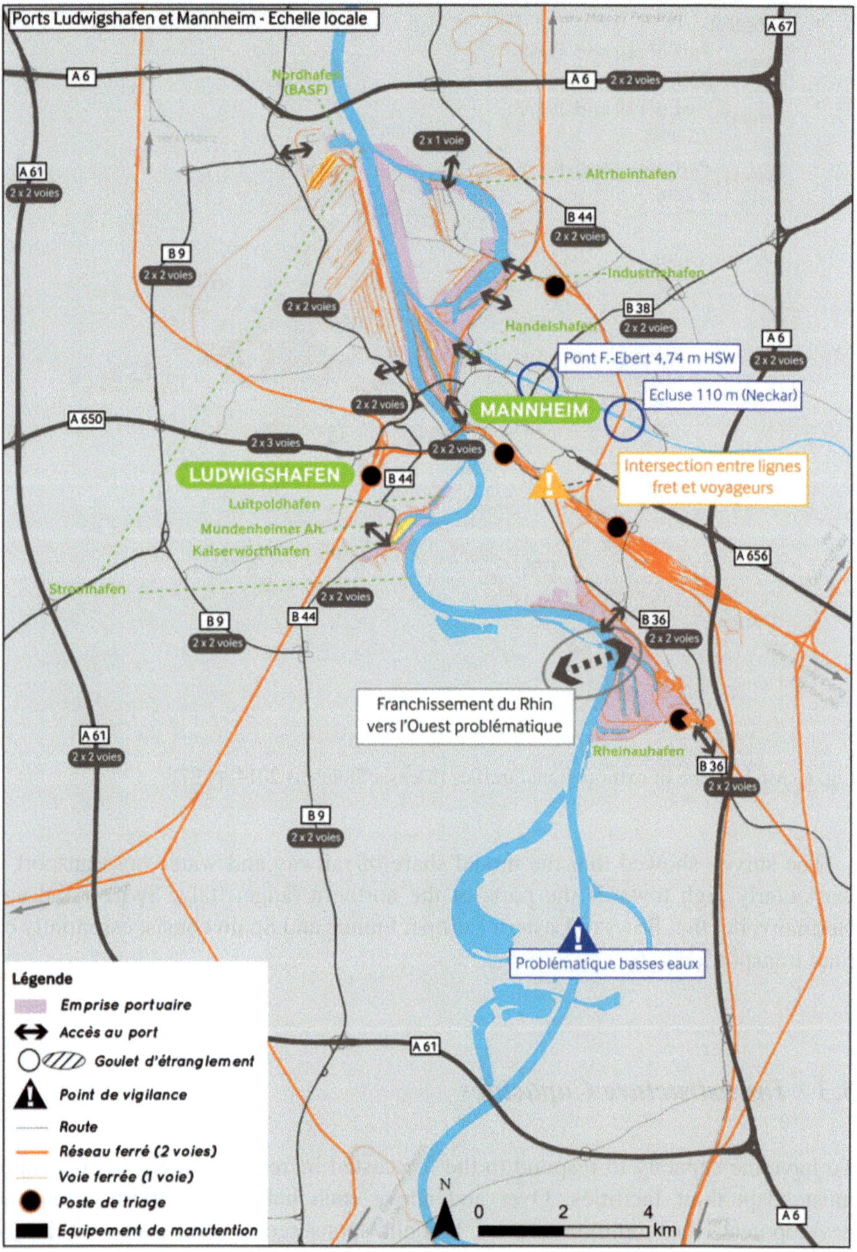

Fig. 7 Local analysis of the ports of Ludwigshafen and Mannheim (Setec-Stratec-ProgTrans 2014, p. 14)

services within the Upper Rhine region in order to identify common projects likely to enhance the territory's logistical performance (Setec-Stratec-ProgTrans 2014).

In a first step, this survey characterised the infrastructures and port facilities at each platform and for all modes using a common methodology. The data generated helped to define residual capacity and development needs. A comparison with the traffic data established by the market survey led to the identification of a series of common investments and actions to be undertaken to increase the overall performance of the system (Fig. 7).

3.4 A Common Masterplan

The results of thinking along these lines have been transcribed in a common master plan for the Upper Rhine. That strategic document is built upon an analysis of needs and projects that meet those needs. In the capacity study, a list of actions has been drawn up, from which a first selection has been made with three conditions: (1) actions lying within the competence of the port, (2) actions of sufficient maturity, (3) actions making sense in a shared vision of the corridor.

The study identifies actions in the field of investments in each platform and cooperation actions between ports. These listed actions have been evaluated within a multi-criteria-analysis, permitting to set up a short list of priority projects (Setec-Stratec-ProgTrans 2014, part II, p. 13).

In a last step, the common masterplan has been built up on the selection of projects, organising them in an investment and a cooperation program covering priority and complementary actions.

I. Investment Program

Priority 1: Secure road access to the ports

Road access to the ports is vital in ensuring the continued existence of and developing their traffic, including waterway and rail (pre- and post-shipping).

- Basel, Strasbourg, Ludwigshafen, Mannheim

Priority 2: Enhancement of the ports' railway access and handling capacity

The quality of the interface between the ports' rail networks and the national rail networks is essential in fully exploiting the ports' trimodality.

In view of the rail development projects for freight on the mainline rail networks and a growth in rail traffic at the ports, the ports' rail networks must be improved or back fitted.

- Basel, Mulhouse, Strasbourg, Kehl, Worth, Karlsruhe

Priority 3: Improvement of the waterway interface

The Rhine already offers a high level of service to its users. A few improvements are nonetheless desirable.

- Basel, Strasbourg, Karlsruhe, Colmar, Weil

Complementary actions

Important investment needs have been identified for several platforms in the field of development of handling capacity for containers and bulk cargo.

II. Cooperation Program

Priority 1: Introduction of shared ICT traffic management platform for the coordination of the fluvial traffic in the inland platforms. The action is required in order to improve the efficiency in the logistic processes and better capacity utilisation.

Priority 2: Constitution of a portal for the common logistic offer

As pointed out within workshops with economic stakeholders, the action is needed to facilitate access to information on the overall logistic offer and benefit to the multimodal transport.

Complementary actions

Development of modal shift towards rail and waterway via the development of strategic partnerships.

The master plan is planned to be used as a road map for the long-term cooperation that points up infrastructure investment and cooperation priorities from a regional point of view and provides all information necessary to and arguments in favour of various funding opportunities.

3.5 *Implementation of the Masterplan and Creation of a Common Governance Structure*

The implementation of the actions listed in the common master plan will take place on two levels: an individual level of each port concerning its infrastructure investments and a regional level for cooperation projects.

This dual approach provides sufficient independence to the port authorities as well as a regional vision for their investments and cooperation opportunities. However, the progressive implementation of individual and common actions will require the solid, well organised cooperation that only a legal entity can provide.

This is why the nine ports plan the creation in 2015 of a legal entity suited to their needs that takes into account the entire range of particularities arising from the trinational context. In parallel, the ports have set up, in early 2015, a proposal for an EU funding program under the title of "Pilot implementation of an Upper Rhine traffic management platform", that closely corresponds to the identified priority for cooperation projects with the masterplan. This project is planned from 2015 to 2018.

4 Summary

Both studies forecast a growth of transshipped cargo along the upper Rhine. To encounter the forecasted increase in flows, the port hubs must adapt their facilities. Only if the traffic modes of water and rail can overcome their bottlenecks, seamless transport chains, alternative to the road, can be strengthened. To answer the current needs inland ports will have to engage in cooperation with other ports and relevant stake holders to develop complementary specializations in certain sectors, e.g. heavy lift cargo containerization etc. Port hubs must adapt their facilities to the needs of further growth and future trends.

At the same time the aimed port development has to be supported through cooperation with local authorities in terms of spacial development. In economic wealthy areas, developing housing and recreational areas together with regulations for environmental protection, e.g. for flood protection, lead to an increasing competition for land use. The availability of space becomes a determining factor for potential development. In this regard cooperation and specialization can have a positive effect. Development can thus be pushed forward through local spacial development concepts, shared infrastructure along the geographical line, networks and joint services.

This form of cooperation cannot be sustained through the effort of every single port, but has to be grounded and resolved on a higher authorized level.

References

CTS-Planco (2013) Schéma portuaire du bassin du Rhin. Strasbourg

CTS-ProgTrans (2014) Etude de marché et flux dans le corridor du Rhin Supérieur dans le cadre du projet RTE-T « Upper Rhine, a connected corridor ». Strasbourg

Drees & Sommer Infra Consult und Entwicklungsmanagement GmbH/PLANCO Consulting GmbH (Dress & Sommer/PLANCO) (in preparation) Master Plan Port.City.Mannheim 2035 +. Stuttgart/Essen

European Commission (2011) White paper on transport. Luxembourg

IVT (2010) Grundlagenuntersuchung zu einem Binnenschifffahrts-und Hafenkonzept Baden-Württemberg. Mannheim

Mensia Conseil-Steinbeis Europa-Zentrum (2014) Rapport de synthèse activité 3 du projet RTE-T « Upper Rhine, a connected corridor ». Strasbourg

Ministerium für Wirtschaft, Energie, Bauen, Wohnen und Verkehr des Landes Nordrhein-Westfalen (MWEBWV) (2010) Binnenhäfen im Spannungsfeld konkurrierender Nutzungsinteressen, Studie der PLANCO in Zusammenarbeit mit ISL. Düsseldorf

Ocean Shipping Consultants (OSC) (2012) North European container port markets to 2025. Ocean Shipping Consultants, Chertsey

Planco Consulting GmbH (PLANCO) (2001) Perspektiven für den Düsseldorfer Hafen. Essen

Planco Consulting GmbH (PLANCO) (2003) Logistikflächen in Hamburg: Zukünftiger Bedarf und Flächenmanagement. Essen

Planco Consulting GmbH (PLANCO) (2008a) Regional- und gesamtwirtschaftliche Bedeutung der Neuss-Düsseldorfer Häfen. Essen

Planco Consulting GmbH (PLANCO) (2008b) Gutachten zur Abschätzung des Flächenbedarfs der Häfen Ludwigshafen/Mannheim. Essen

Planco Consulting GmbH (PLANCO) (2012a) Voruntersuchung zur Hafenentwicklung in Düsseldorf Reisholz. Essen

Planco Consulting GmbH (PLANCO) (2012b) Gutachten zur Erhöhung der Wettbewerbsfähigkeit der Binnenhäfen. Essen

Port of Rotterdam (2011) Port vision 2030. Rotterdam

Setec/Stratec/ProgTrans (2014) Study of capacities and the level of interconnection of the ports platforms. Strasbourg

Manfred Rausch is since 2012 manager of European projects at the Port of Strasbourg. He is in charge of the implementation of the TEN-T Project "Upper Rhine, a connected corridor" initiated by nine inland ports in the Upper Rhine region in order to build up a common strategy. Master of Arts in social and economic sciences at the University of Mulhouse, he is specialized in cross-border cooperation processes.

Nina Marzioch (1983) holds a law degree from the University of Freiburg. After completing the second state law exam at the Regional Court in Mannheim, Nina Marzioch started her professional career as an in-house lawyer for the Port of Mannheim, where she was in charge of two EU funded projects: "Upper Rhine, a connected corridor" and "CODE24". Since November 2014 she works for the Port of Ludwigshafen.

Kerstin Ruppenthal (1976) is since 2009 project leader of urban planning and landscape projects at the planning department of Mannheim. After a degree of urban and environmental planning at the technical University of Kaiserslautern she worked for different planning offices and institutions. In 2010 she co-organized the young academy of urban and regional planning in Mannheim about perspectives in urban planning in a mobile society. Together with the port of Mannheim she is responsible for the case study of inland ports development in Mannheim.

On the Way to a better Interoperability in the Upper Rhine

Results of a Feasibility Study for a Logistics Service Centre Lahr

Markus Ibert and Daniel Halter

Abstract The former military airfield, today the startkLahr Airport & Business Park Raum Lahr with its 600 ha lies in the border triangle of Switzerland, France and Germany. The region of Lahr is situated in the south-west of Germany where the intermodal traffic link of the future trans-European Transport Network is available: rail, waterway, road and airway. The city of Lahr and the development agency IGZ/startkLahr have commissioned a feasibility study, co-financed by the European Union, researching the potential of the combination from area and transport infrastructure and the establishment of a multimodal node in Lahr. This study analyzes the framework, the added value and the impacts of a multimodal logistics service centre in the starkLahr Airport & Business Park Raum Lahr. The analysis focuses on three main issues:

a) Analysis of the market potential including a competitor analysis and an analysis of the potentially qualified area
b) Economic effects in the region
c) Analysis of a qualified operator model, layout model and operational concept

The quantitative analysis comes to the conclusion that the installation of an intermodal logistics terminal in the startkLahr Airport & Business Park is realizable. From the perspective of the economy, the study predicts an added value for the whole region resulting from the terminal operation and the improved regional logistics service.

M. Ibert (✉) • D. Halter
startkLahr Airport & Business Park Raum Lahr, 77933 Lahr, Europastraße 1, Germany
e-mail: markus.ibert@startklahr.biz; daniel.halter@startklahr.biz

H. Drewello, B. Scholl (eds.), *Integrated Spatial and Transport Infrastructure Development*, Contributions to Economics, DOI 10.1007/978-3-319-15708-5_14

1 The Logistics Service Centre Lahr

1.1 Introduction

The Lahr area is located in the Upper Rhine rift valley on one of the central European traffic arteries: Corridor 24 from Rotterdam to Genoa. This region is closely linked in this corridor by the A5 Highway (with its own slip road Lahr), the Rhine port of Kehl, Lahr airport and, in particular, the Rheintalbahn of German Rail.

At the outskirts of the city of Lahr is one of the largest development zones of Baden-Wuerttemberg: the startkLahr Airport & Business Park Raum Lahr. The former military airfield covering an area of 600 ha has been in use and further developed as an industrial zone and business park since the 1990s. One of the main priorities has been to attract the logistics sector to the site.

This combination of multimodal traffic connections, existing development areas and the link to Corridor 24, has provided the impetus for the city of Lahr and the appointed development company for the zone, Industrie- und Gewerbezentrum Raum Lahr GmbH (IGZ), to apply for admission to the EU-Interreg IVb Project CODE24. Upon admission to the programme in the autumn of 2012, the Traffic Planning Office Ernst Basler + Partner, Zurich, was instructed to prepare a feasibility study. This study was to investigate whether a so-called Logistics Service Centre Lahr (subsequently known as LSC Lahr) could be achieved at the site of the startkLahr Airport & Business Park Raum Lahr through a combination of the potential space and the traffic infrastructure. The LSC Lahr consists here of the freight terminal and, in addition, the associated loading bays as well as the adjoining industrial and commercial areas.

The main initial question is based on whether analysis of the achievability of a trans-shipment terminal enables the modal breakdown road-rail and vice versa through a system of unaccompanied combined transport (UCT) and accompanied combined transport ("rolling highway" or Rola).

The feasibility study was made on the basis of the following modular lines of research (Chaumet et al. 2014a, p. 4):

a) Potential and spatial analysis
b) Regional economic classification
c) Transhipment technologies/economic analysis

These three modules were analysed on the basis of several analytical dimensions by the authors of the study.

Apart from a basic presentation of the study concept, their results are also to be presented in this article.

1.2 Study Design

As stated in Sect. 1.1, the feasibility study was split into three independent research modules, each of which was independently processed. The objective of this procedure was to analyse the core topics determined centrally by the authors of the study (see Sect. 1.1) for the implementation of the LSC Lahr. Within these core topics, various research parameters were in turn classified and operationalized for the investigative procedure. These core topics and the research parameters are to be subsequently incorporated.

On the basis of an analysis of the competition, carrying out a potential and area analysis should indicate the market environment in which the LSC Lahr would act, i.e. the specific market potential that exists for the accompanied and also the unaccompanied combined transport. The research parameters concerning the space should enable statements about the basic demands that the setting up of such a facility would place on the potential space of the startkLahr site.

The regional economic grading or classification took place on the basis of an economic analysis. The core aim of this analytical dimension is to show the economic effects, the contributions towards growth of a freight terminal at LSC Lahr. As well the added value for the region.

The third core topic consists of questions about the layout and operation of such a facility and how a suitable operator model could be developed. The research parameters in this connection concern the suitable facility layout and the pertinent appropriate operating concept for the accompanied and unaccompanied combined transport.

1.3 Procedure

The starting point of the following paper is a presentation of the Lahr location and the startkLahr Airport & Business Park Raum Lahr. In this part of the article, the traffic situation and the structure of the land in particular, with regard to those firms that are already established there, are explained. The results of the three central research areas are specified below. The third section focuses on the question about the market potential of a Lahr freight terminal with regard to the setting up of a terminal for unaccompanied combined transport as well as a terminal for accompanied combined transport, the "rolling highway" or Rola. The fourth section of this paper deals with the regional economic aspects of a freight terminal at LSC Lahr. The added value effects that could arise from such a Lahr freight terminal are to be presented here. The analytical section is concluded by a discussion of possible operator concepts. In the concluding section, the central research results of the feasibility study are presented by means of a brief summary. Based on the results of the feasibility study, the further steps and necessary developments for the implementation of a freight terminal at LSC are then discussed.

2 The Location: StartkLahr Airport & Business Park Raum Lahr

2.1 The Location and Structure of the startkLahr Area

Lahr with its 44,000 inhabitants is located in the Ortenau district in the extreme south-west of Baden-Wuerttemberg on the German-French border (see Fig. 1).

The startkLahr Airport & Business Park Raum Lahr is located on the west side of the outskirts of Lahr. The 600 ha site was used by the military until 1994. From 1945, the French forces developed the site into an airfield and the Canadian forces continued to do so from 1967 onwards. Lahr airfield served the Canadian forces as a central European logistics hub and as a worldwide hub through which the vast majority of the military airfreight from Canada to Germany was handled.

From 1995, civilian use of the site began. From early on, those responsible concentrated on logistical usage and focused on the future marketing of the area.

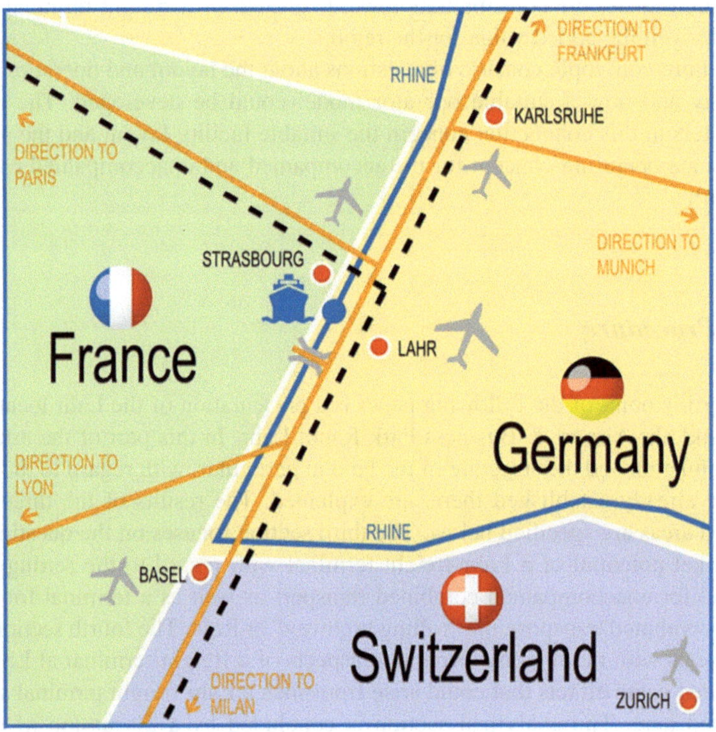

Source: startkLahr/ IGZ GmbH

Fig. 1 Location of the startkLahr site in the border triangle. *Source*: startkLahr/IGZ GmbH

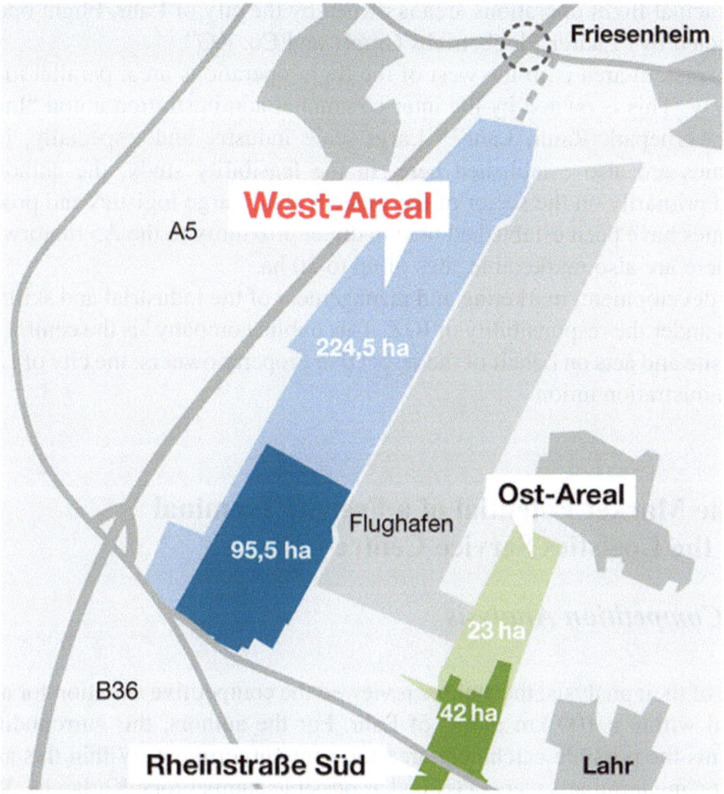

Fig. 2 Organisationally the startkLahr site is split into three areas. *Source*: startkLahr/IGZ GmbH

2.2 The Industrial-Commercial Use of the startkLahr Area

In the meantime, some 155 companies have become established on the site, with over 3500 employees.[1] As well as logistics companies, small and medium sized industrial and production companies and a number of skilled trade and service companies are based here. In the meantime, the location is acknowledged as being an up-and-coming logistics location, which is characterized by its repeated inclusion in the Logistikatlas 2013 of the Fraunhofer-Arbeitsgruppe (cf. Kübler et al. 2013, p. 204) (Fig. 2).

In the eastern area, which is owned by the city of Lahr, there are a number of smaller skilled trade, production and service companies. Here there are still a large number of buildings that were formerly used by the military.

[1] cf. www.startklahr.biz

The actual flight operations area is owned by the city of Lahr. Flight operations are handled by "Lahrer Flugbetriebs GmbH und Co. KG".

The western area is to the west of the flight operations area, parallel to the A5 motorway. This is owned by the inter-communal administration union "Industrie- und Gewerbepark Raum Lahr".[2] Large-scale industry and, especially, logistics companies are also established here. In the feasibility study, the authors have focused primarily on the sector of the western area. Large logistics and production companies have been established here in direct proximity to the A5 motorway exit. Here there are also marketable sites of up to 20 ha.

The development, marketing and management of the industrial and skilled trade areas is under the responsibility of IGZ. This public company[3] is the central contact for the site and acts on behalf of the respective property owners: the city of Lahr and the administration union.

3 The Market Potential of a Freight Terminal in the Logistics Service Centre Lahr

3.1 Competition Analysis

As part of their analysis, the authors reviewed the competitive situation for a freight terminal within a 100 km radius of Lahr. For the authors, this surrounding area represents the possible catchment area for a freight terminal. Within this analysis, the following terminals were classified as possible competitors: Karlsruhe, Weil am Rhein and Basel Kleinhüningen. However, these facilities fall below the intended useful train length of 750 m for the future EU core network for freight transport, which in turn could be achieved in the Lahr freight terminal. By comparison, the Strasbourg terminal, which is also in the vicinity, is primarily focused on the French flow of goods. Due to its distance from Lahr, the terminal in Mannheim is not in any direct competition. The analysis clearly shows that, with regard to the regional competitive situation, there is an adequate market potential for a Lahr freight terminal.

A particularly interesting finding within the competitive analysis was a possible shift of the Rola terminal (accompanied combined transport with a "rolling highway"), currently located in Freiburg, to Lahr.

[2] Members of the special purpose association are: apart from the city of Lahr and the municipality Friesenheim, the Ortenaukreis, the towns of Ettenheim, Mahlberg as well as the municipalities of Meißenheim, Schwanau, Ringsheim, Rust, Kippenheim, Seelbach and Schuttertal.

[3] Proprietors are: City of Lahr, municipality Friesenheim, the administration union "Industrie- und Gewerbepark Raum Lahr" and the Ortenau district, see: www.startklahr.biz.

3.2 *Potential Analysis: Unaccompanied Combined Freight Transport*

When reviewing the potential of a facility that is based on unaccompanied combined freight transport, diverse scenarios were formed on the basis of traffic going north and south. For the traffic going north, a maximum scenario was applied, which presumed that a shift by shuttle, alternatively scheduled trains, would take place. In addition, a forecasting process was employed, which lists the review dimensions freight group (suitability of the freight for combined transport) and catchment area (covering a minimum distance). Within the maximum scenario, the potential for traffic going north was predicted for 2025, with four pairs of trains being forecast per weekday. Based on the estimated ratio of those shipments that were suitable for combined transport, the predicting process produced a factor of two pairs of trains per day.

With regard to the traffic going south, a process was used with which the currently present and future possible modal split ratios were employed. The basis was reached from the analysis of the freight transported over the Alpine region from 2009 (see Table 1), where it was clear that the present modal split ratio is very small without a freight terminal in the Ortenau area. The authors estimate a potential for a scale of approx. 220,000 ton for transport going southwards (Chaumet et al. 2014a, p. 19)

3.3 *Potential Analysis: Accompanied Combined Freight Transport (Rolling Highway)*

As part of the introduction of the NRLA (New Railway Link through the Alps), Swiss transport policy has the objective of transferring a large proportion of the heavy goods traffic crossing the Alps to the railway from 2018 onwards. The goal is to achieve 650,000 trips across the Alps per year.[4] With this in mind, a review needed to be carried out during the study on whether a rolling highway (Rola) terminal in Lahr could contribute towards achieving this transfer goal. Presuming that road haulage traffic through Switzerland would only increase or decrease slightly; the authors of the study predicted a volume of freight amounting to 450,000 shipments per year towards Northern Italy. This is equivalent to around 20 pairs of trains per day.

[4] Bundesgesetz (federal law) über die Verlagerung des alpenquerenden Güterschwerverkehrs von der Strasse auf die Schiene (Güterverlagerungsgesetz, GVVG) of 19 December 2008 (as on 1 January 2010).

Table 1 Transport from the surrounding areas of the Lahr freight terminal going southwards (Chaumet et al. 2014a, p. 19)

Region	Current demand (1000 tons/year)
Northern Italy	356
Southern Italy	144
Total	500

3.4 Layout and Space Requirement

The actual space requirement for the respective utilization of UCT and Rola was determined on the basis of the analyzed potential and an area layout was developed on this basis. For a definition of the area required, a catalogue of basic demands was prepared for the unaccompanied as well as for the accompanied cargo.

The minimum precondition for the loading of unaccompanied cargo is defined as a track length of 750 m. A further precondition concerns the number of tracks below the crane runway or track. It should be two, later four with up to four gantry cranes. Basically, operation should take place with the so-called stand process, for more than four pairs of trains within the flow process.[5] The length of the entry/exit tracks should be 750 m.

According to the feasibility study, a track length of 750 m is adequate for accompanied transport. The concept presumes half-hourly handling for a demand of around 450,000 HGVs per year. This list of requirements results in a track width of 75 m for the unaccompanied transport and a track width of approx. 25 m for the ROLA, which shows a requirement of a total area of 20 ha for the complete facility. This terminal would be one of the few which could provide a full train length of 750 m as an operating length under a crane track.

As the possible location for the terminal within the Airport & Business Park, a zone in the western area has been verified between the A5 and the industrial development area. Under the precondition of a parallel route to the highway for the third and fourth track of the Rheintalbahn, the Lahr freight terminal would be directly next to this line. However, the basic precondition is that the line route is actually placed parallel to the highway, since only with this variant, the startkLahr area, in particular the western area intended for the Lahr freight terminal, would be affected.

[5] The sidings will be repeatedly supplied with train sets during the day.

4 The Regional Economic Effects of a Freight Terminal in the Logistics Service Centre Lahr

4.1 Methodical Aspects

Based on the presumption that the operation of a Lahr freight terminal could have positive economic effects not just for the startkLahr area itself, but also for the whole region, these regional economic effects were analyzed in a second research module.

Consequently, the initial issues be described as follows:

- effects of the Lahr freight terminal for the whole region
- the financial effects of the Lahr freight terminal for the gross added value and for employment in the region
- the contributions to growth of individual active components of the Lahr freight terminal

4.1.1 Effects of the Supply Side

These are the effects that arise from the operation of Lahr freight terminal. A direct effect leads to value creation and employment. In this there needs to be a differentiation between the indirect effect, resulting from the intermediate inputs from the Ortenau district's economy to the operators of the Lahr freight terminal and the induced effect, consisting of the value creation, employment and income effects, which result from the income payments at the Lahr freight terminal.

The authors of the study claim, that through a freight terminal "(...) the demand for goods and services (...) and the income generated by consumption at the terminal (...), processes will be set in motion for the production of goods and for the provision of services (...)" (Chaumet et al. 2014b, p. 8) .

4.1.2 Structural Effects from the Demand Aspect

The potential effects from the use of the Lahr freight terminal are reckoned, when these are more utilized within the framework of the future economic developments (cf. Chaumet et al. 2014a, p. 10)

4.1.3 Location Effects from the Demand Aspect

Location effects are those effects that arise from the characteristics of a site and make a contribution towards the economic productivity of a site. This can include, for example, corporate or residential settlements.

4.2 Value Creating Effects

4.2.1 Initial Situation: Value Creation in the Ortenau District

The basis for the assessment of the value creating effects is an analysis of the actual situation in the Ortenau district as the central starting point. This analysis provides two primary results:

1. With regard to the growth rate compared with Baden-Wuerttemberg, the Ortenau district has economic disadvantages. While 47 % of the regional gross value creation comes from industry and commerce in the Ortenau area, the share in Baden Wuerttemberg is 37 %. However, the activities in this area have less value creation than the activities in the services sector (Chaumet et al. 2014b, p. 15)
2. The services in the Ortenau district are above average in their productivity, however they have a below average share of the regional economy. The authors point out that one element that may be responsible for this situation is a below par infrastructure and the accessibility to purchasing and sales markets.

4.2.2 Value Creation from the Operation of the Lahr Freight Terminal

The direct gross value created (total turnover less advance services) by the operation of the Lahr freight terminal is estimated to be about 5 million euros per annum, by the authors. This assessment is obtained from the hypothesis of an operational concept, whereby this was based on 300 operating days with 12–20 UCT trains and 40 Rola trains per day with 30 full-time workplaces.

4.2.3 Value Creation from Use of the Lahr Freight Terminal

A major objective for the Lahr freight terminal is the provision of additional infrastructure for the region, so as to enable more efficient and more environmentally compatible transport relationships and handling of the flow of freight for the regional economy. A central feature here is the question about the level of transport costs, i.e. the ratio of the transport and logistic costs to the gross added value for the region.

As part of the feasibility study, the level of transport costs for the Ortenau district was compared with those of the Federal German State of Baden-Wuerttemberg (see Fig. 3). It became clear that the level of transport costs for the Ortenau district is 1.1 % higher, i.e. vice versa, that a reduction in this figure would simultaneously result in lower production costs.

The authors set the transport costs for road transport exclusively as the fall-back position and then compared this with the possibilities for combined transport at Lahr freight terminal. This procedure then made it possible to report on the

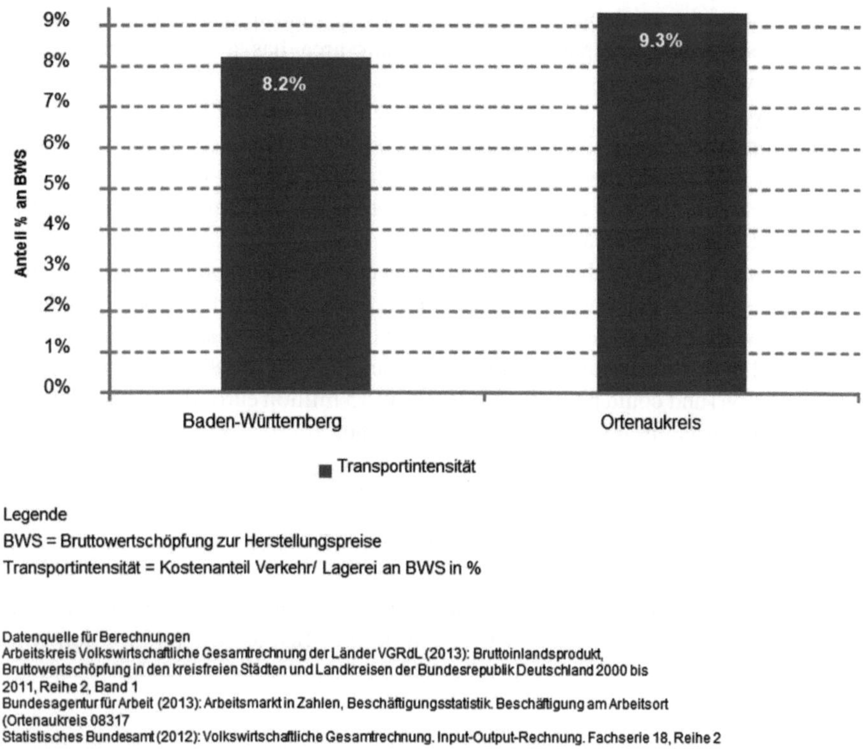

Fig. 3 Transport costs (Chaumet et al. 2014b, p. 20)

economics of such a facility. As part of a comparative analysis the transport costs which arise from exclusive road usage (reference case) were compared with the transport costs which arise from the combined transport with the Lahr freight terminal (plan case). As the basis for the whole economic assessment the production costs were used and not the market prices (cf. Chaumet et al. 2014b, p. 21)

The results show that usage by combined transport and Rola allow the transport costs to fall by 12 million euros. This reduction in costs corresponds to 0.1 % of the regional GNP which is equivalent to the economic performance of 215 employed persons in the Ortenau area (cf. Chaumet et al. 2014b, p. 22)

With regard to the location effects, the marketable industrial and skilled trade areas available in the startkLahr area are reviewed. The possible future benefits for attracting industrial and skilled trade companies can be assigned to the potential for added value. This is the outcome from the fall in transport costs per marketable unit due to moving the freight transport to LSC Lahr, and thus the potential for added value rises. In this, the authors have differentiated between the fall-back position, i.e. area development without the Lahr freight terminal and planned area development with the Lahr freight terminal.

Because of the available development areas for industry and commerce and the already established companies the startkLahr area has a large value creation potential which amounts to 296 million euros per annum. The authors of this study have calculated that with a freight terminal in LLZ Lahr, business and production processes in this industrial and commercial area could be shown to be more cost effective to the amount of 5 million euros per annum.

This is combined with an increase in attractiveness with regard to the use of the land at the startkLahr area.

In the next stage, further investigations were included about the development land that exceeds the startkLahr area. Only those municipalities were included, however, with marketable areas of 10 ha and over. The total added value potential calculated in this regard for the Ortenau district amounts to 981 million euros (fall-back position) and could be increased by about 15 million euros due to the reduction in transport costs through a Lahr freight terminal.

4.2.4 Value Creating Effects for the Ortenau District

When summarizing the effects on the supply, structure and location, the value creating effect for the Ortenau district can be specified as being 45 million euros annually (Fig. 4).

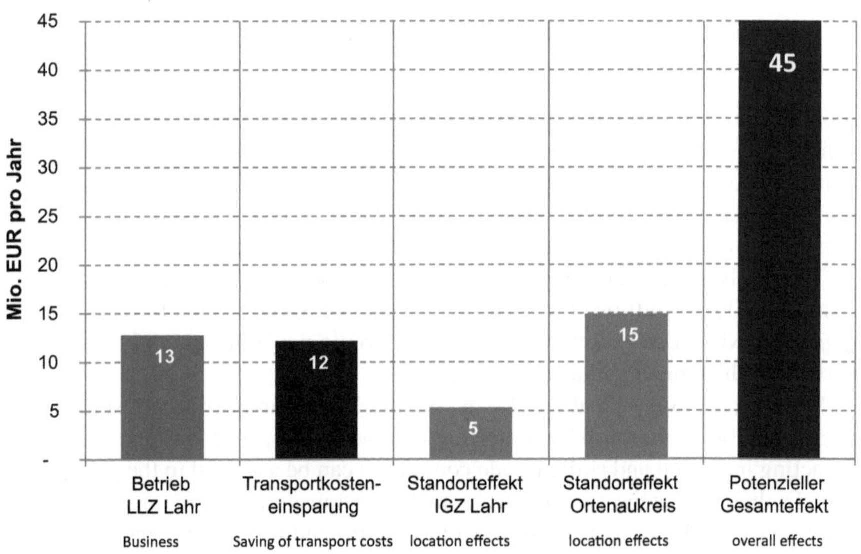

Fig. 4 Regional gross value creation (Chaumet et al. 2014b, p. 26)

5 Operating Concept and Operator Model for a Freight Terminal in the Logistics Service Centre Lahr

5.1 Operational Concepts for the Rolling Highway and for Unaccompanied Combined Transport

The sizing for the rolling highway was methodically prepared on the basis of the estimated loading and unloading times and the associated dwell time at the Lahr freight terminal. A requirement of 3–4 loading and unloading tracks each with a length of 750 m was determined during these calculations.

The sizing of the facility for unaccompanied combined transport needs to be matched to the operating forms of shuttle trains (e.g. Lahr-Rotterdam) or scheduled trains (e.g. Genoa-Lahr-Cologne/Duisburg area Arnheim Rotterdam), since the respective requirements for the facilities differ. For example, the brief stop by scheduled trains presents very high demands on the capability of the crane system.

With a presumed train length of 700 m,[6] the authors predict 5–8 shuttle trains or 4–6 shuttle trains and 2–4 stops by scheduled trains. Based on this, the following tracks would be required:

- Under the gantry crane: 4–5 tracks
- Sidings for UCT trains: 1–2 tracks
- Entry and exit tracks: 1–2 tracks

For exclusive operation with shuttle trains, 3–5 gantry cranes and for shuttle and scheduled trains 4–6 gantry cranes would be required.

5.2 Operator Model

Choosing a suitable operator model for the Lahr freight terminal concentrates primarily on the following aspect which must be defined before the start of the actual project:

- Who is the owner of the site during the construction and operating phase?
- Who is the owner of the terminal infrastructure?
- Who operates the UCT, respectively the Rola terminal?

An additional factor, which has an effect on the operator model selected, is the possibility of dealing with the demand for logistic related areas within the vicinity of the facility and being able to incorporate these in the marketing activities.

[6] The effective length of the set of carriages is 700 m and corresponds to a train length with two engines a train length of 750 m.

Accordingly, there is firstly the possibility of selling or leasing areas of land and to ensure construction and operation by a private company. Alternatively, the area could remain the property of a (partly public) company, which would then also manage the properties. In addition, other hybrids are also possible.

The grant component represents an important aspect of a possible operator model: According to a Federal Administrative Regulation,[7] where various preconditions apply (cf. Chaumet et al. 2014c, p. 24) 80 % of the eligible investment costs for the new building and expansion as well as the extension of the handling facilities can be paid as a non-repayable grant.

6 Summary

The results of a feasibility study on the achievability of a freight terminal at the Logistics Service Centre Lahr (LSC), prepared on behalf of the IGZ Raum Lahr GmbH and co-financed by the EU, was presented in this paper. The diverse aspects were analysed by Ernst Basler + Partner within three independent modules.

The site features of the startkLahr area were presented in detail in a brief overview: multimodal transport connections in combination with a large potential area. The commercial-industrial use of the site is also characterized by a clear logistics-related settlement structure.

By means of a potential and area analysis it was possible to show that with regard to the regional competitive situation there is an adequate market potential for a Lahr freight terminal with a facility for unaccompanied as well as for accompanied combined transport, i.e. for a Rola.

A major characteristic that would open up a good market potential for the Lahr freight terminal is the possibility of handling train lengths in excess of 750 m, which are intended for the future EU core network under the TSI conformity[8] heading.

Analysis of the regional economic effects from operating a Lahr freight terminal and the resultant value creating processes predicted a total value creating effect for the Ortenau district of 45 million euros per annum.

In the third analytical module, the possible operating concepts and their size in relation to the actual area required for Rola and UCT is explained. Within this analytical module, a number of diverse questions are then dealt with, since these will probably be essential when selecting the operator model.

---

[7] "German directive (administrative regulation) on promoting terminals for combined transport by companies that are not state-owned".

[8] cf. Website of the (German) Federal Railway Agency: http://www.eba.bund.de/DE/SubNavi/Recht/GesetzeRegelwerk/TSI/tsi_node.html: "To implement a safe and interoperable, that is a technically compatible European railway system, so-called Technical Specifications for Interoperability (TSI) have been introduced both for high speed traffic as well as for conventional traffic. These TSI include the specifications and test procedures for interoperability components and subsystems."

The feasibility study has verified that basic economic and technical achievability applies for the Lahr freight terminal/LSC. This project has already been presented to a large number of political, union and business representatives by the city of Lahr and by IGZ GmbH. With regard to the Lahr freight terminal/LSC, the Regional Minister for Transport in Baden-Württemberg, Winfried Hermann, has clearly stated that he considers such a project at the Lahr site to be very sensible (cf. Deutsche Verkehrs-Zeitung (DVZ) 07.07.2012)

Of particular importance however is the fact that the still outstanding decision for the location of the third and fourth freight track favours the Autobahn parallel solution. Only with an Autobahn parallel track would the Lahr freight terminal be practical.

References

Bundesgesetz über die Verlagerung des alpenquerenden Güterschwerverkehrs von der Strasse auf die Schiene (Güterverkehrsverlagerungsgesetz, GVVG) of 19 December 2008 (Status on 1 January 2010)

Chaumet R, Buser B, Hofer M (2014a) Güterverkehrsterminal im Logistik-Leistungszentrum Lahr: Potenzialanalyse. Final report of Ernst, Basler + Partner, Zürich

Chaumet R, Buser B, Hofer M (2014b) Güterverkehrsterminal im Logistik-Leistungszentrum Lahr: Prognose der regionalwirtschaftlichen Auswirkungen. Final report of Ernst, Basler + Partner, Zürich

Chaumet R, Buser B, Hofer M (2014c) Güterverkehrsterminal im Logistik-Leistungszentrum Lahr: Anlagenlayout, Betriebskonzept und Betreibermodelle. Final report of Ernst, Basler + Partner, Zürich

Kübler A, Lorenz A, Nehm A, Veres-Homm U (2013) Logistikimmobilien- Markt- und Standorte-Deutschland, Österreich, Schweiz, Belgien und Niederlande. Fraunhofer Verlag, Nürnberg

Since 2005 **Markus Ibert** is the managing director of the IGZ Lahr GmbH, the appointed development company for the former military airfield in Lahr. He is responsible for the development of the whole startkLahr business park. At once he is the director of the Zweckverband "Industrie- und Gewerbepark Raum Lahr", that is the landowner of the biggest part of the startkLahr area. In cooperation with the representatives of the city of Lahr he initiated the project of the Logistics Service Centre Lahr significant and he was one of the main coordinators in the framework of the feasibility study.

Daniel Halter joined the IGZ Raum Lahr GmbH in 2006. He is in charge for the company's real estate management. Since 2012 he is the authorized representative of the company and responsible for the marketing and the technical affairs within the startkLahr business park.

Part IV
Environmental Aspects of Railway Infrastructure Development

Noise Reduction in the Railway Corridor Rotterdam-Genoa: Observations within the Project CODE24

Sebastian Wilske

Abstract The noise emissions of railway traffic, especially of the freight traffic are of the main challenges to gain acceptance for projects to improve the capacity of the corridor—infrastructural projects as well as operational measures. Even decision makers, planners as well as many affected persons have a lack of basic knowledge regarding on the impacts of noise and especially on the ways of function of noise mitigation measures. Hence the decision making procedures have to be improved in the aspects of objectivity, transparency and traceability.

Therefore, CODE24 gave an overview on the approaches in the field of noise protection along the corridor. Some commonalities showed up. Particularly, despite the priority of noise reduction on the source in all legal regulations, in all countries "classic" noise mitigation measures are dominant, especially noise protection walls. The allocation of funds and activities hence does not correspond to the objectively given possibilities for reductions of the impacts of railway noise. The acceptance in the public in many cases is low. The workshops and the survey showed indications, that this can be explained by a lack of knowledge on the on the ways of function of the different noise mitigation measures. CODE24 showed possibilities to strengthen the knowledge basis (toolbox) and especially to give a new and innovative approach (auralisation) to the resulting effects of noise mitigation measures that is much more transparent, traceable and intuitive than conventional ways of providing information, e.g. maps or numerical studies.

All discussions with decision makers and planners within CODE24 showed that noise mitigation concepts in the future have to be considered in an open and fair dialogue with the affected persons and will contain different sets of specific measures that are optimised for the given, from region to region significantly different situations along the corridor. CODE24 showed that all these target groups expect substantial improvements by the application of auralisations in these planning and participation processes in the future. The different bottlenecks along the corridor are starting points for the implementation.

S. Wilske (✉)
Regional Planning Association of the Middle Upper Rhine, Baumeisterstraße 2, 76133 Karlsruhe, Germany
e-mail: sebastian.wilske@region-karlsruhe.de

H. Drewello, B. Scholl (eds.), *Integrated Spatial and Transport Infrastructure Development*, Contributions to Economics, DOI 10.1007/978-3-319-15708-5_15

1 The CODE24 Perspective on the Corridor Rotterdam-Genoa

The corridor development within the project CODE24 follows an integrated approach (Saalbach 2012). This approach is based on a few underlying assumptions which are particularly important for the design of the work package.

1.1 Integrated Perspective

The first core assumption is to consider the corridor not only as an axis of infrastructure but also as a meaningful region to live and to work as well as a region of Europe's cultural heritage. Therefore the target of the corridor development is not a technically optimised infrastructure but rather an equitable development of infrastructure, landscape and settlement (Institute for Spatial and Landscape Development 2013).

1.2 Polycentric Network

The second core assumption is to regard the corridor as a polycentric network of urban nodes, which are mutually linked and which are connected with their hinterland. This intraregional and interregional linking is based on an efficient and comfortable public transport within and between the regions. Two major boundary conditions of spatial development can be concluded at this point.

- On the one hand the needs of freight traffic are not allowed to dominate the infrastructure development; instead the whole infrastructure has to be developed by respecting the needs of freight traffic, short-distance trains and high-speed trains. Therefore, improvements in a single sector should not be at the expense of other transport tasks (Saalbach 2012).
- On the other hand the core areas of settlement are situated now and in the future near to the railroad's infrastructure, which represents an essential feature of this polycentric structure. This strategic concentration is an important planning maxim to support the polycentric network (Scholl 2007).

1.3 Increasing Transport Needs

The corridor's necessity to shoulder upcoming fundamental transport needs constitute the third core assumption. Therefore the main interest of all regions along the corridor is to be able to build the economic development on a capable and reliable link to the transport corridor Rotterdam-Genoa. So the focus on the corridor as an attractive place to live and work does not prevail the development in such a way that infrastructural development is prevented or even transport tasks are displaced to other regions: The freight traffic is seen as a necessary part of the corridor development (Saalbach 2012).

The mentioned core assumptions for the transport-infrastructural and spatial development are not unambiguous: conflicting aims will arise in implementation phases. This leads to subsequent crucial assumption that the acceptance of infrastructural development measures has to be gathered on the regional level, which is the suitable level to find lasting compromises for these conflicting aims. A challenge for the regions is that the people are affected individually whereas the economy profit as a whole.

That is why different facets of railway noise are of decisive importance for the corridor development.

- It is in favour of the regions along the corridor to be linked to an efficient freight traffic corridor and to prepare them for the expected rising transport volume infrastructurally and operationally.
- Noise is a substantial trigger of spatial conflicts following from the proximity of settlements to infrastructure lines. Because of the noise resulting from an unknown operating schedule at the time of the infrastructure's permission, it is only possible to influence the noise volume indirectly. Furthermore, the operating schedule likely changes with the times.
- The acceptance of corridor development projects depends substantially on undermining fears about the tightening of noise problems. So suitable measures have to be named, their mode of action has to be explained and the measures have to be implemented as agreed. In planning and decision processes it is difficult to deal with noise protection issues in a solution-oriented and productive way, mainly because of general difficulties to convey adequately crucial information on these issues.

2 Railway Noise in the Framework of CODE24

The proceeding in the action 6 "planners toolbox for innovative noise protection" was created reflecting the background of these core assumptions. There were mainly three modules: the classification of the already implemented noise protection measures along the corridor, a workshop with experts to find out common grounds and differences about noise protection measures in the different countries with regard to advantages and disadvantages of these measures and testing of an innovative instrument which supports planning and decision processes.

2.1 Common Experiences

The results of the comparative investigation and the workshop with experts can be summarised in three relevant aspects.

Passive noise protection measures dominate at the conflict spots in the countries along the corridor, e. g. noise protection walls and building-based physical measures like noise protection windows. The weaknesses of these solutions are considered in a similar way in all countries. It deals particularly with the visual barrier, with the often only displeasingly solved compromise between cost-efficiency and aesthetics and with its effectiveness limited exclusively to local scale. These results are the basis for new constructions as well as for noise protection measures along existing lines (noise abatement).

There are also common grounds in the field of new constructions. On the one hand the long planning and implementation periods lead to the situation, that projects can no longer give answers to questions regarding the future development, which are asked by involved people today. For example the projects contain increases in capacity and velocity of high-speed passenger transport but no improvement in matters of freight transport are named. In some cases people even suspect negative effects, for example if they are afraid of capacities becoming vacant on existing lines and hence will be used for more freight transport. On the other hand the example of Betuwe-Line shows that new constructions which are developed for rail freight transport cannot unfold their effectivity immediately or only with significant delays. This is particularly caused by missing capacity on further sections of the existing lines and the incompatibility between power, signal systems and the rolling stock.

As a common ground of all examined countries, it seems that the discussion between involved people, planning authorities and policy-makers is extremely dominated by legal limits, technical guidelines and model calculations. But these calculations are ill-suited to give an impression of the expected noise situation and the reduction effect of single projects and packages of measures.

The workshop with experts depicts some notable country-specific features and single measures.

- The Netherlands realised and operates the Betuwe-Line, which represents a new construction focused on rail freight transport. This example illustrates the potential of such a measure with respect to an optimised strategy for noise protection which consists of many complementary single measures. However, the difficulties of such a project became apparent as well. The utilized capacity continues to be relatively low and a significant percentage of rail freight traffic still uses the already existing infrastructure. On the one hand, this is caused by weaknesses of the junctions between the newly built line and the following sections of the corridor. On the other hand, problems occur due to the incompatibility between the used power, the signal system and a significant percentage of the available rolling stock. This example shows once more, that in the face of enormously long implementation periods and the often necessary cross-border coordination of linking infrastructure, the complex interaction between the layout of the line, its technical equipment, the available rolling stock, the timetable and the connections to the existing network result in a seriously delayed occurrence of effects. This fact has to be considered in areas, in which large-scale bypasses are discussed (e.g. Upper Middle Rhine Valley, Jura, northern and southern feeds of the base tunnels). Regularly, you have to assume that people involved on-site won't benefit from the measures in their lifetime and therefore noise protection measures along existing and already operating lines should be implemented as well.

- Examples from Switzerland demonstrate that a compromise between noise reduction and visual barrier effects (knee- and waist-high noise protection walls) is reached more frequently compared to other countries. In addition to this approach the optimisation of the rolling stock is pressed ahead more strongly in comparison to other countries along the corridor (see below for more details).
- Furthermore, in Italy noise issues are discussed with an explicit focus on high speed rail services. One main reason is the minor importance of railway freight traffic in comparison to road transport. That is why, there are less people affected by noise of freight trains, except in the direct catchment areas of the new base tunnels.

2.2 *Common Difficulties*

In addition to these issues, which are more specific for the different countries, all countries along the corridor have in common that they are confronted with substantial difficulties to ensure an adequate provision of information on noise and noise mitigation measures in planning and decision-making processes. It is particularly difficult for the affected persons, planners as well as political decision-makers to receive an adequate insight in these topics by their own intuition. The conventional instruments to give an insight in potential future situations are numerical modeling or comparisons and evaluations of the current and future situation based on simulations. The results are presented in tables and maps. The characteristics of the sounds being crucial for their effects, are for this purpose aggregated into one single number. The same number represents totally different sequences of diverse single events. Hence only a few experts are able to relate the calculated numbers to individual real world listening experiences.

The ex-post assessment of measures causes difficulties due to the fact that most developments take a long time to be implemented so that the effects of different developments interfere with each other. For instance the number of trains is increasing rather gradually than abruptly. Simultaneously, the composition of the rolling stock is changing as well as the speed of the trains, the timetable et cetera. During the construction phases of a measure, there is a entirely different noise situation. This fact makes it difficult to compare the noise situation in advance and after its implementation.

For an objectification of the public discussions and a targeted handling of fears, the conventional instruments are too technocratic. They are inadequate to objectify the often very emotional discussions and to convert them into an almost rational process of compromise seeking and decision-making.

3 "Auralisation" as a New Tool to Support Planning and Decision-Making Processes

Within the project CODE24, a new instrument was developed: it is the so-called "auralisation". "Auralisation" is a software tool which makes existing and planned situations audible. Different sound sources in a specific spatial constellation can be combined with noise reduction measures. This data is calculated on the fly and the resulting noise is made audible. The characteristics of the sound sources, the spatial situation and the noise reduction measures can be modified so that the effects of the changes are made instantly audible. The auditory impression is emphasized instead of the assessment of numbers (Leistner et al. 2013).

The above described basic assumptions and the added value of the approach which was developed in the project CODE24 can be illustrated by using the example of the measure "reducing railway noise at the source by retrofitting the rolling stock". This measure was discussed during the project with various stakeholders: citizens' initiatives, experts in different realms of spatial planning, political decision-makers on the local and regional scale.

For example in a first step the above mentioned long-term developments are important. In the course of the preparation and the implementation of the project CODE24, there was a phase of accelerated economic growth in succession to the crisis of 2008/2009. Hence the public perception was dominated by a recognizable increase of the train numbers. At the same time, regulations were implemented to promote an upgrading of the rolling stock. In Switzerland for instance, a noise bonus is granted for noise reduced wagons (Schweizerische Eidgenossenschaft, Bundesverwaltung 2014). In Germany a noise-related system of track access charges was implemented with the start of the new timetable for 2013 after intense discussions (DB Netz AG 2014). In addition the so-called "ll-block" obtained its permission in 2013. The "ll-block offers" cost advantages for the modernization of existing freight wagons compared to the "k-blocks" already permitted since 2003. Switzerland as well as Germany offer also funding programs with direct subsidies for the retrofitting of existing freight wagons. During the project life span of CODE24 modernized and noise reduced freight trains could be observed on the corridor for the first time.

The discussions with stakeholders mentioned above showed that the majority knew this measure and were informed that the noise of freight trains could be reduced by modifications at the breaking system of the wagons. In addition, many stakeholders also had an idea about the range of possible noise reduction on dB(A).

One has to keep in mind that almost all of the stakeholders were already involved in rail freight noise issues. However several information gaps and misinterpretations appeared.

The stakeholders misinterpreted frequently that the measure would affect the sound of the train when it brakes and not the rolling noise. This misinterpretation emerged in every group of stakeholders named before. Hence many of the affected people, decision-makers and planners thought that the measure's impacts would be limited spatially to those parts of the corridor where the trains reduce their speed frequently, e.g. shortly before stations, freight terminals or shunting yards. For this reason one of the main advantages of the measure, namely reducing the noise in every location along the track, was not taken into account by these stakeholders. Many of the people involved gave a low priority to this measure and argued, that it would not have any effect in the specific part of the corridor they are interested in.

During the project life span of CODE24 noise-reduced freight trains already operated on the corridor. However, many of the stakeholders still did not recognize them. The perception of increasing train numbers and general questions of noise protection defined the discussions. In fact they already recognized unconsciously the less noisy trains in their daily life. The stakeholders were not aware of the relation between the—sporadically—appearance of more silent trains and the implemented measure. The acceptance of the measure "retrofitting rolling stock" increased considerably from the moment on persons became aware of this relation.

Furthermore some people were informed about the supposed noise reduction of the measure. However while discussing, difficulties became apparent to assess the noise reduction in the context of the other sources of noise building up the entire complex noise situation. This fact is crucial if noise mitigation measures convert a noise situation from a situation that is dominated by a single source of noise to a situation that is characterized by a mixture of several sources of noise with a similar volume. This is the case if the so far dominating sound is reduced to a level that it is embedded in the overall noise situation. It might happen as well that the dominating source of noise is decreased to a background sound hardly to identify. It is very difficult to comprehend and assess such drastic changes of a given noise situation. This is crucial in particular in the densely populated areas where the assessment of a noise mitigation measure has to take into account complex noise situations with regard to other infrastructures and the soundscape of the settlement areas.

Another significant aspect represents one main planning principle for an integrated planning of infrastructure and spatial development: It is the bundling of infrastructures. Because of the logarithmic scale of audible sensitivity of the human ear, bundling of infrastructure reduces the impacts of additional sources of noise—"40 dB(A) + 40 dB(A) = 43 dB(A)". Notably the people affected by noise but also political decision-makers refuse the planning principle of bundling. In the course of the discussion, it becomes apparent that the refusal is determined by the difficulties to integrate the mentioned logarithmic correlation into the assessment. Rather the other way round, the assessment is based on a linear scale—"40 dB(A) + 40 dB (A) = 80 dB(A)". These systematically occurring misinterpretations often explain the difficulties in gaining acceptance for worthwhile planning concepts. Once again in such situation, the instrument "auralisation" could be very useful because the overlapping of different sources of noise are presented in a very intuitive way without any calculations rules being contradictory to the intuition of the target groups.

4 Conclusion

The example of the noise mitigation measure reducing railway noise at the source by "retrofitting the rolling stock" clearly shows that misinterpretations and information gaps are not the exception but rather the rule. The mode of functioning and the effectiveness are misjudged systematically. It has to be kept in mind that involved people in the project's discussions have already dealt with noise issues: civic associations, citizens' action committees, local and regional political bodies and experts of local planning authorities.

These misinterpretations lead to a situation that a given noise mitigation measure is not claimed by the affected people and not included in the debates by political decision-makers. In the end, it is not considered as a possible solution by the planners. Consequently, the public does not appreciate the implementation of the measure and does not value the efforts made to find a solution. Hence the implementation does not raise the overall acceptance. Neither for the infrastructure companies nor for the logistic industry incentives exist to implement the measure. On the contrary building noise protection walls promises a more substantial improvement of public acceptance although the effect of this measure is limited to its specific location and has disadvantages like aesthetic deficits and the fragmentation of areas.

To sum up, CODE24 showed that the instrument of auralisation offers substantial advantages:

- It enables unexperienced people to assess noise mitigation measures based on their effects.
- It allows to judge combinations of noise mitigation measures.
- It does not deal with abstract aggregations about the frequency spectrum and the time but with real single noise events being relevant for affected people.
- It offers the possibility to overlay several sources of noise and thereby to assess noise events in the context of complete soundscape.

All the arguments illustrate that the instrument of auralisation has an outstanding potential for more rational, traceable and effective planning and decision-making processes by countervailing typical misinterpretations.

References

DB Netz AG (2014) LaTPS: Förderung der Umrüstung und des Einsatzes leiser Güterwagen. http://fahrweg.dbnetze.com/fahrweg-de/start/produkte/trassen/trassenpreise/latps.html

Institute for Spatial and Landscape Development, ETZ Zurich (ed) (2013) A common strategy for the corridor Rotterdam-Genoa. Institute for Spatial and Landscape Development, ETZ Zurich, Zurich

Leistner P, Naßhan K, Keilhacker P, Hellbrück J (2013) Wirkungsbezogene Gesamtlärmsimulation und –bewertung. Research report on behalf of Landesanstalt für Umwelt, Messungen und Naturschutz Baden-Württemberg, Stuttgart

Saalbach J (2012) Die europäische Entwicklungsachse Rotterdam—Mannheim—Genua. Informationen zur Raumentwicklung 7(8):439–450

Scholl B (2007) Strategien der Raumentwicklung an der Nord-Süd-Transversale für Europa. In: Scholl B (ed) Langfristperspektiven für eine integrierte Raum- und Eisenbahnentwicklung am Hochrhein und Oberrhein. Arbeitsbericht des Instituts für Städtebau und Landesplanung, Universität Karlsruhe (TH). Author, Karlsruhe, pp 11–24

Schweizerische Eidgenossenschaft, Bundesverwaltung (2014) 742.122 Eisenbahn-Netzzugangsverordnung vom 25. November 1998 (NZV). http://www.admin.ch/opc/de/classified-compilation/19983395/index.html

Sebastian Wilske is vice director and chief planner of the regional planning association of the Middle Upper Rhine located in Karlsruhe. After studying civil engineering with a focus on spatial and infrastructural planning he worked as a scientific assistant at the University of Karlsruhe and the ETH Zürich. He did his dissertation on the subject of land use management in densely populated regions. His field professional interests comprises settlement, infrastructural and landscape development as well as land use management on the regional scale. In the project CODE24 he was the action leader for the action "planners toolbox for innovative noise protection."

Management of Ecological Compensation Measures

Claus Peinemann

Abstract CODE24—Corridor Development Rotterdam–Genoa, a project approved under the Strategic Initiatives Framework of the INTERREG IVB program, indicates a future development of the major European north-south axis connecting the Dutch port of Rotterdam and the Italian port of Genoa. Fifty percent of the north-south rail freight is operated along this corridor where 70 million inhabitants are living in this highly populated catchment area. Within the overall project CODE24, one action is focusing on innovative methods and solutions to reduce the negative impact of the corridor's development on the environment. In a European project context, a comparison has been done of planning methods and policies concerning environmental compensation management caused by building and infrastructure projects. Partners from the Netherlands, Germany, Switzerland and Italy were involved in the evaluation of different policies. The main aim of this action is to show how development and infrastructure projects along the corridor could be harmonised with national and European regulations related to nature and environmental protection. The strong linkage between possible impact and compensation measures needs to be more adequately managed in the future. Although legal preconditions for pooling and stocking of real estate and other compensation measures for major infrastructure projects were accomplished, their realisation and the implementation of impact regulations are often ineffective in respect of sustainability.

As a result of questionnaires and interviews for new, innovative and successful approaches for managing ecological compensation measures, the main objective remains to encounter the problem of finding suitable sites for compensation measures. The diversity of existing regulations concerning ecological compensation has been found to be extremely high. Some partner countries currently discuss or even prepare common regulations for ecological compensation. This adoption of a legal framework can be seen as a positive development in nature conservation policy to

Preferred citation: Peinemann, C. (2014). Management of Ecological Compensation Measures; Drewello, H. and B. Scholl (eds.), Integrated Spatial and Transport Infrastructure Development. The Case of the European North-South Corridor Rotterdam-Genoa, Springer-Verlag, Cham.

C. Peinemann (✉)
Verband Region Rhein-Neckar, P 7, 20-21, Mannheim, 68161 Germany
e-mail: Claus.Peinemann@vrrn.de

H. Drewello, B. Scholl (eds.), *Integrated Spatial and Transport Infrastructure Development*, Contributions to Economics, DOI 10.1007/978-3-319-15708-5_16

enhance a wider acceptance and transparency for implementation. A large-scale impact, such as a linear infrastructure development, requires a large-scale compensation. Stakeholders on a regional level should use the opportunity for necessary large-scale nature conservation projects and compensation pool-building. Coherent compensation pool-building, managed within an overall planning concept, will bring forward strategies for coherent sites and effective and sustainable measures that will show visible effects and real habitat improvements.

1 Goal of the Contribution

The project investigates intelligent spatial management solutions, in the sense of compensation management, to secure a sustainable development, thus making planning conditions more effective and implementation more focused. Special attention is given to protect and develop natural resources with consideration for European policies and regulations, such as Natura 2000, the Strategic Environment Assessment, the European Landscape Convention, or the European Water Framework Directive. Further methodologies for compensation areas are being explored with the result that impact regulation can be used as a tool to fulfil local and regional objectives of nature and landscape protection policies as well as economically oriented project development.

Within the CODE24 Corridor, it is necessary to gather appropriate methods and instruments to counterbalance the environmental impact of larger linear infrastructure projects in densely populated and heavily used regions. As this is in a European context, the project shows how Dutch, German, Swiss and Italian stakeholders deal with the problem of not having sufficient space available for environmental impact compensation. Responding to this question allows sharing efficient solutions and detecting future trends. The project, Management of Ecological Compensation Measures, does not want to discuss the existing differences in environmental impact evaluation, the different methods for calculating compensatory needs, nor the obligation to avoid impacts. The focus is on the implementation of required and concrete compensation measures. Gathering experiences, best-practice examples and case studies, discussing common problems that occur during environmental impact compensation and selecting as well as describing several effective instruments ('effective' in a sense of a valuable contribution to the aims of nature protection and of a sustainable treatment of the resource 'space') of the countries involved will enable a transnational exchange and, thus, a benefit for all participating partners.

In this respect, it is also relevant to consider and elaborate new ways of achieving partnership and consensus with a multitude of stakeholders in order to mitigate conflicts about land use.

2 Ecological Compensation

'Ecological compensation may be defined as creating, restoring or enhancing nature qualities in order to counterbalance ecological damage caused by infrastructure projects' (Iuell et al. 2003). This process is of crucial importance with respect to the no net loss principle as an essential policy of the European Biodiversity Strategy 2020 (European Commission 2011). Nevertheless, ecological compensation remains a 'last resort solution, which applies when negative impacts are inevitable since avoidance as well as minimisation measures are exhausted' (Iuell et al. 2003). As described in the guidebook, *COST 341* (Iuell et al. 2003), 'compensatory measures are mainly implemented on a voluntary basis, rooted in agreements between project developers, nature conservation trusts, landowners or other stakeholders.' Seeking suitable land for compensation measures is mostly a time-consuming and exhausting process that often leads to insufficient solutions.

In general, compensation measures should be implemented close to the impacted area. In reality, this is sometimes impossible, since suitable sites are often unavailable. However, the scale for environment compensation is usually the natural region (*Naturraum*) and on this quite large scale, it should be more possible to find adequate sites for the realisation of measures. As Rega (2011) mentions, 'whilst they could be good sites from an ecological point of view, a social question arises, as the benefits of the compensation could be enjoyed by other human communities than the one that is primarily affected.' And of course, a municipality is interested in saving sites to compensate the impact deriving from their own projects. Moreover, an expropriation of property of private citizens or entities is feasible for public use only in exceptional cases.

According to Rundcrantz and Skärbäck (2003) in his article, 'Environmental Compensation in Planning', the term 'compensation' is used in the research to include both the meaning of restoration compensation, which is the environmental compensation for a loss of environmental values in the right functional context, i.e. in-kind compensation, and replacement compensation, which is the environmental compensation for lost environmental values implemented in another functional context, i.e. out-of-kind compensation. The measures can be implemented on site, meaning within the landscape-ecological context of the impacted area or off-site meaning the opposite.

2.1 General Requirements for Ecological Impact Compensation for Infrastructure Projects

The main ecological impact of larger linear infrastructure projects consists of habitat isolation through landscape fragmentation. Landscape fragmentation is one major reason for the loss of biodiversity (Iuell et al. 2003). If avoidance measures are exhausted, impact mitigation and compensation should focus on the reinforcement or (re-)creation of ecological corridor functions. Creating new patches between core habitats, enlarging existing areas of high quality to increase the habitat potential for more species or individuals and the closing of ‘low level’ road networks may represent appropriate compensatory measures. In order to support a sustainable means of impact compensation, the measures should be integrated in local conservation and land use plans. The integration in larger plans protects the compensation sites against future development and ensures a durable functioning of the chosen measures. For the same reason, the subsequent management and monitoring, which have to be included in the overall compensatory plan, should be transferred to compensation agencies or other conservation bodies. In general, compensation sites should preferably require low management input.

2.2 Comparison of the CODE24 Partners

Legal regulations and guidelines, which represent the framework during every environmental impact compensation process, were of major interest. Besides specific national legislation, the focus was also on European directives and conventions that are common to several of the nations involved. Throughout the research, important institutions, organisations, associations and other bodies that are involved in the process of compensation management, were listed. In addition, best-practice examples were researched in all countries involved as a way to follow the application of specific compensation management systems. Relevant experts from all affected countries were identified. Information was gathered using questionnaires and expert interviews (CODE24-Action 5 Final Report 2014).

3 Management of Ecological Compensation Measures in Germany

3.1 Institutions and Stakeholders

The planning of compensation measures is usually realised by an assigned environmental consultancy agency that aligns potential compensation measures with the relevant nature conservation authorities, hence the district or urban municipality. Depending on the kind and size of development, the highest nature conservation authorities are involved as well. The compensation planning has to be authorised by the approving authority. In the case of railway projects, the Federal Railway Authority (EBA) (*Eisenbahnbundesamt*) has to approve the compensation planning.

3.2 Legislation and Regulations

Ecological compensation measures are legally rooted in:

1. The German impact regulation principle (*Eingriffsregelung*)
2. European regulations concerning the protection of habitats and species
3. The Federal Forestry Act

The *Eingriffsregelung* being the German instrument that regulates ecological impact compensation in general, is often translated as 'impact mitigation regulation', 'impact regulation principle' or 'compensation principle'. This legal tool has existed since 1976 and is rooted in par. 13, 14 and 15 of the German Federal Nature and Landscape Conservation Act (BNatSchG). According to the 'the polluter pays' principle, the project developer causing an impact on nature and landscape needs to minimise the unavoidable impact as far as possible and must compensate for any remaining impact.

Since 2010, restoration compensation does not have priority over replacement compensation anymore, but is on the same legal level. In exceptional cases, and as a last resort, monetary compensation is possible. Unlike regulations of other countries (namely Italy, the Netherlands and Switzerland), the German impact regulation principle applies to the total area of the development under consideration. More detailed regulations about eco-accounts, pooling, and trading of compensation measures are fixed by federal state law, e.g. the Hessian *Kompensationsverordnung* (KV) (compensation decree) regulates more specifically how ecological impact compensation should be carried out in the federal state of Hessen. For example, ecological compensation measures should be preferably realised within Nature 2000 sites whilst areas of high value for agricultural land use should be avoided for compensation planning. Furthermore, compensation measures should be bundled as far as possible and as reasonably as possible.

According to Article 6 of the European Fauna-Flora-Habitats Directive of 1992 (92/43/EEC), a plan or a project affecting Natura 2000 sites needs to consider specific measures in order to offset the negative effects of the development and to maintain the overall ecological coherence of the Natura 2000 Network (coherence measures). The German Nature Conservation Act implements this aspect through par. 34 on national nature conservation law.

3.3 Methods and Instruments

In the following section different methods and instruments present objectives to handle questions on environmental compensation policies. Practical approaches should show how compensation can be more effective.

3.3.1 Eco-accounts

The 'Report on best practices for limiting soil sealing and mitigating its effects' by the European Commission (2011) describes the functioning of eco-accounts as follows: 'The eco-account system is based on trading eco-points. Developments requiring nature compensation measures according to the National Nature Conservation Act are charged with eco-points. Developers have to prove that compensation measures of equal value are being carried out somewhere else. Eco-points can be acquired at compensation agencies, which are officially authorised to carry out compensation measures.' Compensation agencies are owners of eco-accounts and can sell eco-points. The price of one eco-point is defined by law, e.g. in the federal state of Hessen, one point has a financial value of 0.35 €. According to Küpfer (2008), in general, an eco-account should be developed out of a landscape plan that covers the entire surface of a municipal district or region. 'The potential of these landscapes within the districts for ecological improvement measures are evaluated. The appropriate and available lots are transferred to a pool. As soon as a measure on one of these lots is realised, it can be transferred to the eco-account and used as a compensatory measure for any impact.'

3.3.2 Compensation Pools

Bundling compensation measures on large and coherent sites forms pools. A project developer can obtain parts of a pool, meaning that the developer himself doesn't need to implement compensation measures. The compensation requirements of numerous developments can be assigned to one single pool. Possible measures are the creation of wetlands or planting hedgerows. Several types of measures can also be combined in one pool. The advantages of compensation pools are mainly:

- Implementation of the impact regulation principle is more efficient
- Spatial improvement of compensation effects through size and/or coherency of areas
- Measures are planned in an ecologically sensible way (not according to available sites or under time pressure)
- Safeguarding of sites at early stage and in a resilient way
- The implementation and monitoring of compensation measures can be assured
- Measures are more sustainable (long-term stewardship)
- Better integration of compensation measures within a larger planning goal, e.g. habitat connectivity
- Creates an impression of worthwhile investment through visible sustainable effects deriving from compensation measures thus leading to higher public acceptance and fewer conflicts, especially with farmers

In general, no legal or procedural changes are necessary to implement a pool. Compensation pools need a responsible body for planning, realising, managing and monitoring the pool areas. Compensation agencies are a suitable service provider for ensuring high-quality compensation pools. Compensation pools may be integrated in regional park concepts, as practiced in the Regional Park Rhein-Main (www.regionalpark-rheinmain.de). Possible compensation measures in regional parks are, e.g. restoration of rivers and river banks, planting alleyways, hedgerows or the creation of green corridors and other biotopes, small bodies of water, etc.

3.3.3 Compensation Agency

Compensation agencies are innovative service providers for nature conservation that offer planning, realisation, safeguarding and management, as well as monitoring compensation measures through the creation of compensation pools. A foundation regulated by public law would be a possible owner of such an agency, for example, the Compensation Agency Brandenburg (www.flaechenagentur.de).

Compensation agencies often provide all-in-one-packages, including the implementation of compensation measures and a long-term-stewardship, e.g. 25 years. If compensation measures require maintenance, land users, e.g. in agriculture or forestry, are often involved in pool management.

The German Assembly of Compensation Agencies (www.verband-flaechenagenturen.de), responsible for lobbying, public relations and the organisation of conferences, published a list of quality standards for compensation agencies that should be retained during the work of agencies such as:

- Ecological improvement (*Naturschutzfachliche Aufwertung*)
- Long-term stewardship and site management (*Langfristige Sicherung von Maßnahmen*)
- Documentation and monitoring
- Integration into regional plans and strategies (Fachliche Abstimmung und Einbindung in übergeordnete Strategien)
- State of the art planning quality

3.3.4 Compensation Register

Par. 17 no. 6 of the German Federal Nature and Landscape Conservation Act demands that competent nature conservation authorities create a central register of compensation measures. The federal state of Baden-Württemberg, for example, implemented this national regulation in federal law by releasing a separate compensation register decree (*Kompensationsverzeichnis-verordnung*), which prescribes the responsibilities and the contents of this register. Restoration and replacement measures and eco-account measures (realised (stocked) compensation measures that haven't yet been assigned to a distinct impact) need to be recorded, always accompanied by information concerning location, type and temporal obligations of the measures and, if existing, the assigned impact. Central tasks and objectives of the compensation register are:

- Enable the verification of the implementation and functioning of planned compensation measures
- Avoid the planning of compensation measures on sites already assigned
- Allow stocking compensation measures
- Avoid the assignation of stocked measures to more than one impact

3.3.5 Integrated Compensation Measures

An integrated compensation measure (ICM) consists of long-term agricultural land use under specific (nature conserving) restrictions whereby nature and habitat qualities should be enhanced, in particular, endangered open land species should benefit from the adaption of farming measures. The farmer changing from intensive to extensive land use is paid for the difference in the yield by an investor. The measures are monitored and accompanied by biologists and/or ecologists and have to be accepted as compensation by the nature conservation authorities. ICMs are a cooperative way of compensation management since the participation of regional stakeholders is required. ICM can consist of, for example:

- Developing a habitat for arable weeds
- Developing a habitat for red kite or hamsters
- Developing flower strips as a habitat for farmland birds, rabbits, partridges, and insects
- Protection of farmland birds
- Organic farming (can be a compensation measure, according to the regulations of the EU)

In contrast to similar agri-environmental measures (AEM) of the European Union, ICM last longer (20–30 years). The specific restrictions for ICMs are defined jointly by authorities and farmers and are thus more adapted to the needs of the farmers who, as a result, show increased willingness to cooperate. Furthermore, unlike AEM, exceptions are possible in an ICM, e.g. application of herbicides, which means more flexibility for the land user.

ICMs create win-win situations: ecological compensation measures are implemented without losing precious ground to agricultural exploitation and, at the same time, the ecological value of the agricultural landscape rises.

3.3.6 Monetary Compensation

In the case of authorised projects having a negative impact on the environment that cannot be avoided, minimised or offset, a monetary compensation may be accepted as a last resort solution (see German Nature Conservation Act par. 15 no. 6). Fixed by the responsible nature conservation authority, the amount corresponds to the average costs of the potential but unrealisable compensation measures, including their planning, maintenance, monitoring and management. If those costs are not assessable, the monetary compensation is calculated based on the duration and intensity of the impact and the benefits derived for the developer.

Monetary compensation is usually paid in advance of the development and should be spent on distinct nature and landscape conservation projects within the same natural region. The nature conservation projects must not depend on other legal obligations.

4 International Comparison

The main principles of ecological compensation are valid for all countries studied: the 'polluter' pays for damage done to the natural environment and needs to follow the hierarchy of (1) avoidance, (2) minimisation and (3) compensation. But, when it comes to comparing the significance of the term 'environmental compensation', some fundamental differences are already noticeable.

Even though all the countries studied have regulations on environmental compensation, the intensity of legal support varies significantly. While Switzerland and Germany have stringent laws and provisions on this topic, Dutch legislation has a medium level and Italian legislation a rather weak support through binding regulations.

However, some fundamental changes have occurred during recent years and now the Netherlands, in particular, has more stringent provisions than 10 years ago. In contrast to that, the reformation of the German Federal Nature and Landscape Conservation Act renders the quite stringent German impact regulation principle more flexible, for example, by permitting ecological compensation measures within the same natural region of the impact or by overriding the former hierarchy of replacement measures prior to restoration measures, unlike in Switzerland, where replacement measures still have legal priority.

4.1 International Workshop

During an international workshop in Mannheim, various experts from Germany, Italy, the Netherlands and Switzerland discussed current methods and perspectives of managing ecological compensation measures. Common constraints of managing ecological impact compensation were described that referred to the densely populated area between Frankfurt a.M. and Mannheim as a case study area. Seven experts with diverse professional backgrounds, ranging from consultants in planning agencies to the developer of large infrastructure projects, gave insights into the ecological compensation management of their countries and presented distinct best-practice examples and innovative solutions to improve the compensation process. The workshop contributed to creating a network of experts who can assist in developing new strategies or improving existing models for ecological impact compensation, with respect to larger linear infrastructure projects (CODE24—Action 5, Final Report 2014).

Even though the Swiss and German legislation about environmental compensation are quite similar, there is a fundamental difference regarding the area where the regulation applies. While the German impact regulation principle is valid for the total surface affected by an impact, the Swiss, and also the Dutch regulations, apply to selected areas. In Italy, the term not only includes compensation of ecological values, but also socio-economic values. So it is quite difficult to study and compare the existing Italian regulations, provisions and habits that refer strictly to ecological compensation. As multiple studies point out (e.g. Rundcrantz and Skärbäck 2003; Jessel 2003; Peters et al. 2002), not only concerning the application area, but also in general, Germany seems to have one of the most stringent and developed legislations about environmental impact compensation.

Nevertheless, it is noticeable that, apart from the diversity of existing legislation, similar problems occur during environmental impact compensation: all the countries involved are currently discussing the creation of a joint methodology for the evaluation of compensatory needs. Increasing problems of public acceptance concerning the impact itself, as well as compensatory obligations, require common rules on a federal level. It is obvious that the acceptance of compensatory obligations will be reduced if the same impact in the Canton of Graubünden and the Canton of Basel, or from a German point of view, in Hessen and in Baden-Württemberg, result in totally different compensatory requirements regarding quantity and quality.

A standard methodology should, of course, be simple in its application and flexible enough to meet all the requirements of the federal states, regions, cantons or provinces. Nevertheless, the main problem is that the implementation of such a framework requires a strong political will and remains a great challenge for the future.

In conclusion, especially in Italy, there seems to be a demand to further develop a fully binding legal framework on impact compensation in a strict ecological sense. The lack of clear rules does not necessarily facilitate the procedure of environmental compensation, as an Italian expert, being involved in the case of the Autostrada Pedemontana Lombarda, underlined.

4.2 Regional Workshop

During a regional workshop in September 2013 in Mannheim, various German experts from Baden-Württemberg and Hessen discussed possible solutions for implementing an efficient compensation management in densely populated and highly used regions, such as the area affected by the future high-speed railway connection Frankfurt–Mannheim. Four experts of different professional backgrounds presented their individual points of view on compensation management and the local issue of a lack of available space.

Professor Dr. Werk is the Head of the Department of Landscape Architecture at the Geisenheim University and Deputy of the Federal Executive Committee of the Bundesverband beruflicher Naturschutz (B.B.N) (Federal Association of Occupational Nature Conservation) and participated in the current evolution of the Bundeseinheitliche Kompensationsverordnung (BKompVO) (German Federal Decree on Environmental Compensation) by writing a statement on the first draft of the decree. In general, a federal compensation decree (BKompV) would be a benefit for all stakeholders involved in compensation management: a consistent framework for the evaluation of inventory and impact, as well as a standardised financial compensation would create uniform conditions and improve the acceptance of ecological compensation. The current discussion reveals numerous cruxes for a nationwide compensation decree. Deviations from the BKompV in favour of distinct industries, e.g. the energy branch, are contentious issues. Federal diversity within the German States obviously hinders a common solution. The primary objective remains: to achieve an agreement that results in an enforceable and stable decree.

Matthias Mähliß from the DB ProjektBau (subsidiary company of the German railway company Deutsche Bahn AG) is responsible for the internal planning and implementation of compensation measures for the future high-speed railway connection Frankfurt–Mannheim. He describes the findings, achievements and expectations deriving from the past 6 years of an intensive planning process. The planned high-speed railway connection Frankfurt–Mannheim requires compensation measures on about 1000 ha. But, southern Hessen does not have any more ground available because the need for space deriving from other realised or planned projects is huge. In addition, the dynamics of the nature conservation regulations represents only one problem of this railway infrastructure project that started in 1993. Besides compensatory needs deriving from the German impact mitigation regulation, compensation tasks result from a more stringent European law, e.g. on species conservation and environmental damage (Directive 2004/35/EC, Umweltschadens-gesetz). Even though compensation management was initiated at an early stage of the project, the lack of available sites for compensation still occurred. This is due to discrepancies between the original compensation planning and the real availability of sites.

Matthias Pollmeier of the Bundesforstbetrieb Schwarzenborn (Federal Forests Schwarzenborn) part of the Bundesanstalt für Immobilienaufgaben (BImA) (Institute for Federal Real Estate) explains possibilities of the so-called 'green conversion'. The Bundesforstbetrieb Schwarzenborn offers ancient military ground for future ecological compensation and thus offers an innovative solution to improve the compensation process. Offering entire properties (up to 250 ha) mainly for compensation and conservation tasks, the Bundesforstbetrieb is especially relevant for developers of larger infrastructure projects. Since over 130 properties are dispersed all over Hessen, compensation sites are usually available close to developments. The organisation is aware of the long-term responsibility for maintaining the target habitats and the related financial risks. Until now, generally positive experiences have been made with the commercialisation of ecological compensation measures. Currently, it is being considered whether to enable project developers to make reservations of unplanned properties.

Gerhard Eppler is the President of the Naturschutzbund (NABU) (Nature and Biodiversity Conservation Union) Hessen, the regional association of one of the largest German organisations. He illustrates that the planned railway infrastructure connection between Frankfurt and Mannheim will pass a region of high ecological value, a so-called 'biodiversity hotspot'. Ecological compensation needs to have a functional, temporal and spatial relationship to the 'harmed' component of nature. From the NABU point of view, compensation planning should make use of the external knowledge of local nature conservation authorities or nature conservation organisations. Species conservation programs or existing regional nature conservation programs should be supported. Often, it can be useful to process conservation instead of planting new trees (afforestation) or some habitats with long lifecycles, forests in this case.

5 Green Infrastructure

Green infrastructure (GI) can be defined as an interconnected network of green space that conserves natural ecosystem values and functions and provides associated benefits to human populations. In this sense, a green infrastructure is the ecological framework needed for environmental, social and economic sustainability. Therefore, it is an approach to open space planning in an effort to combine conservation values and action in concert with land development, growth management and built infrastructure planning as a strategic approach (Benedict and McMahon n.d.).

The Europe 2020 biodiversity strategy emphasises 'that our natural heritage is a major ecological asset which is fundamental to human well-being.' Therefore, 'all Member States should cooperate and coordinate their efforts in order to ensure more effective use of natural resources and avoid net losses in terms of biodiversity and ecosystem services.' Furthermore, the European Commission 'recognises that infrastructure-building, urbanisation, industrialisation and physical intervention in the landscape in general are among the most significant drivers of the fragmentation of ecosystems and habitats.' Consequently, the Commission 'calls on local, regional and national governments (...) to consider these factors, which pose a threat to ecosystems and habitats, in their planning and development projects on both a large and a small scale' (European Commission 2011).

As part of the EU biodiversity strategy, the European Commission will push the full implementation of the Birds and Habitats Directive by developing a green infrastructure (GI) across the EU. A GI is 'a strategically planned network of natural and semi-natural areas with other environmental features designed and managed to deliver a wide range of ecosystem services' (European Commission 2013). Amongst others, designated Natura 2000 sites form the hub of a European GI. The main target of the GI is to reduce ongoing fragmentation of the European landscape as the main reason for the global loss of biodiversity.

Considering those excerpts of the EU 2020 biodiversity strategy, including the Green Infrastructure network, it is obvious that a strategic enhancement of the European transport axis Corridor 24 has to be in line with those strategies by including concepts and solutions for avoiding, minimising and compensating for any negative impact on the natural environment. But the bottleneck situation of Corridor 24 not only applies to transport capacities; the bottleneck also exists from an environmental point of view. An ecological bottleneck is a sustainable project implementation in respect of all existing nature conservation policies, i.e. the European directives concerning habitat and species conservation, becomes an increasingly challenging task—especially in densely populated and intensively used regions. Hence, space-saving mitigation and compensation measures focusing on the reinforcement or (re-)creation of ecological corridor functions, e.g. eco-ducts, wildlife tunnels, are of fundamental importance.

All-in-all, Green Infrastructure plans provide a blueprint for conservation in the same way that long-range transportation plans provide a blueprint for future roads or railway lines. Those plans can create a framework for future growth while also ensuring that significant natural resources will be preserved for future generations. It is even possible to reduce opposition to new development by assuring civic groups and environmental organisations that growth will occur only within a framework of expanded conservation and open space planning (Benedict and McMahon n.d.).

6 Conclusion

In Action 5 of the CODE24 project, a quest to find new, innovative and successful approaches to 'managing ecological compensation measures' was carried out. The main objective remains to deal with the problem of finding suitable sites for compensation measures. Serving as a base of knowledge, a synopsis of current existing regulations, methods and instruments of environmental impact compensation along Corridor 24 was carried out. The diversity of existing regulations concerning ecological compensation was found to be extremely high. Some partner countries are currently discussing or even preparing common regulations for ecological compensation. This adoption of legal frameworks is a positive development in nature conservation regulation and should be enhanced in all partner countries.

Another central task of Action 5 was the identification of factors contributing to a successful and efficient planning and management of ecological compensation measures. In this context, the following recommendations, which come up repeatedly, should be considered during the planning process:

- Focus on large-scale, coherent and ecologically reasonable compensation measures
- Start planning implementation at an early stage and with a long-term view
- Avoid the use of third party land and ensure a long-term safeguard of compensation sites
- Use caution with colliding municipal land-uses and trade-offs within nature conservation interests
- Include local land users at an early stage of compensation planning
- Seeking an early reconciliation (intense communication process) creates win-win situations
- Compensation management should be part of regional land management
- Making use of already existing concepts, plans and strategies, e.g. regional parks, management plans
- Favour low maintenance input for functioning of compensation measures
- Compensation planning should be understood as management tasks
- Create participative processes
- Make use of public relations and image building for a successful external presentation (measures need to be visible and tangible)
- Consider control systems and monitoring concepts in compensation planning

Large-scale impacts require large-scale compensation. This is especially valid for large infrastructure developments. Broad concepts with coherent compensation sites, e.g. compensation pools related to green conversion as they are created and offered by professional service providers (compensation agencies) and usually meet the recommendations listed above. They represent a one-stop solution that allows proceeding with the planning of the original development project. Smaller compensation measures, for example, integrated compensation measures, are especially useful for species conservation-related compensation in open land habitats.

With the focus on large infrastructure projects linking the metropolitan regions Rhein-Main and Rhein-Neckar, green conversion represents a unique opportunity for large-scale nature conservation projects and compensation pool-building. Another large potential for coherent compensation concepts consists of forest-related compensation as offered by the Hessian State Forest. Nevertheless, all compensation planning within the narrow corridor formed by the Rhine and the Odenwald should be part of a regional land management plan that has to be coordinated by a regional partner, e.g. the Regional Association of Rhein-Neckar (VRRN). The latter could adopt and coordinate a regional compensation pool such as that built up in the Bodensee region by the Regional Association Bodensee-Oberschwaben. Regional collaboration in policy networks will grow in importance, in part because no single planning organisation on the local level is usually capable of regulating complex landscape development processes.

The main objective of ecological compensation should be the following, as M. Szaramowicz of the Flächenagentur Brandenburg GmbH summed it up during the international workshop in Mannheim: Coherent sites, effective and sustainable measures, visible effects and real habitat improvement.

Following those standards is a guarantee for successful compensation management and its related public acceptance.

References

Benedict MA, McMahon ET (n.d.) Green infrastructure: smart conservation for the 21st century. Sprawl Watch Clearinghouse, Monograph Series, Washington, DC

European Commission (2011) Communication from the commission to the parliament, the council, the economic and social committee and the committee of the regions. Our life insurance, our natural capital: an EU biodiversity Strategy 2020, Bruxelles

European Commission (2013) Communication from the commission to the European parliament, the council, the European economic and social committee and the committee of the regions. Green Infrastructure (GI) – Enhancing Europe´s Natural Capital, Bruxelles

Gesetz über Naturschutz und Landschaftspflege (Bundesnaturschutzgesetz) (German Federal Nature and Landscape Conservation Act) (2010) Artikel 1 des Gesetzes vom 29.07.2009 (BGBl I S. 2542) in Kraft getreten am 01.03.2010; Stand 01.09.2013 aufgrund Gesetzes vom 06.06.2013 (BGBl I S.1482)

Habitats Directive: Council Directive 92/43/EEC of 21 May 1992 on the conservation of natural habitats and of wild fauna and flora

Iuell B, Bekker HGJ, Cuperus R, Dufek J, Fry G, Hicks C, Hlavác V, Kéller V, Rosell C, Sangwine T, Tørsløv N, Wandall B (eds) (2003) COST 341—Wildlife and traffic: a European handbook for identifying conflicts and designing solutions. KNNV Publisher, Delft, p 176

Jessel B (2003) Die Eingriffsregelung: Bewahrung des Status quo von Natur und Landschaft—mittlerweile ein Alltagsgeschäft?—Schriftenreihe des Deutschen Rates für Landespflege, vol 75, pp 65–75

Küpfer C (2008) The eco-account: a reasonable and functional means to compensate ecological impacts in Germany. Arquitectura e vida. Schriftenreihe des Institut Súperior de Agronomia (ISA), Universidale Téchnica de Lisboa

Peters W, Siewert W, Saramowicz M (2002) Folgenbewältigung von Eingriffe im internationalen Vergleich, vol 82. BfN-Skripten, Bonn

Rega C (2011) SEA and ecological compensation in land use plans. Special Conference on Strategic Environmental Assessment, IAIA SEA Prague

Rundcrantz K, Skärbäck E (2003) Environmental compensation in planning: a review of five different countries with major emphasis on the German system. Eur Environ 13:204–226

www.bodensee-oberschwaben.de

Dr. **Claus Peinemann** works as a planner at Verband Region Rhein-Neckar since 1998. He is responsible for spatial planning on a regional level. Besides regional planning he is involved in landscape development and environmental planning. The focus lies also on regional strategies and networking to enhance regional development and land use management. His university background are studies in landscape architecture and planning in Berlin, Hannover, Edinburgh (GB) and Berkeley (U.S.A.).

Part V
Regions Participating in Development Procedures of European Transport Infrastructure

The Consideration of Local Preferences in Transport Infrastructure Development: Lessons from the Economics of Federalism

Hansjörg Drewello

Abstract The construction of large public infrastructure projects of national importance, such as nuclear power plants, wind farms, electricity, highway or railway lines, regularly leads to mass protests in the population. The main problem is the impact of negative external effects on the people living nearby, which are not taken into consideration during the planning process by the national builders. Democratic coordination processes fail in solving the challenge, for here the problem of 'institutional incongruence' usually arises. This means that the policy makers responsible for the provision of public infrastructure, its users or those affected by it as well as the taxpayers, who finance these services, are not the same people. If the competencies for decision-making, use and financing are separated from one another, then incentives arise to live at the expense of others. The article examines the case of the expansion of the Rheintalbahn on the southern Upper Rhine using the Coase Theorem, and analyses the conditions under which negotiations between the parties involved can lead to an efficient result when building public infrastructure.

1 Introduction

Railway infrastructure planning is a very complex and interdisciplinary challenge. Against a background of scarce resources and limited public budgets, efficiency should be a major objective of political decision-makers and planners. In view of the endless needs and demands of regions as well as a growing lack of infrastructure in times of increasing freight transport, economic rationality requires a selection and prioritization of the most important infrastructure projects. If this is to be done well, the realization of an infrastructure project must also be the object of efficiency

H. Drewello (✉)
University of Applied Sciences Kehl, Kinzigallee 1, 77694 Kehl am Rhein, Germany
e-mail: drewello@hs-kehl.de

H. Drewello, B. Scholl (eds.), *Integrated Spatial and Transport Infrastructure Development*, Contributions to Economics, DOI 10.1007/978-3-319-15708-5_17

considerations. One specific aspect, which is frequently ignored, is the impact of a new or larger transport infrastructure on the regions concerned. In particular, the prevention of negative effects in the region can hardly be successful without the participation of local authorities and the citizens involved. Recent opposition of the German population to major transport infrastructure projects illustrates well the need to implement regional provisos and knowledge into the planning process.

This article focuses on the results of a case study that analyzed the planning process for the enlargement of the Rheintalbahn from two to four tracks in the Upper Rhine Valley. The first part deals with considerations from the economic theory of federalism. Then the German planning process for the implementation of new railway infrastructure is described. Finally, possible changes in infrastructure planning, especially the organization of negotiation procedures, are discussed in order to optimize this process by including regional expertise.

2 The Economic Theory of Federalism

A number of European countries are characterized by a federal structure. Countries such as Belgium, Germany, Austria, Switzerland, Spain or Bosnia-Herzegovina are shaped by a political system in which political power is exercised at different levels. Usually these are the central government level, state government, province or regional governments as well as the local level. The boundaries here between federal and centralized systems are fluid. The structure of all European countries testifies to the development of federal elements, at least to some degree. Even countries like Italy and France, with a rather centralized structure, have gone through processes in the last few decades whereby political power has been decentralized. Beyond that, the Member States of the European Union have given up political power to a "higher level" of jurisdiction.

If there are federal elements in a country, then the question arises as to which political powers are to be exercised at which level. The economic theory of federalism examines "whether there is an ideal federal structure for a political system" (Neumann 1971, p. 493), which public goods should be provided by which level of the federal structure (Oates 1999, p. 1120) and how any existing inefficiencies can be reduced (e.g. Koppel and Lichtblau 2007, 18ff).

Traditional economics assesses the provision of public goods solely on the basis of their costs and benefits. The provision of public goods is only worthwhile from a macro-economic point of view if the resulting benefits exceed the costs. Along with the costs for the utilization of resources, the economic theory of federalism additionally takes so-called "expected external costs" (Buchanan and Tullock 1962, p. 115) or "preference and frustration costs" (Biehl 1994) into consideration. These are caused by an incorrect or insufficient supply of public goods to the population.

2.1 The Provision of Public Goods in Countries with a Federal Structure

The argument for a decentralized provision of public goods is mainly that preferences for public goods and the benefits resulting from them vary from region to region in a heterogeneous society. Smaller units of self-government, therefore, can be more responsive to individual and local preferences than a central government (Oates 1972, p. 54). If decisions on the provision of public goods are made in smaller units, frustration costs tend to decline.

Increasing returns to scale in the production of public goods and the economies of scale associated with them make the central provision of public goods appear advantageous. In addition, there are benefits of joint usage (sharing economies) as purely public goods can be used by other individuals without causing additional costs (Blankart 2007, 62ff.). So-called "resource, decision or organization costs" (Breton and Scott 1977) must also be taken into account. These costs increase with increasing decentralization of decisions (see Fig. 1).

On the horizontal axis, Fig. 1 shows possible decision levels: neighborhoods and urban districts, cities and municipalities, districts and counties, regions, federal and national states, European Union. The curves for the opposing frustration and resource costs result in a total cost curve with a minimum. This minimum reveals information about the type of decentralization to select at the lowest cost. In the example above, the minimum of TC_1 might represent the level of a federal state. Higher frustration costs (FC_2) lead to a total cost minimum at a comparatively decentralized decision level, e.g. a region.

Realistically, this theoretical justification for the provision of public goods at the different levels of a federal structure of government is not sufficient to carry out a concrete assignment of public tasks to individual government levels. Blankart points out two other methods by which this problem can be solved (Blankart 2007, p. 68). On the one hand, the principle of subsidiarity states that the respective public task can always be assigned first to the lowest level of government, e.g. the municipality. If it cannot fulfill the task or can only do so inadequately, then the next higher level is called upon, e.g. the district. The higher-level authority must demonstrate that it can perform the task better. The problem with this method is that higher-ranking government levels are interested in attracting more and more tasks for reasons of power politics. Another method is the practice of direct democracy. Here the political system and administration first decide which level of government performs which public tasks. If this distribution does not meet with the approval of the voters, they can bring about a change by means of a referendum. Both methods can also be combined.

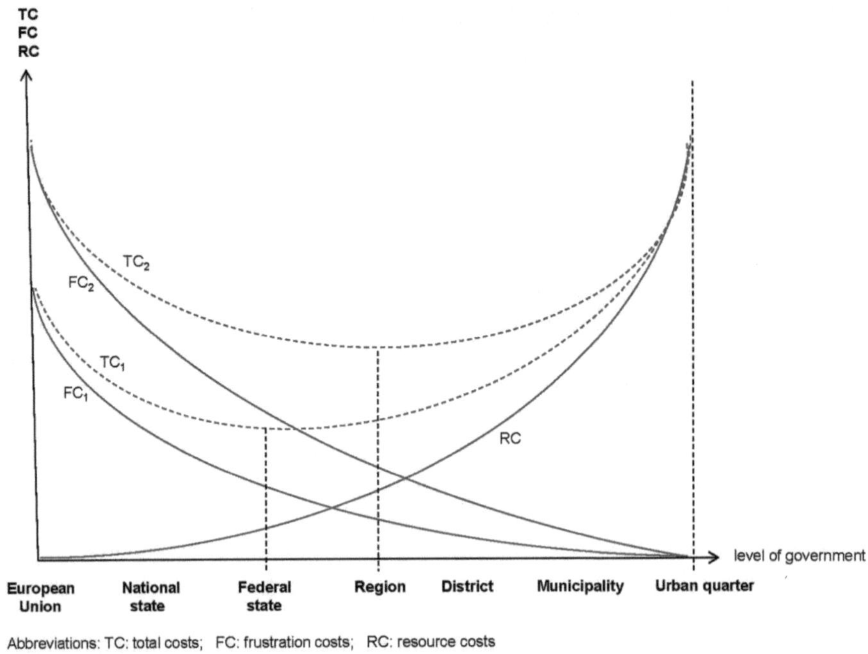

Fig. 1 Basic model of economic federalism. *Source*: Based on Hausner (2005) and Biehl (1994)

2.2 *Institutional Congruence as an Efficiency Criterion for Federal Systems*

Institutional congruence exists if those who make decisions about the provision of public goods, their users (or the people affected) and the taxpayers who fund the supply are located within one area of the federal structure (Blankart 2007, p. 69). If the competencies for decisions, usage and funding are separated, then there are incentives to live at the expense of others. The scope of a public commodity can be larger than the area of responsibility of the decision-makers. This results in spatial externalities (spillovers) in the form of a performance or cost export. In the case of a performance export, one can expect a shortage of the public commodity (spill-out). If the scope of the benefit of the public commodity is smaller than the area of responsibility, then an oversupply can arise (spill-in) (Neumann 1971, p. 506). The supply of public goods will, therefore, be too high or too low under institutional incongruence.

This can be easily understood by looking at the example of public transportation. If the citizens of just one city decide and fund the public transportation provided in a county, then this will turn out to be rather inadequate. If all the citizens in the county take part in the funding but the decision-making authority remains with the city, then incentives are put in place for providing too much. Institutional congruence also results in fiscal equivalence. Each level of the federal system bears the costs of its own decisions.

In federalism, there is often no institutional congruence. Public swimming pools are funded by the individual municipalities, however, they are also used by citizens from outside the municipal boundaries. Decisions on town planning are made in each individual city, but the ensuing costs are often financed through allocations from the state budget. The decision to increase accommodation in nursery schools is made at the federal level. The municipalities must provide the funding. There are many examples of public spending that are subject to institutional incongruence.

In the case of the construction of railway infrastructures, one can also assume that institutional incongruence comes into play. In Germany, decisions about upgrading and new construction of railway lines are made at the federal level. As a general rule, funds for rail projects are provided by the federal budget (cf. Sect. 3.2). Users of the improved infrastructure are companies and people throughout Germany. In that regard, institutional congruence exists. However, the people in the regions where the construction of the infrastructure is implemented are particularly affected. They are confronted with negative externalities, above all in the form of noise but also in land use. These effects correspond to the above-mentioned frustration costs. They are caused by institutional incongruence. If these costs are high enough, they will lead to resistance in the population.

2.3 *The Coase Theorem: A Possible Solution for Institutional Incongruence*

The Coase Theorem (cf. Blankart 2007, p. 73) points to a solution for institutional incongruence. Coase shows that external effects can usually be resolved in an efficient way by the market participants themselves in the course of negotiations (Coase 1960, 5ff.). This is subject to the condition that the negotiations can be conducted without transaction costs and with clear property rights. A simple example based on Coase, which illustrates the situation well, are the external effects in the form of noise, which arise when mowing the lawn. Suppose the two neighbors A and B are not in agreement as to the proper time to mow the lawn. In the summer, A can only mow his lawn on weekends. His neighbor B feels disturbed by the noise. In this situation, one has to consider carefully whether or not the benefit that A has by mowing his lawn on the weekend exceeds the costs of B, which are caused by the noise. Two cases must be distinguished:

1. The benefits of A are lower than the costs of B.

   Suppose A can mow on weekends. By mowing the lawn, he has a benefit of 20 €. This is the price that a gardener would charge per week for mowing the lawn. B suffers damage in the amount of 40 € due to his neighbor's noise. If B now offered his neighbor A 30 € so that the latter dispenses with mowing on weekends, then both would be better off if A took the money offered. Neighbor A would thus have a benefit of 30 € instead of 20 €—for example, he could hire a gardener to mow during the week. B would only have costs of 30 € instead of 40 €.

Suppose that A is not allowed to mow on weekends and B had the right to force A into complying with this rule. A negotiation between A and B with the aim of A being allowed to mow cannot lead to a win-win situation. The result is the situation as before. The efficient solution would be that A does not mow on weekends.

2. The benefits of A are higher than the costs of B.

   Suppose A is not allowed to mow on weekends. He has an additional benefit of 40 € by mowing the lawn. B suffers damage in the amount of 20 € due to the noise of his neighbor. Similar to Case 1, A can now offer B money so he can be allowed to mow on weekends. B will accept the offer if A offers him more money than the costs he has due to the noise. They negotiate a settlement at a price between 40 and 20 €. Neighbor A can mow.

   If A were allowed to mow on weekends, a negotiated solution with the aim of A not mowing would not be reached. The result would again be the efficient solution that A mows on weekends.

This example shows that a legal regulation that assigns one party the responsibility for external costs can only be efficient if this party can deal with the problem at the lowest cost. In the other cases, economic efficiency can also arise through a negotiated solution. Thus, state regulation is not necessary at all to obtain an overall cost-efficient solution.

When applying the above considerations to the situation in the construction of rail infrastructure, one can make the following assertions: The construction of rail infrastructure creates benefits which, simply put, benefits all the citizens of the country. The negative external effects, however, are borne only by the local population. The denser the population is, the higher the frustration costs are. Negotiations between representatives of the government levels concerned are, as follows from the Coase Theorem, a suitable tool for creating an efficient situation.

3 The German Planning Process for Rail Infrastructure: The Case of the Rheintalbahn

The Rheintalbahn (Rhine Valley Railway) is a double-track, electrified mainline railway running through the German federal state of Baden-Württemberg. Its route takes it from Mannheim via Heidelberg, Karlsruhe, Rastatt, Baden-Baden, Offenburg and Freiburg im Breisgau to Basel. It was originally financed and built by the state railway of the Grand Duchy of Baden. The first section to be completed, between Mannheim and Heidelberg, was opened in 1840. By 1855, the line had been finished, a section at a time, as far as Basel (Dumjahn 1984). The line grew in importance for international traffic, particularly after the First World War.

The Rheintalbahn is one of the most important and most heavily trafficked railway lines in Germany as regards both passenger and freight trains. This is due not only to local passenger services, but also, inter alia, to international freight and passenger trains to and from Switzerland and France. All long-distance passenger trains have stops in Mannheim, Karlsruhe, Freiburg and Basel Badischer Bahnhof, and some of them also stop in Heidelberg, Wiesloch-Walldorf, Bruchsal, Rastatt, Baden-Baden and Offenburg. The stations at the two ends of the Rheintalbahn are Mannheim and Basel. These both have marshalling yards, which are among the largest in Europe.

3.1 The Rheintalbahn Expansion

Now, an upgrading and partial new building of the line is intended to increase its capacity quite considerably and to shorten travel and transport times. Once all this work has been completed, there will be four tracks throughout, making it possible to improve the structuring of train movements. As the project from Karlsruhe to Basel is over 182 km, the rail line was divided into 9 sections with a total of 21 sub-sections (Nied et al. 2007, pp. 506–512).

With regard to routing, there are major differences between the Deutsche Bahn Netz AG (German railway infrastructure provider) and the affected municipalities. The Deutsche Bahn favors a route along the existing tracks. The frustration costs in the affected regions and municipalities are manifested in the protest of the residents between Offenburg and Basel against the plan. Various citizens' action committees have been set up. By the end of 2009, a total of 170,000 objections to the expansion of the Rheintalbahn were brought before the Regional Commission in Freiburg.[1] The residents particularly fear high noise levels. They are calling for a new route along the A5 freeway (Fig. 2).

3.2 The Planning Process

The planning process for railway infrastructure measures in Germany can be roughly divided into four planning steps.

1. The basis for the development and upgrading of transport infrastructure is the Federal Transport Infrastructure Plan (FTIP, German: Bundesverkehrswegeplan). It is revised at regular intervals (ca. 10–15 years). In the FTIP, all the project plans for upgrading and new construction of railways, waterways and

[1] Badische Zeitung, 170.000 Einwendungen gegen Ausbau der Rheintalstrecke, 4 December 2009; http://www.badische-zeitung.de/freiburg/170-000-einwendungen-gegen-ausbau-der-rheintalstrecke-23703348.html, accessed 18 June 2015

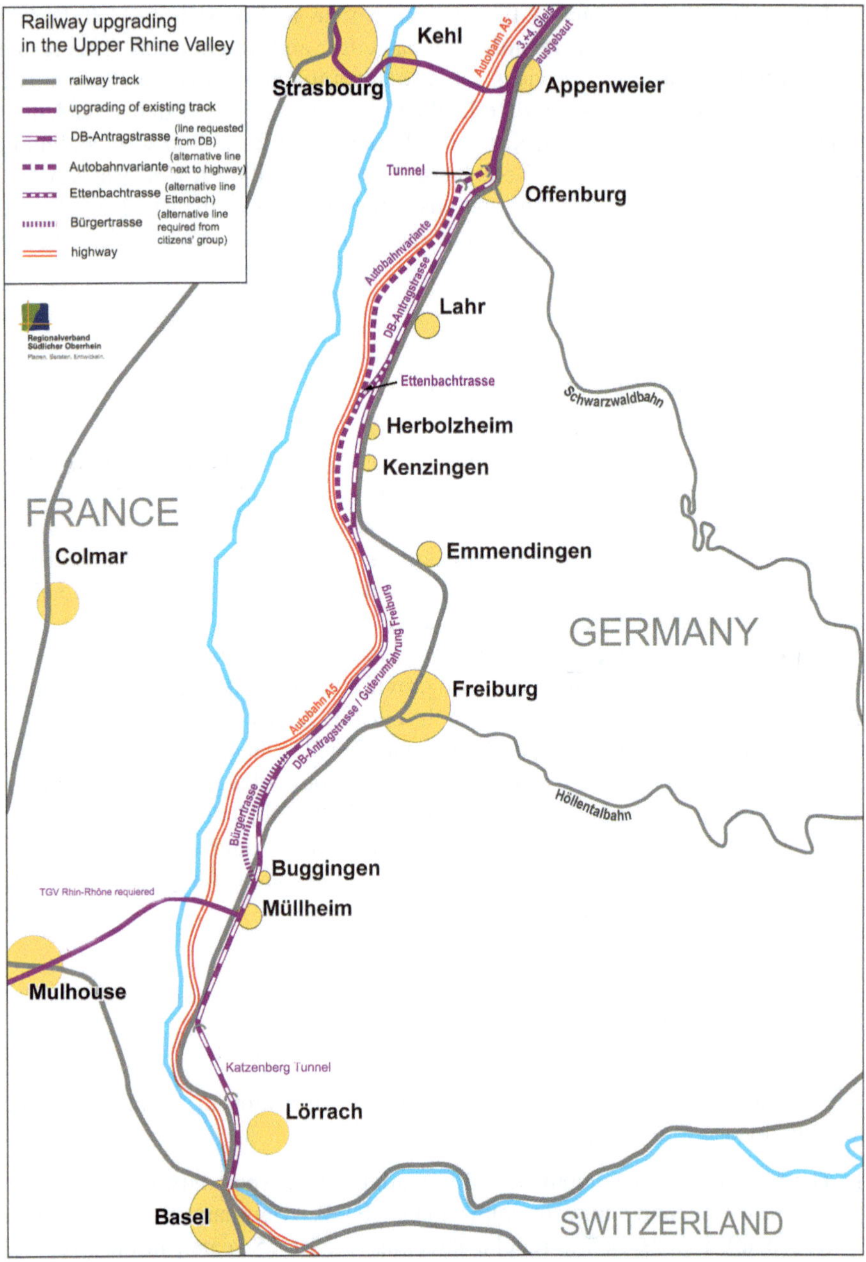

Fig. 2 Alternative railway lines in the Upper Rhine Valley. *Source*: Regional Association Southern Upper Rhine (German: Regionalverband Südlicher Oberrhein, www.rvso.de/de/regional entwicklung/rheintalbahn/Karte-Bahnprojekte-Oberrhein-Basel-Offenburg.pdf, accessed 19 June 2015

highways are specified and prioritized. The Federal Government must show proof that each project is useful and necessary for the overall economy of the country. The necessity is determined by using forecasts for projected traffic volume (Federal Ministry for Transport, Construction and Urban Development 2014, 54ff.). Projects are usually proposed by the states, the DB Netz AG, associations and members of parliament (Federal Ministry for Transport, Construction and Urban Development 2012, p. 10). Citizens and other lobbyists are granted information and consultation rights during the process (ibid, 6). The projects are prioritized into three levels of urgency. A railway infrastructure project can only be brought into the further planning process if it has been classified as very urgent in the FTIP.
2. Very urgent projects, which are to be implemented by decision of the German Parliament, are taken up in the requirement planning of the construction laws. Subsequently, the DB Netz AG, receives a design contract from the federal government. The planning of the rail infrastructure can begin. The DB Netz AG does feasibility studies and draws up different route variations and further planning documents. These planning processes generally do not include a process of citizen involvement.
3. Afterwards, a regional planning procedure (RPP) can be initiated. In accordance with the law, it must be carried out for new rail lines or for substantial alteration of existing routes (§1, Number 9, RoV, German: Raumordnungsverordnung). The Regional Commission (German: Regierungspräsidium) in Freiburg is the higher regional planning authority (§15, Sect. 1, Spatial Planning Act, German: Raumordnungsgesetz, together with §18, Sect. 1, State Planning Act, German: Landesplanungsgesetz). The Regional Commission evaluates the proposal as it relates to regional planning policy and then determines whether it meets regional planning requirements (§2, Sect. 2, Spatial Planning Act). The principle of sustainability plays an important role here. It means that regional planning should create spatial structures that can also serve future generations (Langhagen-Rohrbach 2005, p. 6). Finally, especially for railway projects, one must find the best solution for routing in terms of land use. In a RPP, hearings are conducted with planning authorities, such as municipalities, as well as with public bodies, such as regional associations and nature conservation organizations. For projects with significant effects on the environment, hearings are conducted with the public, who will then be informed about the outcome of the procedure. The result is the presentation of the spatial compatibility of a project in the form of a spatial planning assessment for all of the possible routes. The spatial planning assessment does not bring about any direct legal effects on others. Therefore, the RPP is not binding on the planning approval authority in the subsequent planning approval procedure. For the section of the plan south of Offenburg, no RPP was carried out, as the construction of the third and fourth tracks is intended to be parallel to the existing line (University of Stuttgart et al. 2008, p. 27). Proponents of the route parallel to the freeway are of the opinion that in a RPP, this alternative would clearly have been the more preferable choice. The waiver of a RPP on the part of the Regional Commission

in 1990 as well in 2002 at the repeated request of the German Railway is today viewed as a planning error (DB Project Construction 28 June 2002, p. 4).

4. The planning approval procedure is the approval process for larger infrastructure projects with a concluding decision of approval (official approval of the plans). Here the permissibility of a proposal is determined with regard to the adverse effects on public interests (Hoppe et al. 2011, p. 6). A feature of the planning approval procedure is comprehensive public participation. The possibility for the population to voice objections and conduct hearings with numerous "agencies of public interests" (technical authorities, municipalities, associations, etc.) makes sense because the upgrading and new construction of railway infrastructure involves numerous public and private interests. The process is divided into the following steps: The builder submits the plan to the responsible department. In the subsequent hearing procedure, official comments are obtained. During the following period of public inspection of the documents, the citizens affected can raise objections. After discussions with citizens, builders, approving authorities and other departments, the hearing authority summarizes the results for an approval of the plans. Then the planning approval authority (Federal Railway Authority) issues the approval of the plans. A statement from the Ministry of Transport and Infrastructure in Baden-Württemberg shows how citizens and affected municipalities were involved in the planning approval procedure.[2] In addition to the usual process of conducting hearings, the Regional Commission in Freiburg, in its function as a hearing authority for its area of responsibility, carried out large public information events. Within this framework, citizens were informed about the plans and their rights in the hearing procedure. In addition, tips and suggestions could be found on the website of the Regional Commission.

The entire planning process establishes, within a certain scope, information and consultation rights for the affected residents, municipalities and regions. However, it is not intended that these players should directly influence the planning.

3.3 The Project Advisory Board "Rheintalbahn": An Approach to an Efficient Solution According to the Coase Theorem?

Blankart refers to the possibility of negotiations in the case of institutional incongruence (cf. Sect. 2.3). This raises the question of whether the tool proposed by Coase for internalizing external effects can be applied in the case of the construction of rail infrastructure. Negotiations are called a "form of social interaction"

[2] Statement of the Ministry for Transport and Infrastructure Baden-Württemberg from 18 Dec. 2013, Printed matter 15/4333.

(Thompson and Hastie 1990, p. 99) or a "form of social conflict" (Pruitt 1981) "by which two or more interdependent parties who do not have identical preferences across decision alternatives make joint decisions" (Bazerman and Carroll 1987, p. 252). It is possible that in the negotiation process, a result is obtained which leaves both negotiating partners better off than without the agreement (Voeth and Herbst 2009, p. 5).

Negotiating partners in the upgrading of rail infrastructure are the national and regional level. The aim of negotiations should be, first of all, to examine all aspects of the situation and to provide for transparency. The negotiators of the federal government will try to keep the costs of construction for a predefined performance level of infrastructure as low as possible. In turn, the negotiators of the regional level will work to minimize the resulting negative externalities, in particular, the resulting noise, which will be discussed below.

Ultimately, the negotiation process is a matter of finding out which side can avoid the problem at the lowest cost. One solution might be to reduce the noise as much as possible at its source, e.g. by means of technical requirements for trains passing through. Another solution might be to equip adjacent houses with better soundproofing. Thirdly, the construction of rail infrastructure could take place where the noise does not incur costs. Initially, there is no reason to presume that any one of these solutions is the best one. It is also not absolutely clear whether the residents along the railway line are actually the disadvantaged party because they have to put up with the railway noise in their houses. The railway companies, too, may feel to be at a disadvantage if they are forced by law to implement costly noise abatement measures because people have chosen to live near a railway line.

The key insight from Coase's essay "The Problem of Social Cost" is particularly clear here: externalities are a joint product of the "polluter" and of the "aggrieved party". The main criticism of Coase's Theorem is that a negotiated solution is bound to fail if transaction costs, e.g. caused by lawyers or time-consuming negotiations, are higher than the benefits of an agreement. However, due to the high investment costs for the construction of rail infrastructure, it cannot be expected that the transaction costs of the negotiation will stand in the way of an efficient solution. The negotiations between builders and those affected should, however, take place during the planning process. Once the infrastructure has been built, adjustment measures are sure to cost much more.

In this context, it is worthwhile to take a look at the Project Advisory Board Rheintalbahn, which was established in July 2009. The main objective of this body is the balance of interests between the affected residents and the German Railway or the federal government, respectively. This practice has not yet been institutionalized in German infrastructure planning. Instead, the federal government and the state of Baden-Württemberg agreed to establish this board due to the numerous objections to the enlargement and the intense debates about the correct routing of the rail lines. The board is composed of one representative each from the Federal Ministry of Transport, the Federal Railway Authority and the DB Netz AG. Then there are representatives from the state of Baden-Württemberg, the Regional Commission in Freiburg, the counties of Breisgau-Hochschwarzwald, Ortenau,

Emmendingen and Lörrach, the Regional Associations of Southern Upper Rhine and High Rhine-Lake Constance, and finally, a representative from the umbrella organization of citizens' action committees, the Interest Group for Railway Protest on the Upper and High Rhine (IG Bohr). Individual citizens' action committees also work with the regional working groups, whose task is to support the Project Advisory Board with their expertise and to prepare the board's decisions.

The Project Advisory Board should work towards better planning and more consideration for people and the environment. Regional suggestions for improvement are examined and evaluated. Possible solutions are explored.[3] The alternative plans and the demands from the region concerning the disputed area from Offenburg to Weil am Rhein were formulated in six key demands. At the county level, the Project Advisory Board established three working groups located in different areas. Here, in consultation with local experts, the planning proposal of the German Railway Corporation and the regional suggestions for improvement were examined and evaluated with regard to all relevant aspects. Possible solutions were explored, and all of this was reported to the Project Advisory Board.

The first successes of the negotiations are visible at the current time. According to a press release from the Ministry of Transport and Digital Infrastructure, the Project Advisory Board has negotiated an agreement between the DB Netz and the other lobbyists, who have agreed to the construction of a twin-tube freight-train tunnel near Offenburg.[4] However, there were no statements made with regard to funding.

The method of negotiation between the federal government and regional players, which was applied in the Project Advisory Board Rheintalbahn, offers good opportunities to bring about an efficient solution to the problem of externalities in the expansion of the Rheintalbahn. The fact that this method is possible only because of the "goodwill" of the federal government must be viewed critically. The local players are not entitled to negotiations. In fact, one has the impression that the municipalities and regions can only force the federal level to negotiate by organizing mass protests.

The participation of the DB Netz AG as a part of the German Railway Corporation in the negotiations is also not unproblematic. The German Railway Corporation competes with other railway companies, especially in freight transport. It cannot be ruled out that DB Netz AG uses its influence and expertise during the negotiations to create competitive advantages for the parent company.

Another point of criticism is the lack of public negotiations. Transparency and information are important prerequisites for achieving an efficient solution and for its general acceptance.

---

[3] Statement of the Ministry of Transport and Infrastructure Baden-Württemberg from 18 Dec. 2013, Printed matter 15/4333, p. 7.

[4] Press release of the BMVI from 14 July 2014: Projektbeirat Rheintalbahn tagt in Berlin, No. 054/2014. http://www.bmvi.de/SharedDocs/DE/Pressemitteilungen/2014/054-odenwald-projektbeirat-rheintalbahn.html?linkToOverview=DE%2FPresse%2FPressemitteilungen%2Fpressemitteilungen_node.html%23id133508, accessed 6 August 2014.

4 Conclusion

The realization of railway infrastructure projects involves the risk of negative externalities in the form of noise, vibration and the reduction in value of neighboring properties. The externalities trigger institutional incongruence, a separation of decision-making and funding responsibility on one side and the concerns of the people affected on the other. The economic theory of federalism shows that the economic inefficiency associated with that situation can be prevented by negotiations in accordance with the Coase Theorem.

In the course of the enlargement of the Rheintalbahn, an advisory board with representatives of the federal government and the affected regions was established due to the large number of objections to the project. The process offered the opportunity to manage the planning using local expertise in order to achieve an efficient result for the economy as a whole. However, the planning procedure has some weaknesses, such as the lack of public participation, the participation of the DB Netz in the negotiations as well as the fact that the process has not been institutionalized. The experience gained in the course of the negotiation should be included in a final evaluation process. The evaluation results can serve as a basis for the statutory inclusion of negotiations between the federal government, the region and the municipalities when implementing railway infrastructure.

The establishment of a European Grouping of Territorial Cooperation (EGTC), in which the municipalities and regions along the Rotterdam-Genoa transport corridor join together to pursue common interests (cf. Chap. 19), offers the opportunity to coordinate efforts on a regional level. A strong involvement of local players in the planning of infrastructure for European transport projects should become one of the most important objectives of the EGTC.

References

Bazerman M, Carroll J (1987) Negotiator cognition. In: Cummings L, Staw BM (eds) Research in organizational behavior. JAI Press, Greenwich, pp 247–288

Biehl D (1994) Zur ökonomischen Theorie des Föderalismus: Grundelemente und ihre Anwendung auf die EU-Finanzunion. In: Schneider H, Wessels W (eds) Föderale Union—Europas Zukunft? Analyse, Kontroversen, Perspektiven. C.H. Beck, München

Blankart C (2007) Föderalismus in Deutschland und Europa. Nomos, Baden-Baden

Breton A, Scott A (1977) The assignment problem in federal structures. In: Feldstein M, Inman R (eds) The economics of public services. Macmillan, London, pp 344–357

Buchanan J, Tullock G (1962) The calculus of consent, local foundations of constitutional democracy. University of Michigan Press, Ann Arbor, MI

Bundesministerium für Verkehr, Bau und Stadtentwicklung (eds) (2012) Konzept zur Öffentlichkeitsbeteiligung im Rahmen der Erarbeitung des Bundesverkehrswegeplans 2015. Public Relations booklet of the Ministry, Berlin

Bundesministerium für Verkehr und digitale Infrastruktur (ed) (2014) Grundkonzeption für den Bundesverkehrswegeplan 2015, bedarfsgerecht—transparent—herausfordernd. Public Relations booklet of the Ministry, Bonn

Coase R (1960) The problem of social cost. J Law Econ 3:1–44
Dumjahn H-W (ed) (1984) Handbuch der deutschen Eisenbahnstrecken: Eröffnungsdaten 1835–1935, Streckenlängen, Konzessionen, Eigentumsverhältnisse. Vollständiger unveränderter Nachdruck des unter dem Titel: Die deutschen Eisenbahnen in ihrer Entwicklung 1835–1935. Druckschrift der Deutschen Reichsbahn (eds), Berlin, Reichsdruckerei, 1935. Eigenverlag, Mainz
Hausner K (2005) Die ökonomische Theorie des Föderalismus. Wirtschaftsdienst 1(2005):55–60
Hoppe W, Schlarmann H, Buchner R, Deutsch M (2011) Rechtsschutz bei der Planung von Verkehrsanlagen und anderen Infrastrukturvorhaben – Grundlagen der Planfeststellung, 4th edn. Erich Schmidt Verlag, Berlin
Koppel O, Lichtblau K (2007) Föderalismustheorie: Ökonomische Kriterien für die Konstruktion eines föderalen Systems. In: Institut der deutschen Wirtschaft (ed) Föderalismus in Deutschland Ökonomische Analyse und Reformbedarf. Deutscher Instituts-Verlag, Köln, pp 9–44
Langhagen-Rohrbach C (2005) Raumordnung und Raumplanung. Wissenschaftliche Buchgesellschaft, Darmstadt
Neumann M (1971) Zur ökonomischen Theorie des Föderalismus. In: Kyklos: international review for social sciences, vol 24. Wiley-Black, Oxford, pp 493–510
Nied J, Dassler B, Ziege T (2007) Neu- und Ausbau der Strecke Karlsruhe bis Basel. Eisenbahntechnische Rundschau 56(9):506–512
Oates W (1972) Fiscal federalism. Harcourt-Brace-Jovanovich, New York, NY
Oates W (1999) An essay on fiscal federalism. J Econ Lit 37:1120–1149
Pruitt DG (1981) Negotiation behavior. Academic, New York, NY
Thompson L, Hastie R (1990) Social perception in negotiation. Organ Behav Hum Decis Process 47:98–123
Universität Stuttgart, Verkehrswissenschaftliches Institut, IBK Ingenieur- und Beratungsbüro, Technische Universität Kaiserslautern (eds) (2008) Gutachterliche Stellungnahme zur Bewertung der Variantenuntersuchung des Ausbaus der Rheintalbahn im Abschnitt Offenburg – Riegel. Report on behalf of Regionalverband Südlicher Oberrhein, Stuttgart
Voeth M, Herbst U (2009) Verhandlungsmanagement: Planung, Steuerung und Analyse. Schäffer-Poeschel, Stuttgart

Since 2009 **Hansjörg Drewello** has been a professor of economics, first at the University of Applied Sciences Göttingen and now at the University of Applied Sciences Kehl. He is the German director of the European Competence and Research Centre Cluster Management. His teaching and research deal with regional economics, in particular infrastructure development and the regional economic impact of cluster initiatives in a cross-border context. Until 2009 Hansjörg Drewello was the director of the Department of International Commerce, Transportation, Commerce and Tourism at the Chamber of Commerce and Industry Karlsruhe and a member of its management board. He was the lead partner of the study segment "Freight transport and logistics" of CODE24.

The Extension of the Betuweroute: The Competition of Ideas in the Wesel District

Frank Joneit

Abstract The extension of the Betuweroute is needed for fluent freight transport on rail and also for a rising share of rail in freight-traffic. In addition to that, the impacts of rising freight-traffic on Betuweroute especially by rail transit passing NRW had to be considered carefully so that suggestions for reduction of emissions can be made.

In order to achieve this goal and to activate the needed consensus around the proposal, Regionalverband Rhein-Ruhr (RVR) applied an informal planning procedure in close cooperation with County Wesel. The informal planning procedure should give solutions for an improved handling of rail freight and consider rail specific, spatial and urban aspects.

From the beginning it had been clear that this project was not a singular RVR project but a project that could be carried out in close co-operation with the local authorities only. Two teams of external consultants were chosen to work separately on three exemplary sections as shown on the right map. A team of university graduates ETH Zurich, Politecnico di Torino, University Utrecht took an overview on the whole section and worked on a broader range of themes. Three regional workshops were carried out with regional stakeholder. The results of these workshops, presented in this article, show alternatives for the development of the corridor, which may be more expensive then solutions provided by German Rail.

1 Introduction

To support the interests of the authorities along the railway line Oberhausen–Wesel–Emmerich (as continuation of the Dutch Betuwe route), Ruhr Regional Association and the District of Wesel 24 are partners in the CODE24 project to

F. Joneit, M.A. (✉)
Kampmannstraße 17, 44799 Bochum, Germany

Regionalverband Ruhr, Kronprinzenstraße 6, 45128 Essen, Germany
e-mail: joneit@rvr-online.de

H. Drewello, B. Scholl (eds.), *Integrated Spatial and Transport Infrastructure Development*, Contributions to Economics, DOI 10.1007/978-3-319-15708-5_18

promote the development of the trans-European Transport Corridor 24 Rotterdam–Genoa.

In the Ruhr Regional Association area, German Rail's (DB's) activities are focused on the development of the Oberhausen–Wesel–Emmerich railway line. It is the southern continuation of the Betuwe route in the Netherlands. The route is obtained with a three-pronged expansion, enhanced signaling and measures to increase the speed limit and to obtain the necessary sustainable capacity (DB Projektbau GmbH 2011–2014). The DB plans for the expansion were developed by German Rail (DB AG) that also implemented the legal planning procedure.

With its focus on regional development issues and joint inter-regional strategies, the project aims to strengthen in particular the role of the regional actors along the corridor. Planning tools and tailor-made solutions will be developed in order to eliminate bottlenecks and obstacles for the development and civic participation. Thus, the project supports the development of the railway network and sustainable spatial development. CODE24 is supported program under the INTERREG IVB NWE strategic initiatives.

Today the population along the railway line Oberhausen–Emmerich is affected by considerable noise stress. This is particularly caused by the freight trains. Due to that, extensive active noise control will be an important part of the expansion measures. The measures presented so far by the DB against noise and vibrations are critically assessed by the local authorities and the population affected (Ruhr Regional Association 2011).

The EU supports local authorities and other interested institutions, being affected by the planned extension to take a look at all opportunities, but also at the burdens and risks for the affected areas.

2 Objectives and Methods

The Ruhr Regional Association (RVR) together with the district of Wesel carried out a cooperative process as a competition of ideas between November 2010 and April 2011. So an intensive exchange of information between all stakeholders could be ensured (Ruhr Regional Association 2011).

Three teams of external experts developed different approaches in this period: Two interdisciplinary expert teams

- Schüßler-Plan Engineering mbH from Cologne
- Peter M. Moik transport planning and mobility consultancy based in Dusseldorf.

An international team of postgraduates "CODE24" (ETH Zurich, Politecnico di Torino and Universiteit Utrecht) also worked on this project.

For three typical situations along the Oberhausen–Wesel–Emmerich railway line, exemplary and innovative independent solutions were chosen. These solutions take into account the tension between the maximum noise protection and maximum capacity use on the future three track railway line. The acceptance of the population

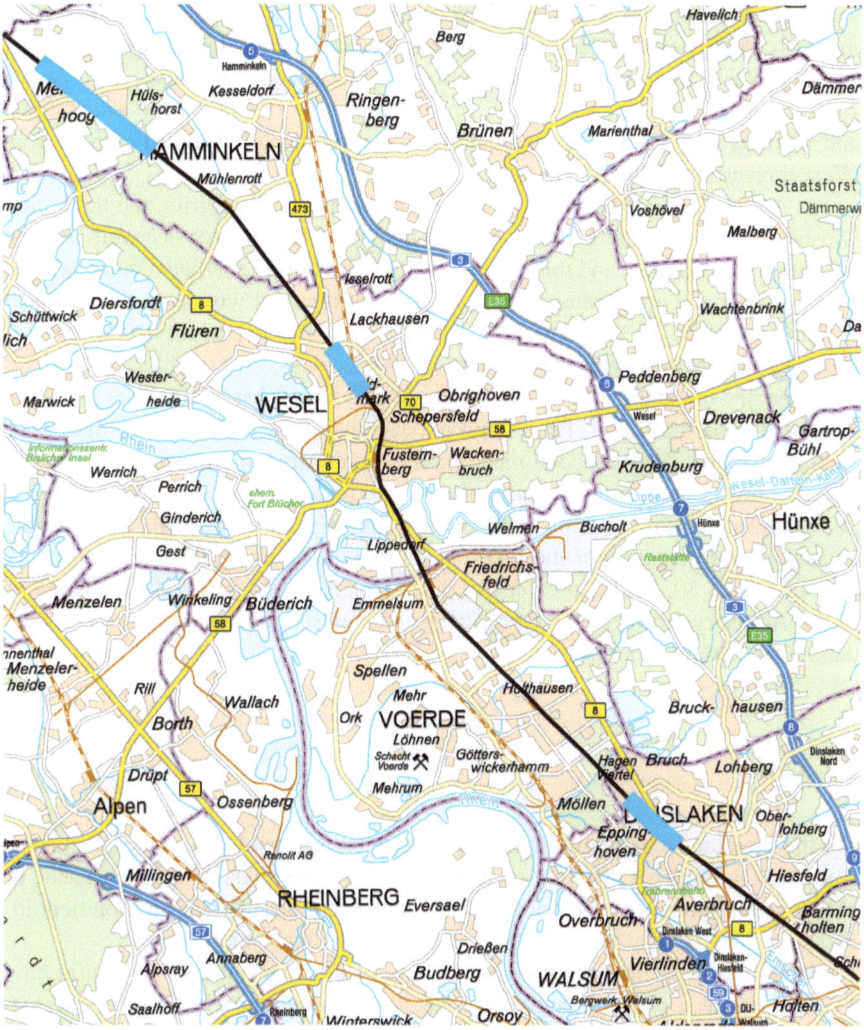

Fig. 1 The railway section Hamminkeln-Wesel-Dinslaken. *Source*: Ruhr Regional Association (RVR), Essen 2010

should be obtained with improved concepts achieving positive response among all affected. After a consultation with the district of Wesel and the affected local authorities one section in each of the towns of Dinslaken, Wesel and Hamminkeln was selected (Fig. 1).

These alternatives should be compared with the design of the DB. Traffic noise reduction measurements should have a higher spatial design quality and at least equal protection than those of the DB AG.

The concepts for the three study sections were examined as part of a standardized sound propagation prediction carried out by a noise measurement expert. The sound propagation prediction was performed by TÜV Nord Systems GmbH, based in Essen. A part of the idea workshop was also an intense discussion and participation process. In several regional workshops discussions took place with experts from transport economics, logistics, science, Chamber of Commerce, cities, towns and districts as well as the regional Passenger Transport Authority Rhine Ruhr (Verkehrsverbund Rhein-Ruhr AöR).

An Advisory Board including representatives of Wesel District, the three cities Dinslaken Wesel, Hamminkeln, and Chamber of Commerce as well as from industry and academia, and the Ruhr Regional Association enabled a more detailed, in-depth thematic and strategic debate. The first regional workshop to launch the project took place on 3 November 2010 and served also to present it. The different actors shared their view of the project.

At the first meeting of the Advisory Board on 8 December 2010, the expert teams presented their first ideas. Hopes and expectations that are placed on the project were exchanged. At the second meeting, the monitoring committee questioned the expert teams to their initial ideas. In the second regional workshop on 16 February 2011 there was a presentation of the first results of the expert teams and the noise expert. The presentation of the final results took place in the third regional workshop on March 30, 2011. An extensive discussion followed during the final meeting of the Advisory Board on 13 April 2011 and was supplemented with recommendations on how to proceed.

3 Task

For the three areas in Hamminkeln-Mehrhoog, Wesel field and Mark Dinslaken a consistent, innovative and cost-conscious concept was to develop. As a basis, the expert teams should consider deepens the three sections and their environment in a first step. The local spatial situations were of particular importance:

Hamminkeln-Mehrhoog is as an example of a settlement area in a rural environment. It has a predominantly open settlement structure with large open areas. Wesel-Feldmark exemplifies a district in a regional center of the metropolitan edge, which is characterized by a dense residential development. The section of Dinslaken serves as an example for an urbanized area.

The present DB plans and municipal urban planning concepts were also analyzed. Proposals for operations management on the future three tracks had to be developed. A multitude of infrastructural and noise related requirements had to be taken into account. These included, for example, the lowest possible generation of pollution and minimizing delay and acceleration. At the same time, a maximum track capacity should allow a high quality of timetable stability and the necessary paths for passenger trains according to the regular interval timetable paying attention to future extensions of services.

The railway line Emmerich–Wesel–Oberhausen is of particular importance for the region due to

- international long-distance passenger transport with the ICE International Amsterdam–Utrecht–Arnhem–Oberhausen–Duisburg–Düsseldorf–Köln–Frankfurt service
- a fast Regional Express service operating hourly
- two regional Rhein-Ruhr services operating hourly each.

Due to the high commuters and travelers volume in this corridor a reliable and punctual rail transport is a necessity. The good transport program plays an important role for the adjoining municipalities as a location factor for residents.

The developed solutions should be prepared and shown that they can be subjected to a subsequent, separate operation of sound propagation prediction.

4 The Results of the Expert Teams

All three teams of experts approached the tasks in a separate way and set different priorities.

4.1 Analysis of the Team of Postgraduates "CODE24"

The postgraduate PhD team took a different look at the actual task described above. It had its own regional access. This team was particularly keen on the regional development prospects within the entire sector. The doctoral student team came to the conclusion that the District of Wesel due to its location between two of the most important and densely populated European metropolitan regions of Randstad and the Metropole Ruhr is an area with a high potential where innovative, high quality and architecturally appropriate solutions should be developed.

Basis for the derivation of development perspectives and approaches of illustrative scenarios are assumptions between growth and stagnation. For the three examination sectors Hamminkeln, Wesel and Dinslaken spatial development opportunities have been identified in the immediate vicinity of the railway stations. The Lippemouth area was seen as a significant potential area of economic development.

4.1.1 Regional Strategy

From the perspective of the doctoral student team local issues and solutions of web development have to be integrated into a regional coordinated development strategy. The future spatial development is heavily dependent on today's decisions on the following topics:

- noise protection in the area,
- development of the railway nodes,
- regional transportation
- future of logistics.

With these elements being identified, some examples will be shown.

4.1.2 The Lippemouth Area

The Lippemouth area is interesting for all modes of transport due to its convenient accessibility among other factors. It is particularly suitable for development of a container terminal as a hub port for ships from Rotterdam. Therefore, the function of the port Emmelsum should be strengthened as intermodal Hub for the tri-modal handling water-rail-road. The Postgraduates team CODE24 proposesd the development of a master plan for this area (Fig. 2).

4.1.3 Extension of the Railway Line and the Railway Nodes

Upgrading of the Oberhausen–Emmerich railway line includes potential for further development of the passenger rail services. These include accelerated regional express trains or the future Rhine-Ruhr-Express services (RRX) that should be extended to Arnhem while serving fewer stations. Thus, the heavier patronized stations would be served only. This could contribute to improved inter-regional accessibility of the corridor and enhance its attractiveness.

According to the Doctorate team the upgrading of the railway node Oberhausen is of high importance. A capacity expansion along the entire route is possible in their view only by the construction of a continuous third track. The train station in Wesel with adjacent hitherto inadequately used areas is seen as a particular potential space for development of a high quality residential district that connects the city with the railway station and additionally serves as an attractive and effective noise control.

4.1.4 Regional Noise Protection

To increase the acceptance of the project and for the benefit of all potentials of development within the area a regional, coordinated noise protection approach should be developed with the participation of all forces. As a further concrete recommendation it is suggested by the doctoral team that the municipalities along the railway line should apply for “Regionale 2019” as “Region Betuwe” (Ruhr Regional Association 2011). In this project “Betuwe” a “noise protection laboratory” should be developed from the perspective of the doctorate team.

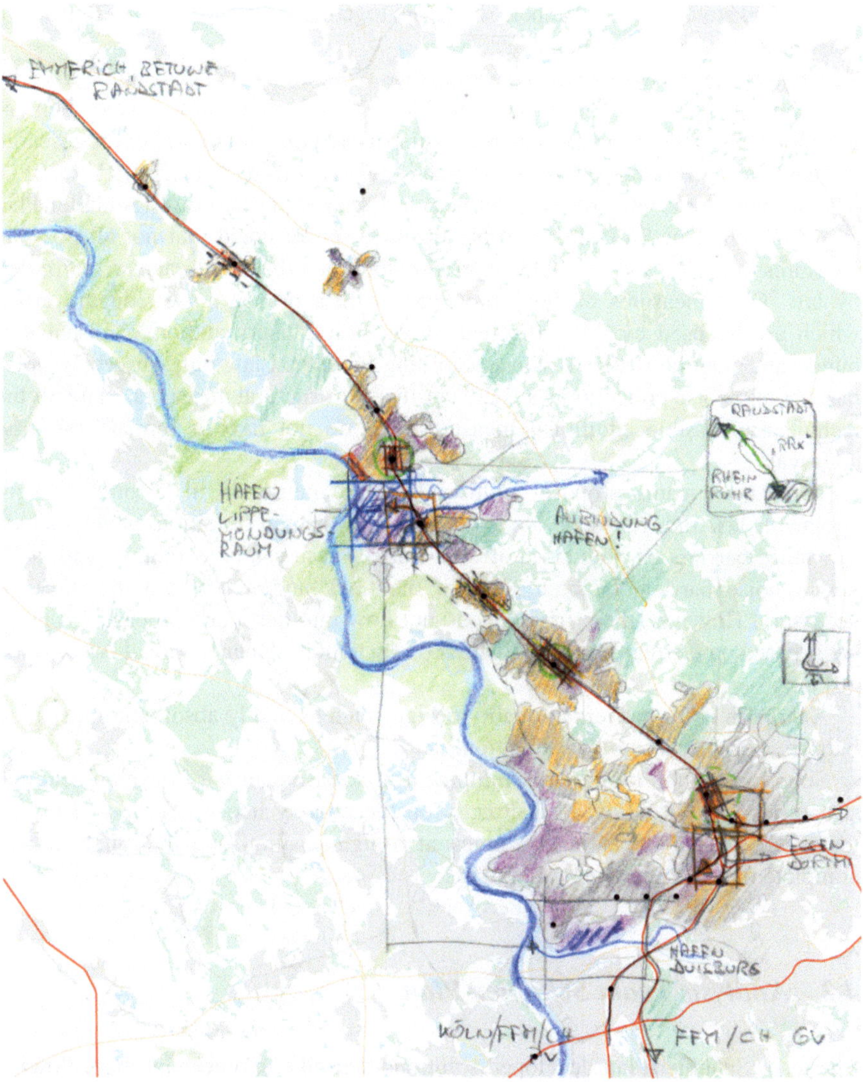

Fig. 2 Illustration of analysis results from the post-graduate team "CODE24", Zurich, Milano, Utrecht 2011

4.1.5 Measures for Noise Abatement

The primary goal for the teams Schüßler-Plan and Peter M. Moik was an integrated overall design for noise control. This was intended to show with urban and architectural measures local custom solutions. In addition, they should also allow ideas for transferability to other areas. For best protection against noise pollution and measures on the track such as foamed track layout, highly elastic rail sleepers, "silent" rail stock and rail dampers should go together with passive noise and active noise barriers along housing areas.

The technically feasible measures to noise reduction have been summarized by the team Schüßler-Plan in a noise toolbox. This includes the development of vehicle-specific measures such as the "whisper break (disk break)" or the research project "Quiet train on real track" (Schüßler-Plan et al. on behalf of RVR 2011). Whilst there are a lot of protective measures against traffic noise (Schüßler-Plan et al. on behalf of RVR 2011), it should however be noted that not all possible measures to enable a sustainable integration into the urban landscape are permitted by law. Only measures of noise protection with a proven noise-reducing effect which are licensed by the competent railway supervisory authority (Eisenbahn-bundesamt) can be realized. Thus, according to current law only absorbing noise barriers and noise protection walls are certified as structural measures. Many of the technically possible additional measures still do not have a formal and legal approval.

Thus, both teams propose solutions that are permitted in accordance with applicable law and are responsive to the respective spatial conditions which include low noise barriers on both sides of each track. These can be combined with half-height noise barriers on both sides of the track and/or between the tracks if necessary. However, these measures do not flow into the sound propagation prognosis based on German sound regulation 03 (Schall03) (Ruhr Regional Association 2011).

Also high noise barriers in quality design, such as highly absorbent transparent barriers were taken into consideration.

Sound insulation by urban development in sensitive areas of development and structurally usable fallow land directly on the railway line was proposed by the teams, too. Ideas for laying the track at trough position or half trough system complete the conceptual considerations.

4.2 Analysis Team Schüßler-Plan

The team Schüßler-Plan developed solutions that take into account the following aspects: The question of noise protection is combined with a search for ways to further develop the adjacent urban space and the future local development. So with regard to the urban space and the embankment of the railway a maximum height of 2.50 m for noise barriers is proposed. Through a combination of track and vehicle-related measures a sufficient noise reduction can be achieved (Schüßler-Plan et al. on behalf of RVR 2011).

In Dinslaken station the construction of a second island platform is provided so that four platform edges are available in the future. For pedestrians, a wide tunnel piercing in the station area is to be created. Thus, the quarters on both sides of the railway line are attractively linked and better *visual links and sightlines* are possible. They are designed in a way to ensure a more comfortable accessibility of the platforms. For the freight yard it is proposed to develop a building on the north side of the railway facilities so that the underlying areas are protected from traffic noise.

In the station area of Wesel-Feldmark are proposed transparent noise barriers. Standard sound-absorbing noise barriers are to be applied in the edge regions. For the railway station two side platforms are provided on the outer track tracks. A tunnel for pedestrians and cyclists shall substitute the existing level crossing. The intention is to create a straight line for having better visual links and thus to a better orientation. A bundling of different traffic structures in the field of Hamminkelner Landstraße and the Emmericher Straße is considered appropriate (Schüßler-Plan et al. on behalf of RVR 2011).

For the station surroundings in **Hamminkeln-Mehrhoog** are the

- street crossing in the course of the Bahnhofstrasse as spatial axis,
- station and its development,
- protection of the residential development in the centre of the planning considerations.

According to the team, the Bahnhofstrasse as a central road in the Hamminkeln-Mehrhoog should be preserved as an axis with its many functions and as a main access to the Mehrhoog Station. A bypass can be avoided. With a road underpass and parallel paths for pedestrians and cyclists, the railway line can be crossed (Schüßler-Plan et al. on behalf of RVR 2011) (Fig. 3).

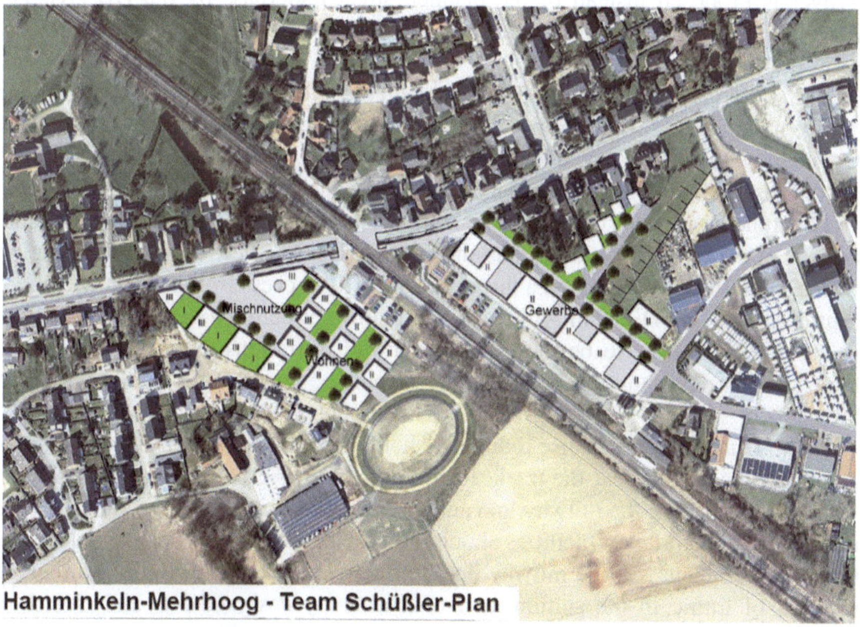

Fig. 3 Train station surroundings in Hamminkeln-Mehrhoog

Along the railway line new commercial buildings are proposed as noise protection on the eastern side while on the western side a closed building with a quiet interior mixed-use is recommended for shielding the existing residential areas from the traffic noise. From the team's point of view it may consist of an attractive centre for Mehrhoog. At the level of the railway overpass Bahnhofstrasse a glass noise barrier is recommended by the team. An island platform and a side platform are planned for the station Hamminkeln-Mehrhoog ensuring three platform edges are available.

4.3 Analysis Team Moik

The solutions proposed by the team Moik aim to provide adequate answers to the station area while ensuring the character of a role model. The team proposes elements that are e. g. combined in different ways, gradually implemented and which can be transferred to other sections along the route. The possibilities of sound insulation by urban development in urban areas while strengthening the internal development are also presented (Ruhr Regional Association 2011).

The concept for **Hamminkeln-Mehrhoog** foresees the lowering of the railway line from about 2.80 m in half trough system as well as a pedestrian and cyclist bridge on today's central location axis Bahnhofstrasse. For motorized traffic a flyover is proposed instead of the planned street underpass by the DB AG. For the noise protection walls and parapets sound absorbing material is provided (Moik et al. 2001).

The station Hamminkeln-Mehrhoog should receive a barrier-free accessible side platform and a central platform, which can be reached via the bridge proposed by the team with a stairway and an elevator. At the station environment, traffic conditions, including an intermodal hub for passenger traffic should be redesigned.

For functional and design optimization in **Wesel-Feldmark** differentiated designs of the crossings and the station are proposed. The new design of the station with a small square shall allow a direct and attractive access from the nearby school centre. Unlike the DB plan with two side platforms both a side platform and an island platform allow to be more flexible in rail operations management. Where high noise barriers cannot be avoided, they should be integrated as good as possible by plants or sophisticated design into the townscape (Moik et al. 2001).

In the **Dinslaken** area noise reduction should be achieved by physical measures and a low noise wall near the track. In addition, a redesign of the station forecourt with buildings along the railway line and a pedestrian tunnel from the station to the north side of the railway facilities should be provided. This will help to reduce the separation efficiency of the railway line. Furthermore, the team proposes a reorganization of traffic in the station area with a newly designed bus and tram hub including a Park and Ride facility as well as a sound insulation construction of offices with underground parking (Moik et al. 2001).

Fig. 4 New design of train station Dinslaken

For the area around the railway station, on the wasteland site east of the station and on the derelict premises of TK-building systems new development could be realized for sound protection (Fig. 4).

5 Noise Propagation Forecast

The analysis developed by the consultant Teams Moik and Schüßler-Plan for the three section designs were subject to a standardized sound propagation forecast for the day and night interval by the noise measurement expert **TÜV Nord Systems GmbH**.

As part of this noise propagation prediction, noise emissions were detected by vehicle type, types of brakes, train lengths, distance rates, types of tracks, bridges, level crossings and curve radii. For the noise level, the distance from the noise source, the directivity and the emission levels are significant. It should be noted that in the calculation of a noise propagation prediction based on German regulation sound 03 (Schall 03) only rolling noise at the top of the rail is taken into consideration. Additional noise sources, such as the slowing down at signals and motor noise at standstill of trains must not be considered (TÜV Nord Systems GmbH on behalf of RVR 2011).

Since the proportion of freight trains is growing, particularly at night, this leads to a significant increase in freight rate and thus to increased emission levels. On the other hand, the limits at night [49 dB (A)] are 10 dB (A) lower than during the day. Therefore, the night time is classified as critical period (TÜV Nord Systems GmbH on behalf of RVR 2011). Because of the critical situation in the night interval, this forecast is of utmost importance.

Table 1 Results of noise analysis

Results	Consultant team Moik	Team Schüßler plan
Dinslaken (TÜV Nord Systems GmbH on behalf of RVR 2011)	In the area north of Dianastraße the selected noise protection walls offer the needed noise reduction capability against the free sound propagation. South of Dianastraße and east of the road in the area of Ursulastraße additional measures such as soundproof windows are necessary in addition to the lower walls chosen. In the area west of the station, the noise-reducing effect of the proposed development is apparent	Because of town development, the noise barriers were limited to a height of 2.50 m. This concept requires additional passive noise protection measures for only a few houses in the southern city area
Wesel-Feldmark (TÜV Nord Systems GmbH on behalf of RVR 2011)	Due to urban planning reasons, the wall height was limited to 4 m. However, additional passive measures are required with this solution	Due to the special situation here on both sides of the railway line 5 m high noise barriers are suggested. Thus, a sufficient noise protection is achieved for most of the adjacent residential areas
Hamminkeln-Mehrhoog (TÜV Nord Systems GmbH on behalf of RVR 2011)	The majority of the residential areas are adequately shielded by the selected sound insulation. Only in the station area on both sides of the road in a 200 m wide sector with residential areas additional measures must be taken into consideration	The proposed noise protection measures and development of the vast space domain can be well shielded. Only in the immediate station surroundings additional passive measures are necessary

The consultants Schüßler-Plan and Peter M. Moik were asked to develop combined solutions in the field of optimal noise protection, good urban spatial compatibility and rising tension. Both consultants have achieved the targets with their overall concepts. Key measures are different heights and differentiated designed noise barriers, additional cultivations and the reduction of noise at the track level (Table 1).

The results of the sound propagation forecasts show that urban solutions with a high level of noise protection can be achieved but may include additional measures to meet the particular nighttime limits.

6 Conclusion: The Findings of the Ideas Workshop

From the perspective of the Monitoring Board, all teams have critically examined the plans of the DB AG and delivered interesting ideas. Both urban design and noise-technical aspects have been incorporated into the expert's suggestions. The planning concepts of the experts show, that it is worthwhile to develop individually tailored measures for each location.

A "tool box" of standard solutions and other complementary elements provides good ideas for further site-appropriate solutions. According to the Advisory Board the entire planning of the expert teams shows viable alternatives to the plans of the DB AG. They illustrate on one hand that location-based solutions are meaningful and sustainable in the long term and on the other hand, the integration of the railway line into a regional master plan including the Lippemouth area would be significant.

Because only new railway constructions or railway upgrades (third track Oberhausen–Wesel–Emmerich) are eligible for noise protection measurements financed by federal government or DB AG, urban planning and noise should be interlinked. An atmosphere of safety at the stations should be ensured through transparency and clarity of the architectural design. The participants in the ideas workshop agreed that innovative solutions should be pursued. However, the question remains whether the cities can participate in financing the additional costs of sophisticated solutions.

There is agreement by all participants in the project that the results of the planning ideas from the workshop are of high value for local authorities in the context of future discussions with the DB AG and the pending plan approval process. The municipalities have to discuss the solutions of the ideas workshop in depth and take a stand. The Ruhr Regional Association will continue to support the local authorities in the further process.

References

DB NETZE, DB Projektbau GmbH (2011) Unterlagen für ein Verfahren gem. § 18 AEG

Moik PM et al on behalf of RVR (2001) Ideenwerkstatt Fortsetzung der Betuwe-Linie Abschnitt Oberhausen – Wesel- Emmerich in der Entwicklungsachse Rotterdam – Genua. Project report, Düsseldorf

Ruhr Regional Association (RVR) (2011) Fortsetzung Betuwe-Route (CODE24) Ideenwerkstatt Kreis Wesel. Project-report, Essen

Schüßler-Plan et al on behalf of RVR (2011) Abschlußbericht Ideenwerktatt Betuwe-Linie. Final project report, Köln, p 15

TÜV Nord Systems GmbH on behalf of RVR (2011) Gutachten Geräuschimmissionen durch Schienenverkehr auf der Bahnstrecke "Oberhausen – Emmerich" (Sogenannte Betuwe-Linie) nach dreigleisigen Ausbau bei zwei alternativen Maßnahmenvarianten. Survey report, Essen

Frank Joneit works since 1996 at the Ruhr Regional Associations department for regional development. He studied political sciences, English and history at Ruhr-Universität Bochum. His work includes the management and implementation of communication processes with politics and public institutions as well as local authorities. He also works on the implementation of competitive procedures in interdisciplinary teams. He had also been lecturer at the University of Applied Sciences in Bochum for the topic of public transport in the early years 2000.

European Grouping of Territorial Cooperation: Giving a Voice to Local and Regional Government in European Infrastructure Development

Jörg Saalbach

Abstract Since the Rhine–Alpine Corridor can be considered a spatially coherent axis with a multitude of common interests and interrelationships between the individual regional areas of the Corridor, there is a wide scope for cooperation. It was already clear at the beginning of the strategic initiative CODE24 that not all of the complex challenges of this major European axis could be solved in the 5-year project period. Therefore, a solution for continuing cooperation beyond the funding period of CODE24 had to be found. Consequently, the preparation of an appropriate legal form for further cooperation became one major objective of the partnership.

The establishment of a European Grouping of Territorial Cooperation (EGTC) became the aim as a means to enable the regions, cities and other stakeholders of the Rhine–Alpine Corridor area to continue their cooperation, to speak with one voice and to keep on developing the Corridor jointly. The EGTC is now called 0 the Interregional Alliance for the Rhine–Alpine Corridor and was established during CODE24's final conference in November 2014 in Mannheim.

1 The Need for Further Cooperation

The INTERREG Project CODE24, Corridor Development Rotterdam–Genoa, was aimed at an integrated joint approach towards the future development of this major European axis and intended to interconnect economic development and spatial, transport and environmental planning. To pursue these goals, facilitate transnational cooperation between partners along the axis and manage the complex challenges of corridor development, it was decided to establish an organisation: the European Grouping of Territorial Cooperation (EGTC).

It was obvious from the beginning of the cooperation project CODE24 that it would be impossible to deal with all challenges the development of such a major corridor implies and to achieve solutions for the reoccurring problems within the restricted period of the INTERREG project. For this reason, the objective of

J. Saalbach (✉)
Verband Region Rhein-Neckar, P7, 20-21, 68161 Mannheim, Germany
e-mail: joerg.saalbach@egtc-rhine-alpine.eu

H. Drewello, B. Scholl (eds.), *Integrated Spatial and Transport Infrastructure Development*, Contributions to Economics, DOI 10.1007/978-3-319-15708-5_19

establishing a durable framework for the future cooperation was already part of the project application for CODE24.[1]

There is an apparent need for continued cooperation in order to discuss and solve development issues of this European core network corridor. Project partners of CODE24 and the future members of the EGTC expressed their will to create a basis for their cooperation that aims at coordinating the local and regional level of the corridor's stakeholders. At the same time, it was vital to continue the collaboration of this bottom-up consortium and even enlarge it by including new members.

2 What Kind of Legal Form for a Transnational Cooperation?

The decision had to be made in which form this fruitful cooperation within the project partnership of CODE24 could be sustained. Basically, one can distinguish between informal and formal kinds of cooperation; both categories again offer a variety of cooperation forms. Therefore, it was necessary to scrutinise different forms of cooperation that might be applied for the continuation of the partners' cooperation in the future.

The leading criterion for this choice has always been the identification of a legal form that assures the seriousness of intent and improves the visibility of the partnership. Although an informal type of organisation without a legal background is easier and quicker to achieve, formal types are obviously more appropriate for the tasks the partners envisage for the future. In considering the formal types of cooperation, there was one important issue that was finally decisive, i.e. to avoid applying a national form of cooperation. For example, any association that follows a national legislation was considered inappropriate, as the various partners would all have different regulations. In fact, the predominant objective was to clearly demonstrate the European and transnational character of cooperation in the Rhine–Alpine Corridor. Consequently a European, rather than a national legal form, was considered the most suitable and appropriate.

After taking the pros and cons of possible forms of cooperation into account, the EGTC was finally judged to be the most applicable. This new European instrument thus became the legal form of cooperation, as preferred by the partners. The intention of the EU, already evident from the title of the group, was to facilitate territorial cohesion and territorial cooperation, thus fitting perfectly into the future objectives of the CODE24 partnership.

Albeit the establishment of an EGTC was more complicated than other working groups, considering the interventions of all the different approval authorities along the Corridor, the EGTC still remains the most adequate legal form for the kind of cooperation the partners had in mind.

[1] cf. CODE24 Application for funding from the INTERREG IVB Programme, Northwest Europe.

3 The European Grouping of Territorial Cooperation

The European Grouping of Territorial Cooperation was established on 5 July 2006 by Regulation (EC) 1082/2006 of the European Parliament and Council and came into force on 1 August 2006.[2] The European Union considered it helpful to provide a community-wide European form of cooperation, tailor-made for public authorities at all levels of government. In order to foster cooperation across borders, there was a need for a legal form directly legally effective in all Member States of the European Union. So far, legal cross-border organisations, based for example on the Madrid Convention,[3] have always referred back to national law. Up to now, 53 EGTCs have been created, making concrete use of the EU regulation.[4] In most cases, these EGTCs are involved in cross-border cooperation, i.e. cooperation on a rather small scale between neighbouring areas on two sides of a border. Only a few of the existing EGTCs incorporate large-scale areas.

The European Union encourages the foundation of more EGTCs and supports interested organisations in making use of the respective EU regulation. The Committee of the Regions (COR) plays a central role by providing support and an EGTC platform.[5]

4 The Interregional Alliance for the Rhine–Alpine Corridor

The European Grouping of Territorial Cooperation's on the Interregional Alliance for the Rhine–Alpine Corridor has the main aim to continue the strategic initiative of the INTERREG Project CODE24 in order to secure a long-term partnership and cooperation beyond the limited INTERREG project period.[6] The legal documents needed to create an EGTC, i.e. the Convention and the Statutes, clearly define the future objectives and tasks of the EGTC.

[2] Regulation (EC) No. 1082/2006 of the European Parliament and the Council of 5 July 2006 on a European Grouping of Territorial Cooperation (EGTC). This regulation has meanwhile been amended by Regulation (EU) No. 1302/2013 of the European Parliament and of the Council on 17 December 2013.

[3] European Outline Convention on Trans-frontier Cooperation between Territorial Communities or Authorities, 1980.

[4] cf. https://portal.cor.europa.eu/egtc/en-US/Register/Pages/welcome.aspx. Accessed 18 Nov 2014.

[5] cf. https://portal.cor.europa.eu/egtc/en-US/Pages/welcome.aspx. Accessed 18 Nov 2014.

[6] cf. Draft Convention and Statutes for the EGTC Interregional Alliance for the Rhine–Alpine Corridor, April 2014.

4.1 Objectives and Tasks of the New EGTC

The general objective of this EGTC is to jointly strengthen and coordinate an integrated territorial development along the multimodal Rhine–Alpine Corridor from a regional and local perspective. The objectives and tasks were deliberately formulated in a rather general form in order to maintain a certain level of flexibility and allow room for variation in the scope of action for future activities. This will also help avoid the necessity of amending the Convention for a new issue that might appear in the future. The partnership of CODE24 has always considered its scope to be broader than pure transport or even freight transport matters. In fact, a broader approach seemed more suitable when taking into account the integrated and multidimensional perspective of regional development.

At this time, it was determined that certain achievements and concrete measures resulting from the cooperation during CODE24 should be continued and updated regularly by the EGTC. This concerns, for example, the Corridor Information System (CIS), an Internet-based tool that was developed as one major and tangible result of CODE24. The EGTC should host and update the CIS in future.

Another important and useful tool of CODE24 is the Mobile Exhibition. This exhibition consists of various modules that can be combined based on the target audience and available exhibition space. The Mobile Exhibition had been a very useful way to inform the public, as well as experts, about the content and results of CODE24. It is thus very reasonable for the EGTC to further utilise this communication tool and keep it updated.

The objectives and tasks of the EGTC were formulated against the background of the CODE24 project. The tasks encompass five main topics that clearly define the fields of action for this EGTC. In more detail, the tasks are as follows[7]:

1. Combining and focusing the joint interests of its members towards national, European and infrastructure institutions. Within this task, two kinds of activities are foreseen:
   - The organisation and implementation of joint lobbying activities for the development of the Rhine–Alpine Corridor from a bottom-up perspective
   - The representation of the EGTC members in the EU Rhine–Alpine Corridor Forum
2. The evolution of the joint development strategy for the multimodal Rhine–Alpine Corridor. The task comprises the continuation of the most important functional aspects of the cooperation and contains two kinds of activities:
   - The coordination of regional development in the Rhine–Alpine Corridor by taking into account local and regional perspectives
   - The consideration of transport infrastructure projects and land use conflicts along the Rhine–Alpine Corridor

[7] cf. Draft Convention of the EGTC "Interregional Alliance for the Rhine–Alpine Corridor", Article 4.

3. Directing funds to Corridor-related activities and projects. One goal of the EGTC consists of the ambition to generate new projects that fit into the context of developing the Rhine–Alpine Corridor and trying to obtain EU funding for these operations. Therefore, the two subtasks are:

   - Supplying information to EGTC members about funding opportunities for corridor related projects
   - Applying for new EU-funded projects and for joint management of EU funds.

4. Provide a central platform for mutual information, an exchange of experience and an opportunity to meet. It was proven during the implementation of project CODE24 that providing opportunities for meetings and exchange are a vital basis for cooperation. At the same time, information flow is a prerequisite for sound decisions. Consequently, the EGTC will facilitate meetings and opinion-making through the following activities:

   - Organising meetings of members
   - Ensuring information transfer
   - Taking charge of the Corridor Information System

5. Improving the visibility and promotion of the Corridor. Public relations, communication and promotion are key for successful progress and for informing the public and target audiences about the objectives and measures concerning the development of the Rhine–Alpine Corridor from a local and regional point of view. The following activities are meant to support this objective:

   - Organisation of corridor events (congresses, workshops, etc.)
   - Elaboration and distribution of publications (newsletters, leaflets, brochures)
   - Maintenance of the Website www.code-24.eu, which was developed as part of the CODE24 project
   - Taking over and maintaining the Mobile Exhibition

Once the EGTC becomes operational, it will follow up and fulfil these tasks. Of course, it might happen that certain tasks will become more important than others; such things can change over time. In case new tasks do arise, there is always an option to amend the convention and statutes according to current needs.

4.2 Members of the EGTC

Basically, the partners of CODE24 are foreseen as the members of the EGTC. So far, the following organisations have officially declared their interest in joining the EGTC:

- Province Gelderland
- Region Köln/Bonn e.V.
- City of Mannheim
- Regional Association of Rhine-Neckar
- City of Karlsruhe

- Technological Region Karlsruhe GbR
- Regional Association of the Middle and Upper Rhine
- City of Lahr
- Uniontrasporti, Milan
- Rotterdam Port Authority
- Region Piedmont
- Regional Association Southern Upper Rhine
- Duisburger Hafen AG

Invitations to join and the internal requirements for becoming a member are ongoing in many of the organisations along the Corridor. The goal is to officially create the EGTC with 10–20 members.

According to EU regulations,[8] additional members who are situated in the Corridor area can join the EGTC upon application with approval of the Assembly. New members are also welcome to join the EGTC as it is bound to grow and be open to new members. Since the EGTC explicitly represents organisations from the local and regional levels, cities and regions are the focus for recruiting further members. This also includes organisations from non-member states of the EU, e.g. Switzerland. It is the aim of the EGTC to attract members from Switzerland, which holds a central position within the Rhine–Alpine Corridor and is of strategic relevance for coordinating the future development of the Corridor, thus, including partners from Swiss regions and cities is important.

4.3 *Formal Matters*

According to EU regulations, certain formalities have to be respected. This concerns the obligatory definition of the envisaged organs, liability concerns, the legal seat of the EGTC and other issues. However, compared to other regulations, the EGTC regulation may be judged as rather lean, containing only a minimum of formal obligations. In the founding documents, the following organs are foreseen for the EGTC: are foreseen: the Assembly, the Director and the Managing Committee.

Mannheim/Baden-Württemberg has been considered for the legal seat because it is in a central location within the Corridor area. The EGTC office also has a plan to be hosted at the premises of the Regional Association of Rhine-Neckar (Verband Region Rhein-Neckar), which was also the CODE24 lead partner.

As stated in the EU regulation, liability matters must be defined.[9] The draft convention contains the following formulation:

[8] Article 3, paragraph 1, subparagraph 1 of Regulation (EG) No 1082/2006 as amended by Regulation (EU) No 1302/2013.

[9] cf. Regulation (EC) No 1082/2006 of the European Parliament and the Council of 5 July 2006 on a European Grouping of Territorial Cooperation (EGTC). This regulation has meanwhile been amended by Regulation (EU) No 1302/2013 of the European Parliament and of the Council on 17 December 2013.

> The EGTC shall be liable for all its debts. To the extent that the assets of the EGTC are insufficient to meet its liabilities, its members shall be liable in equal parts for its debts irrespective of the nature of those debts.[10]

Of course, it would have been preferable to be a 'limited' organisation, but the EU regulation, respectively the national regulation, does not allow this. Then again, the realistic financial risk, considering the tasks and operations of the EGTC, is low. Moreover, the annual budget, which has to be agreed upon by the members of the EGTC, does not permit any financial operations of a size that could become hazardous to the EGTC budget.

4.4 Budget

The EGTC budget is an important element and must be defined by the assembly of the members. According to the principle of not building up a new bureaucracy, the budget is planned to be modest. Although, depending on the further development and success of the EGTC, it might be expanded. As incorporated in the Convention and Statutes, the budget consists mainly of the annual fees of its members. As stated in the tasks of the EGTC, EU funding for new projects that have to be generated will enlarge the budget. These additional financial means will have to be co-financed through proportional contributions of those members who participate in such projects.

The calculation of the annual fee has been kept simple: The total amount of the budget is divided into equal parts between the members. It was judged as not advisable to create a complex fee system and define different contribution rates, taking into account, for example, size of population or financial power of a member and distinguishing categories for different scales. Furthermore, the distinction of contribution rates would also affect the voting rules. Logically, contribution rates would need to be proportional to the voting rules, which would create different levels of voting power between the members. Therefore, the proposal was made to provide one vote per member and all members have the same level of contribution.

5 The Procedure to Create an EGTC

In order to create an EGTC and become a member, certain obligations have to be respected. In short, founding documents have to be drafted and potential members need official approval first. Since this European legal form is quite innovative and thus interested parties are probably not familiar with its application, there are

[10] cf. Draft Convention of the EGTC "Interregional Alliance for the Rhine–Alpine Corridor", Article 13.

support lines and manuals on offer.[11] These documents and advice desks[12] are very useful when setting up an EGTC.

5.1 The Formal Documents: Convention and Statutes

The legal documents have to be drafted and must include a defined number of obligatory issues, which are listed in the respective articles of the EU regulation.[13] Two documents are compulsory before applying to found an EGTC: the convention and the statutes. Both documents have to be jointly developed by the future members of the EGTC. Customarily, one partner formulates the first draft as a basis for discussion among the other potential members. For CODE24, the lead partner, the Regional Association of Rhine-Neckar, prepared the first draft. After various rounds of consultation, when the two documents have been amended and serve as a basis for internal decision-making, then the potential members can submit the formal application for approval.

5.2 Application and Approval for Membership

Based on the carefully crafted drafts of the convention and the statutes, every organisation willing to become a member of the future EGTC is obliged to get approval from its respective competent board. Usually these boards request a copy of the two documents in their own language.

Once this decision has been taken, every organisation willing to join the EGTC is obliged to officially apply to the respective national approval authority for permission to become a member of the EGTC, attaching the decision of its own board. The national approval authorities will inspect the draft documents, taking the EU regulation into consideration, as well as the status and competence of the respective organisation wanting to join.

---

[11] cf. Leitfaden zur Gründung eines EVTZ für Akteure der transnationalen Zusammenarbeit, Februar 2014.

[12] http://www.interact-eu.net/egtc_setup/egtc_setup/69/52. Accessed 19 Nov 2014.

[13] cf. Regulation (EC) No 1082/2006 of the European Parliament and the Council of 5 July 2006 on a European Grouping of Territorial Cooperation (EGTC). This regulation has meanwhile been amended by Regulation (EU) No 1302/2013 of the European Parliament and of the Council on 17 December 2013.

5.3 *Foundation of the EGTC*

Only those organisations that have successfully completed this process and have received official approval to become a member of the EGTC may then participate at the founding meeting of the EGTC. At this meeting, members have to discuss and decide on the final version of the convention and statutes. Each signature on the convention confirms the membership.

All in all, this process can take a long time, depending on the meeting schedules of the boards that can decide upon membership in the EGTC. The national approval authorities need to decide within 6 months after the application has been submitted. If pertinent comments are received from one or more approval authorities, the entire procedure has to start all over again. The amended version of the convention and the statutes taking the relevant remarks from the approval authority into account will need to be reconsidered by all potential member organisations.

6 Planned Activities of the EGTC

The approach to planning activities has been delineated in the Convention as follows:

- The new EGTC will capitalise on the achievements and results from project CODE24, pursue the activities of project CODE24, and continue with relevant actions that are useful for the future development of the Rhine–Alpine Corridor.
- The new EGTC will not create a bureaucracy. It intends to organise work with a compact and lean structure. Therefore, only a small secretarial office is envisaged, enabling the EGTC to act jointly and implement the decisions of the EGTC Assembly.
- The EGTC plans to define new projects of common interest. These projects shall be submitted for co-financing from the appropriate EU funding schemes, such as INTERREG or CEF.
- The EGTC also intends to become a full member of the Corridor Forum created by the European Commission for each of the nine TEN-T Core Network Corridors. These fora are bound to play a consultative role for supporting the European Union by defining the corridor policy for each corridor. These fora consist of a variety of stakeholders involved in the development of the newly defined core network corridors. It is therefore an objective for the EGTC to be part of this body and represent the interests of the local and regional levels in the Rhine–Alpine Corridor.
- In order to raise awareness for the Rhine–Alpine Corridor specifically from a local-regional angle, public relation measures play an important role. Thus, public events, press conferences and communication activities will be organised.

It will be very useful to have a legal entity available when implementing the activities mentioned above. Obviously, it makes a difference when acting for the sake of the Corridor's development from a local and regional perspective, when a proper legal organisation with an assembly of members, a chair and a competent secretariat with a director can act on behalf of all its members. Compared with the short-term project consortium, if a European Grouping of Territorial Cooperation has a proper legal form, and actually, a European legal form, then its acceptance will bring considerable positive acknowledgement.

7 Conclusion

The intention to establish an EGTC in order to continue the fruitful cooperation of the CODE24 partnership and also allow new partners to join is an ambitious undertaking. In principle, the new legal instrument 'European Grouping of Territorial Cooperation' is a useful and appropriate tool. Its practical application, though, is a challenge. The bureaucratic barriers still impede the creation of an EGTC. It causes frustration and irritation amongst organisations willing to join when they find themselves confronted with strict regulations and procedures that they are forced to execute in order to gain approval to finally become a member.

Nevertheless, these obstacles can be surmounted and it is worthwhile to achieve the goal of establishing an EGTC. Once the official foundation is completed, the work and implementation of the defined tasks of the EGTC can start.

The new EGTC Interregional Alliance for the Rhine–Alpine Corridor will then give a strong voice to local and regional governments by being member of the EGTC along the Corridor and will enable them to significantly influence the development of the Corridor.

Jörg Saalbach earned his Master's Degree in geography before starting his post-graduate studies of Spatial Planning at the ETH Zurich. He started his professional career as a regional planner in South-West Germany. For 15 years he worked for the French-German cross border co-operation including the management of INTERREG A programmes and projects. Since 2007 he is Representative of European Affairs for the Regional Planning Authority Rhein-Neckar which again acts as Lead Partner of the INTERREG IVB-NEW project "CODE24". He was appointed as Director of the Interregional Alliance for the Rhine-Alpine Corridor EGTC in April 2015.

Conclusion: Perspective on the Future of the Rhine–Alpine Corridor

The preceding contributions clarify and define the challenging tasks of spatial and transport development in the most important North–South corridor of Europe and its multifaceted problems.

The contributions in this publication also make it clear that cooperation towards a solution in the assignment of tasks lies well within the mutual interests, because it would stimulate economic development and strengthen cohesion between Europe's north and south. One of the biggest challenges will be to bring the estimated growth in transport volume, mainly in freight transport, into harmony with the needs of the more than 70 million people in the settlement areas.

The development of a corridor is like putting a wooden interlocking puzzle together. Many actors have contributed their puzzle pieces to the whole structure. Leading thoughts on spatial and transport development are (1) to connect the future settlement development in the catchment area with efficient railway stops and stations in public transport and (2) to transfer freight transport from road to rail and the Rhine River waterway. The first requires a consistent use of richly available land reserves, as presented in this publication, and the last demands a consistent availability of capable, cross-border and reliable rail traffic for passengers as well as freight. Measured against the planned harbour investments for the northern range, this should be feasible in the next 10–20 years.

If the road-to-rail transfer cannot be realised, then implementing any strategic approach of spatial and transport development carries a very high risk that with excessive freight transport on the roads and an unreliable railway operation, settlement development will be driven further onto greenfield areas and is then suddenly burdened with more cars, leading ultimately to a negative spiral with more road traffic and more traffic congestion. In this scenario, the value-creation potential of such important corridors for the economic performance of Central Europe could never be fulfilled, which would also be a distinct disadvantage for other regions of Europe.

H. Drewello, B. Scholl (eds.), *Integrated Spatial and Transport Infrastructure Development*, Contributions to Economics, DOI 10.1007/978-3-319-15708-5

In order to be able to achieve the desired effects according to (1) and (2) in the mid to long term, the following essential knowledge is of central importance, in our opinion, for the future of the process:

- In order to maintain an overview of events, in future, regular mutual evaluations should be conducted for the entire corridor space to protect the status of the mutual knowledge achieved, as well as to recognise where future focal points can be set. These could be special themes, but mainly they will be areas singled out for their importance.
- During the work on CODE24, a corridor formation system was started. It should be further developed and brought up to date. It should also be tested to find out whether the synergies organised by the EU will work within the framework of the corridor platform.
- Of course, the possible technological advances of noise reduction in goods transport should be used, as well as the increase in efficiency in goods transport logistics through consistent, useful modern logistics, for example, through furnishing or constructing tri-modal equipment for the transfer of freight.
- The corridor space is a very dynamic area. Therefore, a thorough and simultaneous execution of all tasks between Rotterdam–Antwerp and Genoa is not possible. A strategy requires defining the focus points and determining the areas of concentration. However, deciding the targets is only possible when they are based on qualified overviews, such as those made available during the CODE24 process. And, of course, these overviews should be updated regularly. A maxim to remember: Anyone who tries to coordinate everything ends up coordinating nothing.
- As the work on CODE24 has shown, there are several locations where a continuing cooperation could be even more intensive between the important actors at the various levels of local government and, based on the tasks, the important actors within special processes for infrastructures. Such cooperative processes are manageable and when measured against the investments in question, the costs are also very manageable as they lie in the parts-per-thousand range of the respective expenditures of the infrastructures under consideration.

At the final conference on the strategic EU Project CODE24 in Autumn 2014, the Regional Rhein–Neckar Association took the lead in establishing a European special purpose association to ensure that the cooperation, which had proven reliable during the CODE24 Project, will be stabilised and intensified after the conversion to the Rhine–Alpine Corridor. This is an encouraging sign! We wish much success to the participants of the European Group on Territorial Cohesion (EGTC) and the actors responsible for further spatial and transport development.

We suggest that a report on the progress, experiences and findings during the further development of the Alpine–Rhine Corridor, as well as other corridors of EU importance, should be presented in a few years time at an event such as an international conference on corridor development. A large-scale, integrated European spatial and transport development will confer a further—and much needed—stimulus to future plans.

MIX
Papier aus verantwortungsvollen Quellen
Paper from responsible sources
FSC® C105338

If you have any concerns about our products,
you can contact us on
ProductSafety@springernature.com

In case Publisher is established outside the EU,
the EU authorized representative is:
Springer Nature Customer Service Center GmbH
Europaplatz 3, 69115 Heidelberg, Germany

Printed by Libri Plureos GmbH
in Hamburg, Germany